METHODOLOGICAL SURVEYS IN BIOANALYSIS OF DRUGS

(Former series title: Methodological Surveys in Biochemistry and Analysis)

Series Editor: Eric Reid
Guildford Academic Associates
72 The Chase
Guildford GU2 5UL, United Kingdom

This series was divided into Subseries A: Analysis, and B: Biochemistry up to Volume 22. Enquiries concerning Volume 1–18 should be sent to the above address.

Recent Titles

Volume 15(B):	Investigation and Exploitation of Antibody Combining Sites Edited by Eric Reid, G.M.W. Cook, and D.J. Morré
Volume 16(A):	Bioactive Analytes, Including CNS Drugs, Peptides, and Enantiomers Edited by Eric Reid, Bryan Scales, and Ian D. Wilson
Volume 17(B):	Cells, Membranes, and Disease, Including Renal Edited by Eric Reid, G.M.W. Cook, and J.P. Luzio
Volume 18(A):	Bioanalysis of Drugs and Metabolites, Especially Anti-inflammatory and Cardiovascular Edited by Eric Reid, J.D. Robinson, and Ian D. Wilson
Volume 19(B):	Biochemical Approaches to Cellular Calcium Edited by Eric Reid, G.M.W. Cook, and J.P. Luzio
Volume 20(A):	Analysis for Drugs and Metabolites, Including Anti-infective Agents Edited by Eric Reid and Ian D. Wilson
Volume 21(B):	Cell Signalling: Experimental Strategies Edited by Eric Reid, G.M.W. Cook, and J.P. Luzio
Volume 22(B):	Bioanalytical Approaches for Drugs, Including Anti-asthmatics and Metabolites Edited by Eric Reid and Ian D. Wilson
Volume 23:	Biofluid and Tissue Analysis for Drugs, Including Hypolipidaemics Edited by Eric Reid, Howard M. Hill, and Ian D. Wilson
Volume 24:	Biofluid Assay for Peptide-related and Other Drugs Edited by Eric Reid, Howard M. Hill, and Ian D. Wilson

How to obtain future titles on publication

A standing order plan is available for this series. A standing order will bring delivery for each new volume immediately upon publication. For future information, please write to:

Turpin Distribution Services Ltd
Blackhorse Road
Letchworth
Herts. SG6 1HN
Telephone: Letchworth (01462) 672555

Biofluid Assay for Peptide-related and Other Drugs

Methodological Surveys in Bioanalysis of Drugs, Volume 24

Biofluid Assay for Peptide-related and Other Drugs

Edited by

Eric Reid
Guildford Academic Associates, Guildford, United Kingdom

Howard M. Hill
Corning Hazleton (Europe), Harrogate, United Kingdom

Ian D. Wilson
Zeneca Pharmaceuticals, Macclesfield, United Kingdom

THE ROYAL SOCIETY OF CHEMISTRY
Information Services

Based on the proceedings of the 11th International Bioanalytical Forum mounted by the Forum Syndicate of the Chromatographic Society, entitled Biofluid Assay for Peptide-related and Other Drugs, held 29 August to 1 September 1995, in Guildford, UK.

Special Publication No. 185

ISBN 0-85404-741-7

A catalogue record for this book is available from the British Library.

Published by The Royal Society of Chemistry,
Thomas Graham House, Science Park, Milton Road,
Cambridge CB4 4WF, UK

Printed by Athenaeum Press Ltd, Gateshead, Tyne & Wear

Senior Editor's Preface

with some abbreviations and argued 'MS shorthand' policies

This volume reflects the continuing tradition of strong editing to achieve clear yet compact presentations that give adequate detail which, however, does not devour page space.- Validation evidence, whilst important, may not need multiple Tables such as prevail in other publications which, moreover, may have Figs. with unrectified miniscule lettering. Putting instrument-generated diagrams into legible and uncluttered form with elimination of 'junk', and curing faintness, is a growing burden now that fewer diagrams are hand-drawn.- Authors please note! Faintness marred some material in the previous volume (printers may use a camera without 'intelligent' adjustments). The present book may fulfil more consistently the aim of good appearance with fair homogeneity that embraces abbreviations.

The Bioanalytical Forum series, 21 years on.- The 1975 Forum was a pioneer venture to encourage problem-discussing amongst bioanalysts (mainly company-based), whose timidity made the Forum hard to set up. Its success has since been emulated by other meetings, and it is now common for company staff to 'go public' with their methods; but the series still flourishes as evidenced by the 1995 Forum (11th, at the usual University of Surrey venue) on which this book (not mere 'Proceedings') is based. For a decade the series was under 'Guildford Academic Associates' auspices. Now, with the same team and aided by an endowment, a 'Forum Syndicate' of the Chromatographic Society has responsibility. The 1997 Forum dates are 2-5 September.

Some features of the book.- The inclusion of Discussion remarks seems to be commendable rather than idiosyncratic, as hopefully applies also to insertions by this Editor. These include an Assay Compendium, based on therapeutic classes but now with diminished coverage of therapeutic and toxicological monitoring (for which our publisher now produces Abstracts); the emphasis is on more sensitive assays and on automation. Good indexing remains paramount, with the usual Analyte Index based on some chemical features which analytes not listed may also possess. In the past this Editor's Preface has alluded to some 'dislikes'; disfavoured vague terms include '2-D'/'multi-D' and 'hyphenated'. A traditional book feature - 'Special topic' - here comprises peptide-type analytes (as in Vol. 16), too diverse for comprehensive coverage.

Acknowledgements.- Dr D. Stephenson was Co-organizer for the 1995 Forum, to which SmithKline Beecham made an appreciated donation. Drs D. Thomas and M.V. Doig helped settle editorial policies for mass-spectrometry (MS) 'shorthand' [see overleaf]. For the benefit of fellow-bioanalysts, busy authors 'made time' to produce publication texts. Ackowledgements for allowing use of items already published elsewhere appear within the text.

Alertings to readers and other analysts and library staff

Abbreviations, especially for MS.- Settling policies for the book was prompted by diversity amongst Forum Abstracts, e.g.:
LC/MS/MS LC/MS-MS LC-MS/MS LC-MS-MS HPLC-MS/MS HPLC-MS-MS
as reflected in Journals, wherein guidance to authors is scant.
Shorthand now adopted, acceptable in some MS circles *(including* IUPAC *?)*, employs **hyphens**, not 'slashes' // unless 'options' are signified (e.g. FTICR/MS). A further policy [some individuals like it] saves multiple hyphens: **parenthetical sub-descriptions** (), e.g. for thermospray, or atmospheric pressure ionization, chemical (c) or electrospray (ESP/ESI; variant: ionspray, ISP). #Thus: LC-MS(APcI) or merely LC-MS(API) [*not* LC-API-MS], maybe amplifying the 'c' approach: n (*not* N)/p = negative/positive ion • as in *examples:* LC-MS-MS(APncI) LC-MS-MS(TSP) and similarly • electron impact, GC-MS(EI); fast atom bombardment, TLC-MS(FAB) Also: SIM = selected ion monitoring, or MRM if multiple reaction.

Other abbreviations.- HPLC prefixed NP/RP, normal/reverse(d) phase [ISRP, internal surface]. Detection: fluor = fluorimetric - ex[citation] & em[ission] nm maybe stated; EC = electrochemical [*any support for* ECh*?*]; **not** ECD, = electron-capture detection as in GC-ECD. FID = flame ionization; NPD = nitrogen-phosphorus.

Ab, antibody (mAb, monoclonal)
ELISA, Enzyme-linked immunosorbent assay
IA, immunoassay (RIA, radio-)
i.s., internal standard
LLOQ/LOD, limit of quantification (lower)/detection
BSA, bovine serum albumin
CE, capillary electrophoresis
C.V., coefficient of variation
i.v., intravenous; s.c., subcutanous; i.m., intramuscular
QC/QA, quality control/assurance
SPE, solid-phase extraction

In some arts.: MeCN, acetonitrile; PCA/TCA, perchloric/trichloroacetic acid; TEA, triethylamine; PK/PD, pharmaco-kinetics/dynamics. Temperatures ° are Celsius. 'Multi-D' *disfavoured (unclear term).*

Bibliographic points affecting Libraries besides readers.-

- Now ***A*** and ***B*** distinction (specify! vol. nos. shared): *J. Chromatog.*
- Dual-title muddle (synonyms): *HRC* or *J. High Resolu. Chromatog.*
- Sloppy variants: *Pharmac./Pharmacol. Endocr./Endocrin./Endocrinol.*
- Vol. 23 of present series: erroneous 'Cumulative Index' phrase on spine (was in Vol. 22!).- *Mask over! (by tape?)*
- Earlier vols.- *#Analysis* subseries: 5, 7, 10 & even nos. up to 22 - all focused on drugs, as were many arts. in *Biochemistry* subseries. #Former series title (affects shelving or procurement; info. wanted*?*): *Methodological Surveys in Biochemistry and Analysis.*

72 The Chase, Guildford GU2 5UL, U.K.
Fax: (0)1483-565324

ERIC REID
Forum Syndicate Sec.

CONTENTS

Besides main contributions there are 'NOTES' (short articles) and 'COMMENTS'. The suffix **'n'** as in #A-7n distinguishes NOTES from main articles. COMMENTS (put at end of Section) comprise Forum discussion remarks and Editor's 'annotations'.

'n' as in #C-5n signifies a Note

Contents

'n' as in #E-4n signifies a Note

List of Authors with p. nos.

Primary author

D.J. Anderson: (i) 78-81, (ii) 183-193
Cleveland State Univ., OH

C.J. Bailey: 40-47
Zeneca Pharm'ls, Alderley Pk.

M. Bayliss: 141
Astra Charnwood, Loughborough

M. Bentley: 327-328
AS FOR Hill

C.D. Bevan: 299, & see Tiller
Glaxo Wellcome, Stevenage

K. Borner: 147-150
Klin. Chem./Biochem. Inst., Berlin

S. Braggio: 158
Glaxo Res., Verona, Italy

D. Browne: 254-256, & see Robotham; AS FOR Hill

J.E.C. Burnett: 333-335, & see Hill (i); AS FOR Hill

G.C. DiDonato: 57-58
Bristol-Myers Squibb Res. Inst., Princeton, NJ

T.A. Enos: 126-127, & see James, Simmonds; AS FOR James

J.D. Gilbert: 65-77
Merck Res., West Point, PA

K. Grob: 246-253
Kantonales Labor, Zürich

D. Gygax: 11-19
Ciba-Geigy, Basel

P. Heizmann: 304-311
Hoffmann-La Roche, Basel

H.M. Hill: (i) 48-56, (ii) 223-224, (iii) 316-326, & see Burnett, Thomas
Corning Hazleton, Harrogate

S-H. Hsu: 35-39
Rhone-Poulenc Rorer Central Res., Collegeville, PA

A.J. Hutt: 154-157
Kings Coll., Chelsea, London

C.A. James: 82-88, & see Enos
Pharmacia & Upjohn, Crawley

Co-authors, with relevant name to be consulted in left column

M. Abermann - Nemes
J.A. Allanson - Pleasance
A. Arroyo - DiDonato
D. Ayres - Wring
S.J. Bacon - Bailey
R.J. Barnaby - Braggio
G. Barnard - Venn
K. Bijmholdt-Sierat - van Amsterdam
E. Bojti - Nemes
J. Borlak - Bentley
E. Borner - K. Borner
L. Botta - Gygax
S.C. Bridges - Westwood
N.K. Burton - James
J.L. Byard - Seymour

M. Churchill - Mason, Moore
S.D. Clarke - Hill (i), Thomas
C.S. Creaser - Seymour
S. Dayal - Wring
V. de Biasi - Hutt
J.J. Dervan - Enos
M. Ehrat - Gygax
J.D. Ellis - Gilbert
S. Feely - Seymour
C. Fernández-Metzler - Gilbert
B. Ferraiolo - Mohler
I.J. Fraser - Pleasance
J.L. Furness - Bayliss

W.A. Galloway - Mason, Moore
C. Gooijer - Lingeman
R. Gora - Heizmann
P. Graf - Gygax
S-J. Groves - Lough
P. Guntz - Nuez

M. Hail - DiDonato
E.L. Hand - Gilbert
G. Harrison - Wring
H. Hartwig - Borner
M.J. Hemsley - Noctor (i)
E. Houghton - Seymour
E. Jochem - van Amsterdam
A.E. Jones - Pleasance
I. Klebovich - Nemes
J. Körner - Heizmann
P. Kwasowski - Stevenson

S.F. Lane - Tiller
R. Laplanche - Nuez
B. Law - Lough
J. Lee - Hsu
M. Lee - DiDonato
G.PhD. Leeflang - van Amsterdam
G. Lefèvre - Gygax
M. Lemaire - Nuez
Y. Liu - Anderson (i)
H. Lode - Borner
J.M. Long - Enos

P.V. Macrae - Venn
J. McGowan - Moore
D.A. McLoughlin - Gilbert
S. Maier - Heizmann
J. Maltas - Lough
A.J.G. Mank - Lingeman
H. Marsh - Seymour
P. Martin - Wilson
P. Maude - Robotham
C.M. Middlekoop - van Amsterdam
K.G. Miller - Poole
M.J. Mills - Lough
W. Morden - Wilson
I.M. Mutton - Bevan, Tiller

Primary author

H. Lingeman: 275-287
Free Univ., Amsterdam

W.J. Lough: 142-146
Univ. of Sunderland

C. Mason: 137-140, &
see Moore - AS FOR Moore

M.A. Mohler: 3-10
R.W. Johnson Pharm. Res. Inst., Raritan, NJ

J. Moore: 89-93, & see Mason
British Biotech, Oxford

K.B. Nemes: 103-104
Egis Pharm'ls, Budapest

T.A.G. Noctor: (i) 151-153, (ii) 213-214, (iii) 257-261, & see Hill (i) & (ii), Thomas
Corning Hazleton, Harrogate

C. Nuez: 96-102
Sandoz Pharma, Basel

S. Pleasance: 118-125
Glaxo-Wellcome, Beckenham

C.F. Poole: 194-208
Imperial Coll., London SW7

L. Robotham: 312-315
- AS FOR Noctor

M.A. Seymour: 215-218
- AS FOR Westwood

R.J. Simmonds: 128-136, & see Enos, James
Pharmacia & Upjohn, Crawley

D. Stevenson: 219-222
Univ. of Surrey, Guildford

M.R. Taylor: 288-298, & see Westwood; AS FOR Westwood

D. Thomas: 230-238, & see Hill (i) - AS FOR Noctor

P.R. Tiller: 329-332
- AS FOR Bevan

P.H. van Amsterdam: 105-117
Solvay Duphar, Weesp, Belgium

R.F. Venn: 20-34
Pfizer Central Res., Sandwich

S.A. Westwood: 209-212, & see Taylor
Horseracing Forensic Lab., Newmarket

continued from left-hand column

I.D. Wilson: 239-245
Zeneca Pharm'ls, Alderley Pk.

S.A. Wring: 262-274
Glaxo-Wellcome, Triangle Pk., NC

Co-authors, with relevant name to be consulted in left column

T.V. Olah - Gilbert
K. O'Neill - Bayliss
P. Oroszlan - Gygax
G. Padbury - James
D. Perrett - Taylor
C. Pfister - Gygax
S.K. Poole - C.F. Poole
S. Pratt - Hill (ii)
J.R. Preston - Bayliss

B. Rashid - Stevenson
S.A. Rees - James
K. Reeve - Seymour
L. Rich - Hsu
S. Richards - Moore
C.J. Roberts - Seymour
J.N. Robson - Hill (i), Noctor (ii), Thomas
R.J. Ruane - Bailey

P. Sadra - Pleasance
M. Saeed - Lough
K.C. Saunders - Venn
F.C. Schwende - James
E. Sharp - Noctor (i)
M. Shepherd - Browne
D.S. Siebert - Poole
P. Solanki - Mason

W. Speed - Enos, Simmonds
N. Stevens - Hutt
J. Stubbs - Mohler
J. Taberner - Wilson
S.C. Tan - Hutt
S. Taylor - DiDonato
P. Teale - Seymour
K. Tennant - Hill (i), (ii) & (iii), Bentley, Browne
T. Thompson - Burnett

D.J.K. van der Stel - van Amsterdam
H.J.C. van Schie - van Amsterdam
T. Wangsa - Hsu
B. Warrack - DiDonato
S.A. White - Hill (iii)
D. Wilkinson - Bayliss
V.J.C. Willson - Bayliss
S.A. Wood - James
K. Woodward - Taylor, Westwood
V. Wroblewski - Mohler
A.S. Yuan - Gilbert

Section #A

PEPTIDE-TYPE AGENTS

The suffix '**n**' as in #A-7n distinguishes a 'NOTE' from main articles. COMMENTS (starting on p. 59) comprise Forum discussion remarks and Editor's 'annotations'.

Overleaf: some general literature (not bioanalytical)

Some general literature [see also ref. list in art. #A-1]

'Amino Acids, Peptides and Proteins', in the RSC Specialist Periodical Reports series; Vol. 26 (1995) includes surveys of peptide hormones and cyclic/modified/conjugated peptides.

In **Adv. Drug Res.**- Dutta, A.S. (1991) **21**, 147-286: 'Design and Therapeutic Potential of Peptides'. McMartin, C. (1992) **23**, 41-106: 'Pharmacokinetics of Peptides and Proteins...'. Fauchère, J-L. & Thurieau, C. (1992) **23**, 128-159: 'Evaluation of the Stability of Peptides and Proteins as a Tool in Peptide Drug Design'.

Entries in Sect. #ABC (p. 161) reflect the wide **therapeutic span** (prefix ° denotes a peptide-type drug).

Therapeutic usefulness may blossom for some gene-coded agents:- 'Antimicrobial Peptides' [Ciba Foundation Symp. #186 (1994); ed. J. Marsh & J.A. Goode; Wiley, Chichester, 283 pp.]. They are of diverse occurrence and chemical type, e.g. toad-skin magainins.

For hirudin (cf. art. #A-2), a review on its clinical potential gives good chemical and 'bio' background: Johnson, P.H. (1994) Annu. Rev. Med. **45**, 145-177.

'The Comparative Physiology of Regulatory Peptides' (1989; ed. S. Holmgren), Chapman & Hall, London, 392 pp.

'Cell Signalling: Experimental Strategies' [Vol. 21 (1991) of PRESENT SERIES, ed. E. Reid *et al.*] embraces IL's and other cytokines (p. 32), PA (p. 323), etc.; see p. vii and Index. Also it has a notable review (LeVine & Brown, with 85 refs.; not specially for peptide ligands) on receptor investigation as featured too, with guidance on binding assays, in Vol. 13 (1984; Plenum), 'Investigation of Membrane-located Receptors'. The calcitonin theme of the Vol. 13 art. by J.M. Moseley & co-authors is informative in respect of Hsu's art., #A-4 below.

#A-1

UNIQUE ISSUES AND ANALYTICAL METHODS FOR PHARMACOKINETIC AND METABOLISM STUDIES WITH PROTEIN/PEPTIDE PHARMACEUTICALS

Marjorie A. Mohler†, ¤Victor Wroblewski, John Stubbs & Bobbe Ferraiolo

R.W. Johnson Pharmaceutical Research Institute,
P.O. Box 300, Route 202, Raritan, NJ 08669, U.S.A.

and ¤Bayer Corporation, West Haven, CT 06516, U.S.A.

Analytical methods in disposition studies for biotechnology products range from the straightforward to the elaborate. Many protein-analysis methods do not positively identify the analyte. IA's and bioassays are often used. IA's are relatively specific and easy to perform, while* in vitro *bioassays are more difficult but illuminate biological effects. Radiolabelled proteins can be advantageous in respect of detection and analysis of the molecular form of the protein; but many limitations exist. Along with these methods, additional analysis by MS, chromatography or electrophoresis can help. MS can aid protein analyte identification, while HPLC in IE, RP and size exclusion modes may help in isolating protein products from complex biological matrices. PAGE is a notably powerful and much used technique for protein mixtures.*

PK parameters vary widely for proteins, maybe due to minor molecular variations; the profile of decline partly depends on assay sensitivity and sampling duration. Metabolism reflects proteolysis and possibly adding-on of groups. Degradation at extravascular injection sites may affect bioavailability, for which assay limitations may give artefactual results. Complexing to binding proteins may affect metabolism and/or clearance, and hence activity and toxicity. As the complexes are usually not covalent, their study may need non-denaturing analytical methods; disposition results may suffer if analysis does not distinguish bound and unbound forms. Studies by approaches now outlined show the usefulness for safety evaluation of the data obtainable.

This article aims to survey the unique issues and various analytical methods that are encountered in protein/peptide disposition studies (as more fully discussed elsewhere: [1, 2]).

†addressee for any correspondence

**Abbreviations* [others in later footnotes & Preface): Ab, antibody (mAb, monoclonal); IA, immunoassay; ELISA, enzyme-linked immunosorbent assay; IE, ion-exchange; PAGE, polyacrylamide gel electrophoresis; PK, pharmacokinetic(s). 'Bioassay' may be enzymic.

The issues that need to be addressed in protein disposition studies* are similar to those for conventional small-molecule drugs. The primary aims are to find the fate of the active parent, to identify metabolites and binding proteins, and to determine the mechanisms of clearance. It is also important to identify the sites of catabolism and to correlate the drug's concentration with the pharmacological effect. In metabolism studies a major difference between conventional and protein drugs in biological matrices lies in the analytical methods used to isolate and measure the drug. With conventional drugs, endogenous proteins are removed by precipitation or extraction procedures so that methods such as HPLC or GC can be applied. With protein drugs such isolation approaches are precluded.

IA's and bioassays (sometimes enzymic) are often employed in protein PK and metabolism studies [3]. IA's are relatively specific and easy to perform, while *in vitro* and *in vivo* bioassays are more difficult but provide information regarding biological effects. Radiolabelled proteins can be useful in detecting and analyzing the drug molecular form. Complementary analyses can be of value.- MS can significantly aid identification of the protein analyte, while IE, RP or size-exclusion chromatography serves to separate out proteins from biofluids and electrophoresis to resolve and size-characterize the protein species of interest [4]. It is often useful to employ a combination of analytical techniques for the sake of reliable conclusions.

Antibodies (Ab's) in relation to analysis

As a much used IA approach, ELISA's are sensitive, do not employ radioactive reagents and can be completely automated†. The main drawback of IA's is lack of positive identification of the analyte, e.g. its exact biochemical form or sequence. There can also be interferences in IA's from diverse endogenous or exogenous materials including Ab's, binding proteins, metabolites and non-specific interferants from the biological matrix [5]. Ab's can also have a major impact on quantitation of the administered protein. In pre-clinical studies with recombinant human tissue factor, the presence of Ab's to the factor interfered with ELISA quantification of plasma levels. Conversely, recombinant tissue factor in plasma interfered in quantitation of tissue factor Ab's using a similar assay method.

A unique issue encountered in protein disposition studies is the production of Ab's. Their presence in an animal model with a human protein drug can be reckoned on; but Ab production has also been observed in clinical studies [6-9]. The immunogenicity of a protein is governed by its features including primary/secondary/tertiary structure. Stepping-up of dosing amounts, frequency or duration increases the cumulative dose,

*In contexts such as this, 'protein' embraces small polypeptides.
†For automated ELISA see S.A. Westwood's art., #D-3.

so making an immunogenic response more probable. The route of administration can also affect the apparent immunogenicity of a protein drug. Thus, in a Phase I clinical trial of human IL-2* where the cytokine was administered to patients either s.c. or i.v., Ab titres to IL-2 were detected regardless of route, but were highest with s.c. injections [10]. Disease states can also affect Ab production. Thus, after treatment with INF-α, patients with renal cell carcinoma on average produced Ab's within 2 months, whereas patients with hairy cell leukemia did not produce Ab's till nearly one year after treatment started.

The presence of Ab's can confound interpretation of the results of metabolism and PK studies by affecting distribution, metabolism or excretion, or can affect efficacy studies by inhibiting the biological activity of the protein drug. Thus, if circulating Ab's are present the clearance of recombinant human proteins/peptides may be decreased (rCD4, t-PA, INF-α) or increased (mAb's to INF-α) [10-12]. Interference with efficacy can occur if the Ab is of neutralizing, not non-neutralizing, type, as usually identified by an *in vitro* bioassay wherein the biological activity of the protein is suppressed. Although *in vitro* bioassays are easier to perform than *in vivo* bioassays, neutralizing Ab's identified by an indirect *in vitro* assay may not predict neutralization of a desired pharmacological effect [10]. Thus, where uterine cell cultures are exposed to an Ab combined with recombinant human relaxin and the expected rise in intracellular cAMP is inhibited, the Ab is defined as neutralizing. However, it is unknown whether these Ab's behave the same *in vivo* as *in vitro*. Hence *in vivo* bioassays may be more accurate in identifying neutralizing Ab's. Since induction of Ab's may markedly affect the outcome of protein disposition studies, the presence or absence of Ab's to the protein drug must be carefully monitored to fully understand the results.

Glycosylation aspects.- While Ab production can influence clearance of a protein, the degree of glycosylation can also affect clearance. The presence of terminal sialic acid on the oligosaccharide chains protects many serum glycoproteins from clearance by asialoglycoprotein receptors on hepatocytes that recognize terminal galactose. Winkelhake & Nicolson [13] found that non-denatured radioiodinated desialylated IgG Ab's were taken up in the liver and had shortened half-lives. Moreover, they showed that removal of galactose residues (exposed by desialylation) restored the survival characteristics to normal and eliminated liver localization. These later observations led to the conclusion that uptake by asialoglycoprotein receptors does not have a general role in IgG metabolism; possibly

---

**Further abbreviations:* IL-2, interleukin-2; INF-α, interferon-α; t-PA, tissue plasminogen activator; i.v., intravenous; s.c., subcutaneous. Prefix r = recombinant. '*In vitro*' embraces *ex-vivo*.

IgG's have too few exposed galactose residues to participate in asialoglycoprotein receptor-mediated catabolism [14]. Studies with human rt-PA demonstrated that its removal from the blood by the liver depends significantly on the nature of the oligosaccharide structure on the molecule [15]. Other studies have shown, by liver subfractionation, that rt-PA distributes 55% in parenchymal cells, 40% in endothelial cells and 6% in Kupffer cells [16]. The parenchymal-cell association of rt-PA was not mediated by a carbohydrate-specific receptor; however, the association with endothelial cells could be inhibited 80% by ovalbumin (a mannose-terminated glycoprotein).

Studies with radiolabelled proteins

An alternative to relying exclusively on IA's or bioassays in protein disposition studies involves using radiolabelled proteins. While these are advantageous for detection, numerous pitfalls must be avoided. If the protein contains a suitable amino acid such as tyrosine or lysine, an external label such as ^{125}I may be used [17-19]. If this is not possible, the protein may be internally labelled by growing the production cell line in the presence of amino acids labelled with ^{3}H, ^{14}C or ^{35}S [19, 20]. Once the labelling method has been selected, it must be optimized and the resulting labelled protein fully characterized to be sure that it is essentially identical in physico-chemical and biological properties to the unlabelled protein [21]. This full characterization is best done in the appropriate biofluid or tissues *in vitro* to gain insight into the behaviour of the radiolabelled protein and to begin to develop the analytical methods that will be used for samples from *in vivo* studies. Such analytical methods for characterization include TCA precipitation, electrophoresis, immunoaffinity purification and HPLC.

Mass balance/tissue distribution studies employ radiolabelled proteins/peptides routinely. When tissue distribution was determined in mice for ^{125}I-TNF and ^{3}H-TNF, both were found to have accumulated mainly in the liver [19]. After 4 h, however, the stomach also accumulated a significant amount of ^{125}I radioactivity - but not of ^{3}H, suggesting that the gastric accumulation was secondary to dehalogenation and release of free iodide, which is known to accumulate in thyroid, G-I tract and skin. A major limitation to using radiolabelled proteins/peptides to study disposition is re-utilization of the labelled amino acid in the replacement synthesis of endogenous protein such as albumin. An advantage of having a tyrosine residue as ^{125}I-labelling site is that no tRNA exists for ^{125}I- tyrosine, so there can hardly be labelled protein unrelated to the parent.

Whole-body autoradiography is also gaining popularity for protein-disposition studies. WBA has added a visual and quantitative tool for understanding the relative distribution of radiolabelled proteins/peptides in the intact animal, using a computer-linked gamma camera for digitized monitoring [22]. A limitation of the method is the choice of radionuclide label: if beta-emitting, the particles penetrate tissues only slightly and are undetectable externally. Hence a gamma photon energy of ~150 KeV is ideal [23]. A second WBA method entails whole-carcass sectioning after radiolabel dosing, then quantitative autoradiography [24]; it allows use of beta-emitters. Within-organ discrimination is feasible: thus ^{125}I-r-huHGF* was mainly localized, 5 min after i.v. dosing, in kidney cortex rather than medulla [25]. Checks on disposition results may include: pre-treatment with NaI or dehalogenase inhibitors; distribution of NaI or labelled amino acid; a chase with excess unlabelled protein; repeat with a second label; TCA-precipitable radioactivity; SDS-PAGE, HPLC, bioassay.

Protein-drug metabolism, and implications of protein binding

There are numerous reports of binding proteins for protein therapeutics, both stimulatory or inhibitory in pharmacodynamic respects and maybe specific to species and disease states [26]. They may greatly influence disposition profiles and present, in metabolic studies, some unique issues and analytical challenges. Depending on the specificity of the Ab employed, an IA may not detect protein drugs complexed to specific or non-specific binding proteins [5]. Where concentration values for bound and unbound drug are each wanted, analytical methods that do not disrupt protein-protein complexes are required (review: [26]).

Protein drugs give rise to metabolites, as for conventional drugs but harder to investigate due to the many potential sites of metabolism, their potency and, for most protein drugs, their physico-chemistry; moreover, many of the available analytical methods entail relatively non-specific identification [27]. Peripheral metabolism of protein drugs predominantly involves hydrolysis of the amino acid backbone by a combination of exo- and endo-proteinases [28]. Information obtained through the IA and immuno-extraction techniques depends on the ability of an Ab to recognise proteolytic fragments. Similarly, information from use of radiolabelled tracers is limited by both the type of label and the position(s) within the primary sequence which are labelled. Due to the limitations of the analytical methods currently available, the complexity of protein degradation *in vivo* is not well understood.

The enzyme systems involved in the metabolism of dosed exogenous proteins have not been well described [29]. Studies with purified enzymes, subcellular fractions, whole cells or isolated organs have helped in defining proteolytic mechanisms

---

**Further abbreviations*.- huHGF, human hepatocyte growth factor; TNF, tumour necrosis factor. CE, capillary electrophoresis; WBA, whole-body autoradiography.

and cleavage products that may be important *in vivo*. However, the physiological relevance of *in vitro* information to a protein administered to an animal or human is not easy to establish experimentally. Complete assessment of the metabolism of a protein *in vivo* requires a combination of approaches. Amongst important considerations are the selection of an appropriate radiolabelled tracer, characterization of multiple antisera that will detect different portions of the protein sensitively and specifically, and ability to separate the parent molecule from degradation products.

Human insulin metabolism has been studied more extensively than for any other peptide hormone (review: [30]). Insulin binding to the insulin receptor is thought to be the essential step in insulin clearance from the circulation with subsequent degradation of the molecule [31]. The major enzyme involved in the cellular metabolism of insulin is the insulin protease. Although extensive work on insulin metabolism *in vitro* has been carried out, with biochemical characterization of the enzymes involved and products formed, there is only limited information regarding the metabolism of insulin *in vivo*. Metabolic studies have also been done for human relaxin [32], human ANF [33] and human INF-γ [34].

Human relaxin-2 (r-hRlx-2) studies illustrate the need for multiple analytical methods to characterize the disposition of a protein therapeutic agent. The molecule comprises an A chain 24 amino acids in length and a 33-amino acid B chain. After an i.v. dose of 0.5 mg/kg to rhesus monkeys, hRlx-2 was isolated from plasma by an mAb affinity purification method; the rRlx-2 thus purified was then subjected to RP-HPLC, and changes in the B chain were observed. The column fractions were then submitted to MS for identification. At 60 min after i.v. administration, $<70\%$ of the recovered immunoreactive rRlx-2 was intact B chain. The remainder of the immunoreactive rRlx-2 was accounted for by losses of 1, 4 or 6 amino acids from the C-terminus.

CONCLUDING REMARKS

Current analytical methods are not entirely suitable for routinely analyzing recombinant proteins in biological matrices. Radiolabelled protein may have limited stability, and reincorporation of amino acids labelled with ^{3}H, ^{14}C or ^{35}S into new proteins/peptides has been known to occur. Since proteins contain numerous antigenic combining sites for Ab's, IA's will not always provide positive identification of the analyte; for correct interpretation of results a combination of techniques is often required. The extent of characterization necessary in protein metabolism studies needs to be settled on a case-by-case basis. The rationale for doing more conventional metabolism studies from a regulatory or safety viewpoint may be debatable in cases where the protein agent is identical to the endogenous protein and is being used only for replacement therapy. However, conventional metabolism studies may be of greater concern for second-generation molecules that have been substantially altered from the endogenous form of the molecule.

Future goals for protein disposition studies will need to focus on 'growth areas' of analytical method development, notably HPLC with tandem MS, automated extraction methods and CE. While LC-MS-MS is becoming established for bioanalysis of low mol. wt. drugs and for protein characterization and mol. wt. determination in purified systems, it is still in the evolutionary stage for proteins in biological matrices - for which CE is an emerging and perhaps more promising approach, especially CE-MS.

References († denotes a complementary ref. added by Editor)

1. Ferraiolo, B.L. & Mohler M.A. (1992) in *Protein Pharmacokinetics and Metabolism* (Ferraiolo, B.L., Mohler, M.A. & Gloff, C.A., eds.; 292 pp.), Plenum, New York, pp. 1-33 *(Goals and Analytical Methodologies for Protein Disposition Studies).*
2. Ferraiolo, B.L. & Mohler, M.A. (1992) in *Handbook of Experimental Pharmacology*, Vol. ~100±, *Pharmacokinetics of Drugs*, Springer, Berlin, pp. 99-123. (Eds.: P. Welling & L. Balant.)
3. Gosling, J.P. (1990) *Clin. Chem. 36,* 1408-1427.
4. MacNamara, E.M. & Whicher, J.T. (1990) *Electrophoresis 11,* 376-381.
5. Chen, A.B., Baker, D.L. & Ferraiolo, B.L. (1991) †*Pharmacokinet. Pharmacodyn. 3,* 53-71; also appears (same pp.) in a book *Peptides, Peptoids and Proteins* (1991; Garzone, P.D. *et al.* eds.), Harvey Whitney, Cincinnati, OH.
6. Pringle, P.J., Hindmarsh, P.C., DeSilvio, L., Kurtz, A.B. & Brook, C.G. (1989 *J. Endocrinology 121,* 193-199.
7. Steis, R.G., Smith, J.W., Urba, W.J., Clark, J.W., Itri, L.M., Evans, L.M., Schoenberger, C. & Longo, D.L. (1988) *N. Eng. J. Med. 318,* 1409-1413.
8. van Haeften, T.W. (1989) *Diabetes Care 12,* 679-688.
9. Gray, R.S., Cowan, P., di Mario, U., Elton, R.A., Clarke, B.F. & Duncan, L.J. (1985) *Br. Med. J. 290,* 1687-1691.
10. Working, P.K. (1992) in *same book as for* 1., pp. 73-93 *(Potential Effects of Antibody Induction by Protein Drugs).*
11. Larsen, S.M., Brown, J.P., Wright, P.W., Carrasquillo, J.A., Hellstrom, I. & Hellstrom, K.E. (1983) *J. Nucl. Med. 24,* 123-129.
12. Pimm, M.V., Perkin, A.C., Armitage, N.C. & Baldwin, R.W. (1985) *J. Nucl. Med. 26,* 1011-1023.
13. Winkelhake, J.L. & Nicolson, G.L. (1976) *J. Biol. Chem. 251,* 1074-1080.
14. Mariani, G., Ferrante, L. & Rescigno, A. †(1991) in *NATO ASI Ser., Ser. A 221,* 189-202; & (1986) *Radioaktiv Isotope im Klinik und Forschung*, Egerman, Vienna, 741 pp. *(Kinetic Modeling for the Analysis of Digital Images from...).*

15. Hotchkiss, A., Refino, C.J., Leonard, C.K., O'Connor, J.V., Crowley, C., McCabe, J., Tate, K., Nakamura, G., Powers, D., Levinson, A., Mohler, M. & Spellman, M. (1988) *Thromb. Haemost. 60,* 255-261.
16. Bakhit, C., Lewis, D., Billings, R. & Malfroy, B. (1987) *J. Biol. Chem. 262,* 8716-8720.
17. Regoeczi, E. (1987) *Iodine-Labeled Plasma Proteins,* Vol. 1, CRC Press, Boca Raton, FL, 102 pp.; cf. †*Am. J. Physiol. 252,* R457-R461.
18. Woltanski, K.P., Besch, W., Keilacker, H., Ziegler, M. & Kohnert, K.D. (1990) *Exp. Clin. Endocrinology 95,* 39-46.
19. Ferraiolo, B.L., Moore, J.A., Crase, D., Gribling, P., Wilking, H. & Baughman, R.A. (1988) *Drug Metab. Disp. 16,* 270-275.
20. Cossum, P.A., Dwyer, K.A., Roth, M., Chen, S.A., Moffat, B., Vandlen, R. & Ferraiolo, B.L. (1992) *Pharm. Res. 9,* 415-420.
21. Bennett, H.P. & McMartin, C. (1979) *Pharmacol. Rev. 30,* 247-292.
22. Mariani, G. & Strober, W. (1990) in *Receptors and the Action of Antibodies* (Metzger, H., ed.), American Society of Microbiology, Washington D.C., pp. 94-177 *(Immunoglobulin Metabolism).*
23. Chilton, H.M. & Witcofski, R.L. (1986) in *Nuclear Pharmacy.* Lea & Febiger, Philadelphia, pp. 1-35. *(An Introduction to the Clinical Application of Radiopharmaceuticals).*
24. Johnston, R.F., Pickett, S.C. & Barker, D.L. (1990) *Electrophoresis 11,* 355-360.
25. Zioncheck, T.F., Richardson, L., DeGuzman, G.G., Modi, N.B., Hansen, S.E. & Godowski, P.J. (1994) *Endocrinology 134,* 1879-1887.
26. Mohler, M.A., Cook, J.E. & Baumann, G. (1992) in *same book as* 1., pp. 35-71. *(Binding Proteins of Protein Therapeutics).*
27. Moore, J.M. & Wroblewski, V.J. (1992) in *same book as* 1., pp. 93-126 *(Pharmacokinetics and Metabolism of Protein Hormones).*
28. Bond, J.S. & Butler, P.E. (1987) *Annu. Rev. Biochem. 56,* 333-[364.
29. Harris, R.B. (1989) *Arch. Biochem. Biophys. 275,* 315-333.
30. Duckworth, W.C. (1988) *Endocrin. Rev. 9,* 319-344.
31. Terris, S. & Steiner, D.F. (1975) *J. Biol. Chem. 250,* 8389-8394.
32. Ferraiolo, B.L., Winslow, J., Laramee, G., Celniker, A. & Johnston, P. (1991) *Pharm. Res. 8,* 1032-1038.
33. Koehn, J.A., Norman, J.A., Jones, B.N., LeSueur, L., Sakane, Y. & Ghai, R.D. (1987) *J. Biol. Chem. 262,* 11623-11627.
34. Ferraiolo, B.L., Mohler, M.A., Cook, J.E., Chen, A., Reed, B., O'Connor, J. & Keck, R. (1990) *Pharm. Res. 7,* S-46 (Abstract BIOTEC 2028).

#A-2

IMMUNO AND OTHER 'BIOLOGICAL' APPROACHES IN PHARMACOKINETICS

[1]Daniel Gygax†, [1]Luigi Botta, [2]Markus Ehrat, [1]Peter Graf, [3]Gilbert Lefèvre, [2]Peter Oroszlan & [1]Christian Pfister

[1,3]Dept. of Bioanalytics & Pharmacokinetics [2]Corporate Analytical Research

all Ciba-Geigy Ltd.:
[1,2]Basle, Switzerland; [3]Rueil Malmaison, France

For the evaluation and interpretation of PK data, reliable quantitative determinations are a requirement that can be met only by well characterized and fully validated analytical methods. To this end, an approach based on an integrated and automated FOBIA system for IA's is being established, with consideration herein of this system's performance characteristics when used in monitoring recombinant hirudin, a highly potent and selective inhibitor of human thrombin. The PK assessment of biotechnology drugs shows peculiarities which are related to mol. wt., the integrity of the complex 3-D structure and the immunogenic potential.*

During a 3-month i.v. tolerability study in dogs, some dogs developed Ab's against r-hirudin. The hirudin-Ab complex accumulated in plasma, and the apparent hirudin plasma concentrations were hence much higher than expected from single-dose kinetics. Hirudin captured by Ab's showed an extended ½-life and the complex was still pharmacologically active as demonstrated by the observed increase in thrombin time. This illustrates that only appropriate analytical methods allow adequate monitoring and PK characterization of biotechnology drugs in biological samples.

This overview discusses immunoanalytics as a tool for drug monitoring, and considers PK assessment of a recombinant DNA-derived therapeutic protein, r-hirudin (CGP 39 393). Due to size and charge, r-hirudin is eliminated mainly by glomerular filtration. But only a fraction of the hirudin dose seems to be reabsorbed at the proximal tubule by luminal endocytosis and hydrolyzed by lysosomal enzymes, leaving ~50% of the dose to be excreted in the urine. Thus, renal clearance of r-hirudin in the absence of renal insufficiency appears to depend primarily on glomerular filtration rate.

---

†addressee for any correspondence
**Abbreviations include* [others in Preface]: (m)Ab, (monoclonal) antibody; FOBIA, fibre optic biospecific interaction analysis; IA, immunoassay; PK, pharmacokinetic; r-, recombinant-.
Some abbreviations have been introduced editorially.

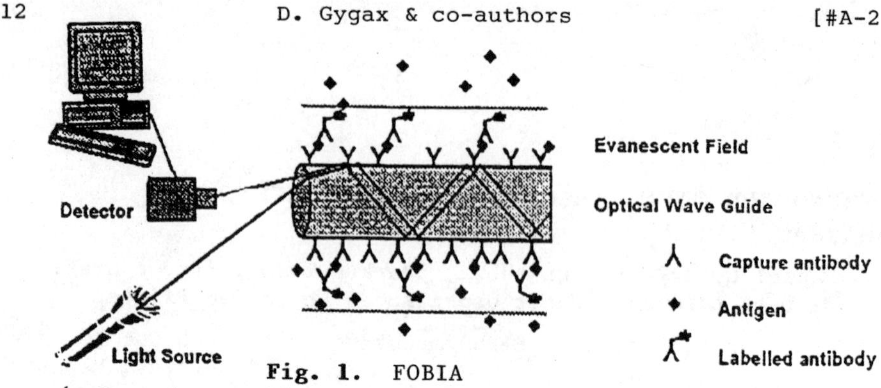

Fig. 1. FOBIA (schematic; working vol. injected = 1.5 ml). Typically the 70 µl flow cell with a pre-silanized fibre (optical wave guide) is filled with capture Ab solution (1 mg/ml); after a 2 h incubation non-specific sites are blocked with BSA for 30 min.

USE OF AN IMMUNOSENSOR FOR DRUG MONITORING

Reliable values for drugs and metabolites, requisite for evaluating and interpreting PK data and for monitoring, demand well characterized methods with full validation [1] that must include specificity, sensitivity, precision and sampling [2]. Such features interrelate with laboratory productivity where saving time is now a paramount objective for drug development.

To meet the productivity demand an IA concept based on an integrated and automated fibre optic biospecific interaction analysis system is being implemented. FOBIA (Fig. 1) comprises three modules: a precise sample and liquid handling module; an evanescent wave fibre optic immunosensor module; and a software module for data reduction and IA optimization [3]. The sensing approach allows real-time monitoring of biospecific interaction such as antigen-Ab binding and so one can characterize affinity by extracting on/off rates and equilibrium constant of binding (Fig. 2).

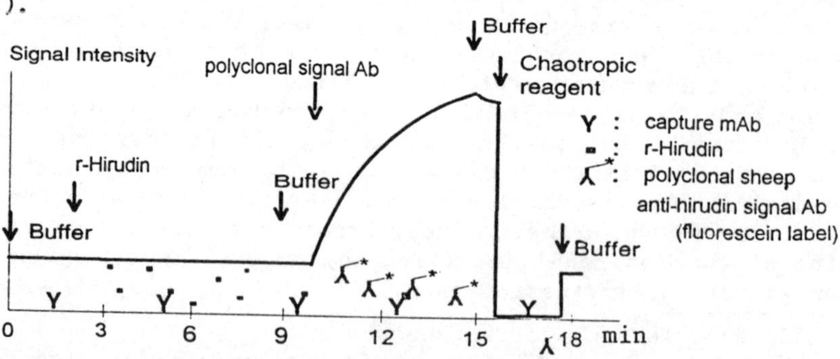

Fig. 2. Regenerable r-hirudin sensor. Emission signals (V) accompanying a cycle of the sandwich r-hirudin IA on FOBIA. Continuous flow over the fibre sensor (1 ml/min) throughout the assay. *Arrows* indicate changes of supplied reagent solutions.

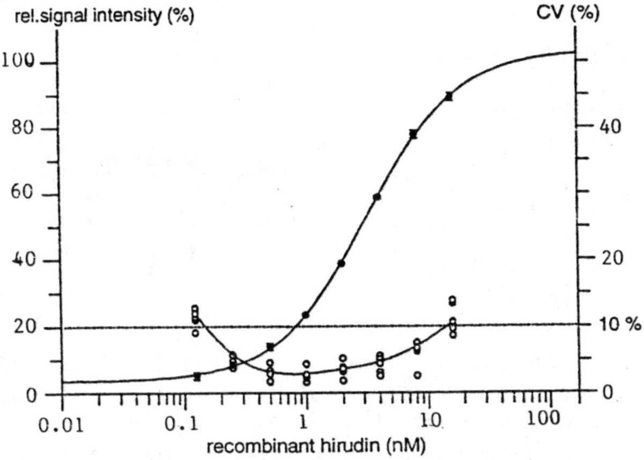

Fig. 3. FOBIA standard curve and fibre-to-fibre precision profile for recombinant hirudin. Within 10% C.V. the estimated working range for 10-fold diluted working solution is 0.12 to 16 nM.

Computer-aided assay development (CAAD) can simulate an IA's optimal performance. Based on experimentally determined antigen-Ab binding constants, by a sequence of steps tracer concentration, sensitivity and assay dynamics can be optimized efficiently and afterwards validated by experiment. Shifting the optimization work from the experiment to the computer may substantially decrease the time used for assay development.

The performance of FOBIA underwent a rigorous test when r-hirudin plasma concentrations from a large clinical trial were measured. At 10% precision the working range was between 0.12 and 16 nM, the lower limit representing the assay sensitivity (Fig. 3). FOBIA allows dilution of plasma samples up to ×1000 without deviation from linear relationship between found and given concentrations. To evaluate recovery and precision, fibre-to-fibre data (n = 7) were collected during 5 consecutive days with a freshly prepared fibre each day. The mean recovery was 108.5 ±6.1% (S.D.). The precision being good (C.V. <10%), singlicate assays sufficed for plasma samples. FOBIA, as a fully automated and computer controlled system, can be set for serial analyses, each taking only 18.5 min. The working range being wide, most samples could be assayed directly, avoiding unnecessary dilution steps. Hence a high weekly throughput of up to 360 plasma samples using the same fibre was achievable.

IMMUNOANALYTICS IN PK STUDY OF RECOMBINANT HIRUDIN

For biotechnology drugs the PK parameters are generally as for small organic compounds: clearance, volume of distribution, half-life, AUC, maximum concentration, minimum concentration and fraction of dose absorbed [4]. The kinetics of a drug is inherently interrelated with its structure. For proteins the critical determinants are mol. wt. and conformation. The mol. wt. of a therapeutic protein is in the range 3-160 kDa. The integrity of a protein in the body is governed by the spatial arrangement of amino acids and carbohydrates.

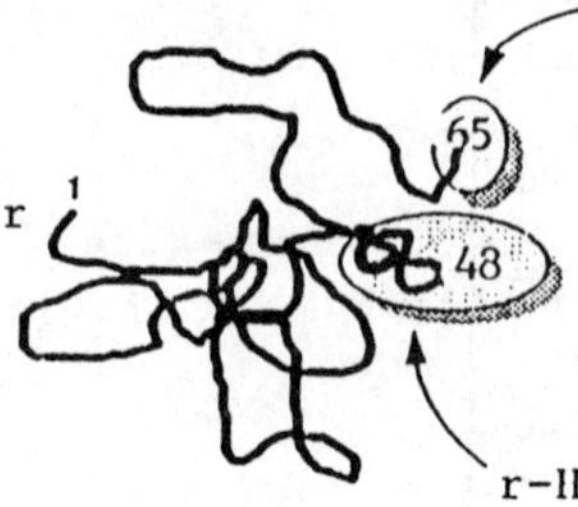

Fig. 4. A mouse mAb collecting free and metabolically changed r-hirudin was used for measuring hirudin in biological samples. The mAb was used for analyses by FOBIA and ELISA.

Native hirudin is produced by the salivary glands of the leech *Hirudo medicinalis,* whereas r-desulphatohirudin is produced in yeast [5]. Recombinant hirudin is a 7 kDa single-chain polypeptide consisting of 65 amino acids with 3 -S-S-bridges; 'desulphato' means that the r-form lacks the sulphate at tyr-63 that arises by post-translational modification of the native hirudin. Hirudins are known to inhibit thrombin very strongly and selectively; the K_i of r-hirudin is 2.1×10^{-13} M. Hence r-hirudin will be developed as an anticoagulant drug for arterial and venous thrombosis.

Quantitation of hirudin

Analyses for r-hirudin were originally done by an **ELISA** which, like FOBIA, is a structural assay [6]. A solid phase is coated with mouse anti-hirudin mAb's so as to capture hirudin. Then r-hirudin is added as standard, QC or sample. Finally, polyclonal sheep Ab's conjugated to alkaline phosphatase are used as signal Ab's. The working range derived from the precision profile (concentrations within 10% C.V.) was, for the working solution, 0.03 nM (LLOQ†) to 0.3 nM. For undiluted samples there was a linear correlation up to 265 nM (1:1100 for assay) of r-hirudin. Throughput was ~100-120 samples per week.

The capture mAb recognises an epitope between residues 41 and 47 on the surface of r-hirudin (Fig. 4). This recognition is dependent on the conformation of the epitope which is also involved in the binding of thrombin; so thrombin-bound hirudin is not captured by the mAb. Since free, but not necessarily unchanged, hirudin is measured by the ELISA, C-terminus truncated hirudins are also recognized.

'TCA', the thrombin chromogenic assay [7], is a second approach, directed to the biological function of hirudin, which binds to thrombin added in excess to the assay system. The hydrolyzing activity of the residual thrombin is then related to the hirudin concentration. The more hirudin there is in the sample, the less is the availability of active thrombin and the less product is formed. The assay is specific for free hirudin. Truncated r-hirudin (tyr-63) inhibits thrombin activity *in vivo* but not *in vitro;* hence the concentration of functional hirudin speciies is underestimated.

† lower limit of quantification

Fig. 5. Structural *versus* functional assay: correlation between ELISA and TCA for r-hirudin in plasma of healthy volunteers (i.v. infusion, 0.3 mg/kg per h, for 6 h).

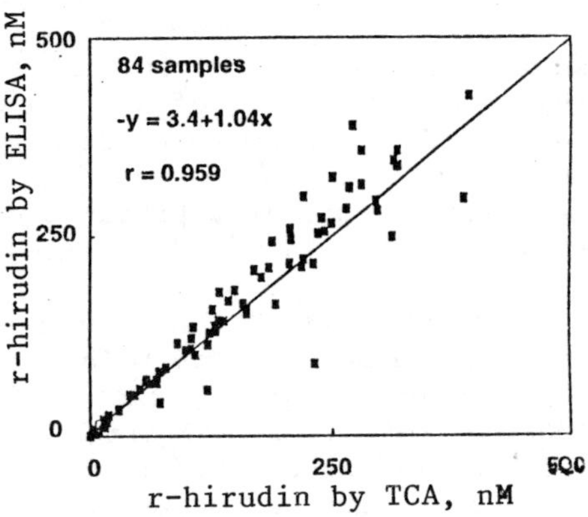

The structural (ELISA) and functional (TCA) assays were found to correlate very well (Fig. 5). We conclude that the conformational integrity of hirudin in the body, as indicated by Ab recognition, reflects biological activity, i.e. both methods measure the same active molecules (hirudin and potential metabolites). Thus, importantly, PK parameters derived from IA data also characterize the active molecule(s) in the plasma.

IN VIVO HANDLING OF INJECTED HIRUDIN

The plasma concentration-time profile of r-hirudin after an i.v. bolus injection of 0.3 mg/kg to healthy volunteers shows a multi-exponential decay; ~90% of the hirudin disappears from the systemic circulation within 2 h. This suggests either that elimination is much faster than distribution or that the volume of distribution is very large. That of the central compartment, estimated from plasma concentrations immediately after injection, is ~5 L, which roughly equals the plasma volume. That of central and potential peripheral compartments at steady-state is only ~3 times larger, 17 L. This volume corresponds roughly to the extracellular space, viz. plasma, the fluid outside the cells of various tissues, and lymph. These relatively low volumes of distribution may indicate that the rapid initial drop in hirudin concentration is caused primarily by elimination.

For elimination of **proteins** [8-10] the general routes are renal filtration, hepatic extraction, and degradation or complexation - the latter processes being widespread and mediated by specific enzymes, binding proteins or receptors. Thus Ig's, tissue plasminogen activator and asialoglycoproteins are very efficiently eliminated by specific receptors in the liver.

Depending on mol. wt. and charge, proteins are filtered at the glomerulus. Those with mol. wt. <20-50 kDa are easily filtered from the systemic circulation at the glomerulus and afterwards reabsorbed by cells of the proximal tubule. Reabsorption is very efficient and therefore proteins are normally not excreted in the urine. Subsequent to reabsorption the proteins are degraded to smaller peptides and individual amino acids. Only these basic building blocks re-enter the circulation.

An additional clearance mechanism has been described for insulin, epidermal growth factor and atrial natriuretic factor. These proteins are extracted from peritubular capillaries by active transport, secretion, degradation by peptidases, or receptor-mediated clearance. Reabsorption as well as this extraction may be regarded as metabolic processes aimed at providing the body with basic building blocks for biosynthesis.

Hirudin.- After administration to healthy volunteers, 40-50% of a dose of r-hirudin is found in the urine. The renal clearance is ~80 ml/min. The difference between creatinine clearance and renal clearance is assumed to account for the tubular reabsorption of r-hirudin. This process seems not to operate very efficiently for hirudin. The hindered reabsorption of hirudin may partially be explained by its low i.e.p., 3.7. The difference between herudin's systemic clearance and creatinine clearance thus represents the non-renal clearance.

Reabsorption of hirudin in the tubules was investigated in rats by Markwardt & co-authors [11]. They showed that radio-labelled hirudin accumulated in the kidney.

Our current understanding of hirudin clearance is based on the notion that the total clearance is the sum of renal clearance, the metabolic clearance which is essentially reabsorption and non-renal clearance. What kind of elimination processes contribute to the non-renal clearance is not yet known. Elimination from peritubular capillaries and elimination *via* thrombin are processes which are under discussion. Recently, data were published showing that hirudin administered as a complex with thrombin is largely degraded in kidney and liver [12].

Immunogenic potential, maybe with an undesirable immune response, resides in parenterally administered therapeutic proteins, depending on factors such as size, structure, protein denaturation, aggregation, purity, formulation, species and route of administration. The undesirable response may be manifested by allergic reactions or tissue damage [13]. Another aspect, pertinent to this overview, is the elicitation of IgG. These Ab's may be of neutralizing or binding nature and hence alter the PK and dynamics of a therapeutic protein.

Fig. 6. Architecture of an IA to detect Ab's elicited against r-hirudin administered to dogs.

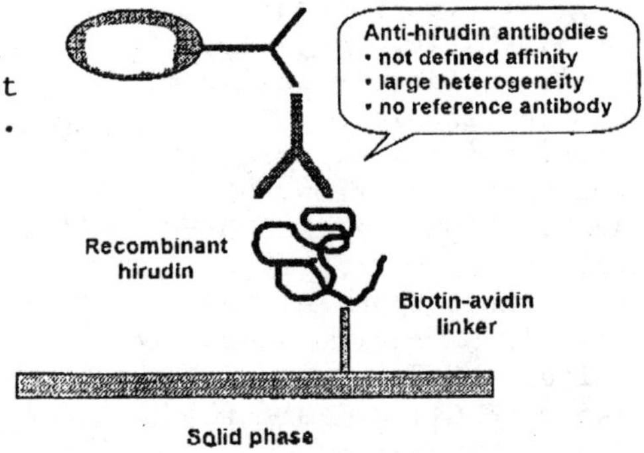

Ab's to hirudin.- Extensive studies in animals and man during pre-clinical and clinical development revealed that r-hirudin has hardly any immunogenicity. Only some dogs in an i.v. tolerability study manifested anti-hirudin Ab's.

The design of an assay to detect anti-hirudin Ab's is based on the capability of hirudin attached to a solid phase to collect anti-hirudin Ab's in plasma (Fig. 6). Quantitative determination of these Ab's is not easy, because anti-hirudin Ab's have no defined affinity, show marked heterogeneity and, accordingly, a reference Ab is by definition not available. As a further complication, especially in repeated dose studies, circulating hirudin may cause false positive results.

After single daily i.v. injections of 25 mg/kg r-hirudin to dogs during a 3-month tolerability study, some of the dogs showed elevated and persistent plasma hirudin concentrations. 'Trough' values on day 44 after a single dose and apparent half-life, and Ab titres on day 25 were as follows.-

Dog 404:	hirudin 6700 nM	with half-life 21 h;	Ab (O.D.) 20
Dog 520:	360 nM	5 h;	9
Dog 538:	30 nM	4 h;	0.5

Whereas dog 538 was non-responsive as identified by the above-mentioned ELISA, dogs 404 and 520 were immune-responsive and evidently had relatively high hirudin concentrations (half-life being prolonged in Dog 404) and Ab titres. Blood coagulation was completely inhibited in all samples from responsive dogs.

If the hirudin plasma concentration was elevated because of elicitation of anti-hirudin Ab's and subsequent formation of immune complexes, the latter should be chromatographically separable from free hirudin. Chromatograms of day 44 plasma samples taken 4 h after hirudin administration showed, for the immune-responsive dogs, two nicely resolved peaks (Fig. 7). The high mol. wt. peak corresponds to the immune complex and the low mol. wt. peak to hirudin. In the non-responsive dog only the hirudin could be detected.

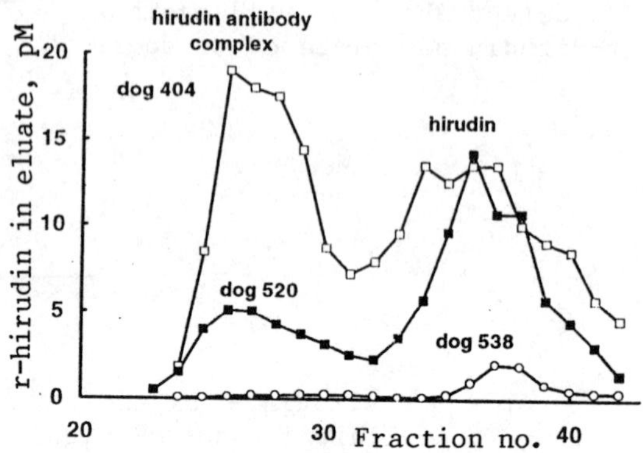

Fig. 7. Gel-permeation on plasma (0.4 ml, ×10): r-hirudin levels in 1.6 ml eluate fractions. For nature of the experiment, and interpretation, SEE TEXT. Sephacryl S 300-HR column (45 × 1.5 cm), equilibrated and eluted with 10 mM sodium phosphate buffer, pH 7.4, containing 146 mM NaCl; 0.8 ml/min flow rate (peristaltic pump).

Altered hirudin PK in the immune-responsive dogs can be explained by the presence of hirudin-binding Ab's. The anti-hirudin Ab complexes have mol. wt. >150 kDa and are therefore not filtered at the glomerulus but are retained in circulation. The kinetics of hirudin is now determined by the kinetics of the immune complexes, which are probably eliminated from the circulation by receptor-mediated hepatic extraction [14]. The Ab's are of non-neutralizing nature; hence the immune complexes still inhibit blood coagulation. Consequently, the complex can be regarded as a hirudin depot, resulting in an extended period of time in which blood is not coagulable.

CONCLUDING COMMENTS

The PK assessment of biotechnology drugs depends greatly on their structure and conformation. This overview has sought to show that these features translate into processes such as: hindered distribution, urinary excretion, tubular degradation, receptor- or cell-mediated elimination and immunogenicity. Consequently only appropriate analytical methods allow adequate monitoring and PK characterization of biotechnology drugs in biological materials.

References [largely review-type; hence titles given]

1. Shah, V.P., Midha, K.K. *& 10 co-authors* (1992) *J. Pharm. Sci. 81,* 309-312. Analytical methods validation: bioavailability, bioequivalence and pharmacokinetic studies.
2. Valcárcel, M. & Rios, A. (1993) *Anal. Chem. 65,* 781A-787A. The hierarchy and relationships of analytical properties.
3. Oroszlan, P., Thommen, Ch., Wehrli, M., Duveneck, G. & Ehrat, M. (1993) *Anal. Meth. Instrumentation 1,* 43-51. Automated optical sensing system for biochemical assays.... [cf.] ELISA.
4. Gloff, C.A. & Benet, L.Z. (1990) *Adv. Drug Delivery Rev. 4,* 359-386. Pharmacokinetics and protein therapeutics.

5. Märki, W.I. & Wallis, R.B (1990) *Thromb. Haemost. 64,* 344-348. The anticoagulant and antithrombotic properties of hirudins.
6. Schläppi, J.M., Vekemans, S., Rink, H. & Chang, J.Y. (1990) *Eur. J. Biochem. 188,* 463-470. Preparation of monoclonal antibodies to hirudin and hirudin peptides.
7. Spannagl, M., Bichler, J., Birg, A., Lill, H. & Schramm, W. (1991) *Blood Coag. Fibrinolysis 2,* 121-127. Chromogenic substrate assay for determination of hirudin in plasma.
8. McMartin, C. (1992) *Adv. Drug Res. 22,* 39-106. Pharmacokinetics of peptides and proteins: opportunities and challenges.
9. Bocci, V. (1990) *Adv. Drug Delivery Rev. 4,* 149-169. Catabolism of therapeutic proteins with implications for drug delivery.
10. Maack, T., Park, C.H. & Camargo, M.J.F. (1992) in *The Kidney: Physiology and Pathophysiology* (Seldin, D.W. & Giebisch, G., eds.), Raven, New York, pp. 3005-3038. Renal filtration, transport and metabolism of proteins.
11. Markwardt, F., Fink, G., Kaise, B., Klöcking, H.P., Nowak, G., Richter, M. & Stürzenbecher, J. (1988) *Pharmazie 43,* 202-207. Pharmacological survey of recombinant hirudin.
12. Bichler, J., Baynes, J.W. & Thorpe, S.R. (1993) *Biochem. J. 296,* 771-776. Catabolism of hirudin and thrombin-hirudin complexes in the rat.
13. Konrad, M. (1989) *TIBTECH 7,* 175-179. Immunogenicity of proteins administered to humans for therapeutic purposes.
14. Working, P.K. (1992) in *Protein Pharmacokinetics and Metabolism* (Ferraiolo, B.L., Mohler, M.A. & Gloff, C.A., eds.), Plenum, New York, pp. 73-92. Potential effects of Ab production by protein drugs.

#A-3

ANALYTICAL STRATEGIES FOR A NEW PEPTIDE DRUG

Richard F. Venn†, ¤Geoff Barnard, Paul V. Macrae & Kenneth C. Saunders

Drug Metabolism Dept., Pfizer Central Research, Sandwich CT13 9NJ, U.K. | ¤Diagnostics Research, Endocrine Unit, Duthie Bldg., Southampton General Hospital, Southampton SO9 4XY

A sensitive and selective assay was needed for the clinical stage of developing UK-X, a Pfizer dual enzyme inhibitor given orally with doses as low as 25 mg and expected to have low bioavailability. A method for plasma with a 0.5 ng/ml LLOQ was developed using HPLC-MS-MS(APcI) after extraction and derivatization with BF_3/methanol; but throughput was lower than that needed for further studies. The molecule being peptide-like and not metabolized, we developed an IA ('DELFIA'; plasma and urine LLOQ's 0.2 and 5 ng/ml) - a 96-well plate-based competitive IA - with time-resolved fluorescence as end-point and a europium label introduced using isothiocyanate chemistry. This IA has immense throughput; the MS method was used to validate its selectivity.*

While this IA was under development, alternative approaches were tried including HPLC-MS-MS with or, later, without prior derivatization, and HPLC-EC. All provided the sensitivity and selectivity required; but only HPLC-MS-MS(ISP) gave sufficient robustness and speed, meeting our precision/accuracy criteria.

UK-X is a Pfizer dual enzyme inhibitor under development (Fig. 1). For supporting the clinical stage, a sensitive and selective assay was judged to be required, since initial doses as low as 25 mg were planned and low bioavailability was anticipated. For prelinical studies a bioassay had been developed, based on UK-X inhibition of a rat peptidase[θ]; but its LLOQ*

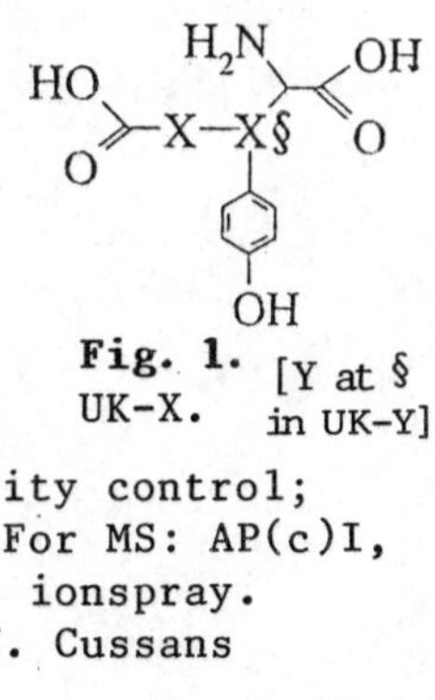

Fig. 1. UK-X. [Y at § in UK-Y]

†addressee for any correspondence
Abbreviations: Ab, antibody; AmAc, ammonium acetate; EC, electrochemical (detection); Eu, europium; FMOC, 9-fluorenylmethylcarbonyl chloride; IA, immunoassay; i.s., internal standard [UK-Y was the one used]; LLOQ, lower limit of quantification; LOD, limit of detection; MTBE, methyl tert-butyl ether; PPD, pneumococcal polysaccharide derivative; QC, quality control; TEA, triethylamine; TFA, trifluoroacetic acid. For MS: AP(c)I, atmospheric pressure (chemical) ionization; ISP, ionspray.
θpersonal communication from a colleague, Dr N.J. Cussans

was ~10 ng/ml; preliminary studies in man showed this to be inadequate. Thus there was an urgent need for a more sensitive assay, probably sub-ng/ml, to continue to support clinical development. Alternative approaches such as HPLC-EC, HPLC-fluorescence with pre-column derivatization and HPLC-MS(API) were investigated, as well as raising Ab's for an IA approach. Initially an assay involving derivatization and MS(API) was developed and validated. This was rapidly developed (<2 weeks) to a sensitivity of 0.5 ng/ml, relying on our in-house experience of applying MS(API) to bioanalytical problems [1]. This method was used in early clinical studies, but whilst sufficiently sensitive its throughput was low. Thus, ultimately, a dissociation-enhanced lanthanide fluorescence IA (DELFIA®) for UK-X in human plasma and urine was also developed. The plasma LLOQ was 0.20 ng/ml, and for urine we validated the method to 5 ng/ml. The assay's validated dynamic range is >2 orders of magnitude, the method is robust, and using the 96-well format for batch-analysis a high throughput has been achieved which adequately supports the current clinical program.

The DELFIA system relies on the fluorescence of lanthanide chelates as an end-point [2]. For UK-X, a competitive IA (96-well system) was set up by using plates coated with anti-rabbit Ab (IgG). UK-X in standards, QC's and unknowns competes with Eu-labelled UK-X for the rAS binding sites. After equilibrium is reached and the plates washed, their bound label is released with a low-pH solution which forms a new chelate of the lanthanide in a different micellar environment, greatly enhancing the fluorescence yield on UV irradiation. This chelate (usually a β-lactone) absorbs UV light and transfers the absorbed energy by inter-system crossing and intramolecular transfer to the lanthanide; in particular, the excited micellar ion emits long-lived fluorescence which has a very large effective Stoke's shift (e.g. λ_{exc} at 340 nm, λ_{em} 614 nm) which can be distinguished from background interference which is much shorter-lived.

Each well is read for 1 sec (1000 separate readings), and a 96-well plate can be read in <2 min which is very advantageous compared with conventional RIA. The dynamic range in a DELFIA system is also large (up to 10^6 for a non-competitive assay), and the system has potential advantages over conventional EIA: better sensitivity, lower interference, greater dynamic range and ease of use. This article describes the methods finally validated and discusses their relative merits, and also outlines the alternative methods and the reasons for their failure.

MATERIALS AND METHODS

UK-X, UK-Y (i.s.) and other 'UK' compounds were obtained in-house from Compound Control. All reagents used were AnalaR grade or better. DELFIA® products - plate-washer, plate-shaker, Research Fluorimeter #1234, assay buffer, plate-wash concentrate and enhance solution - together with anti-rabbit (goat)

IgG microtitration strips (plates of 8 ×12 wells) were obtained from Wallac EG & G Berthold UK (Milton Keynes, U.K.). Wallac Oy (Turku, Finland; see below) prepared Eu-labelled UK-X and supplied heavy-metal-reduced BSA, 75 mg/ml. Rabbit L anti-UK-X antiserum, bleed date 29.3.93, came from King's Diagnostics (Bournemouth; see later). Human plasma was outdated stock from the Blood Transfusion Service or else was collected locally from volunteers in Li-heparin tubes and transferred after centrifugation into vials for storage at -20°. Other sources (U.K.) were: methanol 205 Super Purity and acetonitrile 195, both HPLC grade: Romil Chemicals (Loughborough); MTBE (TBME; HPLC grade) and pH 7 phosphate buffer solution: FSA Supplies (Loughborough) 99% sodium azide, TFA, AmAc (Gold Label) and BF_3, 12% w/w in methanol: Aldrich (Gillingham, Dorset); TRIS, acetic acid, TEA and HCl: Merck-BDH (Poole); SPE extraction cartridges (BondElut 3 ml Certify II and 1 ml 100 mg C-18): Jones Chromatography (Hengoed, Mid Glam.). Water (18.2 MΩ) came from a Milli-Q-plus system (Millipore-Waters, Watford). An IBM-compatible microcomputer was used to collect fluorimeter output data which were subsequently analyzed using the program Multi-Calc (Wallac UK).

HLPC-MS-MS(APcI)

HPLC and MS entailed use of the following: Shimadzu LC-6A pump from Dyson (Newcastle-upon-Tyne) with an ISS 200 PE autoinjector (Perkin-Elmer, Beaconsfield); mobile phase (1 ml/min), passed twice under vacuum through a 50 mm GF/F filter (Whatman, Maidstone): 80% (v/v) aqueous methanol containing 20 mM AmAc and 5 mM TEA; PE column: 3 µm C-18 30 × 4.6 mm (3×3CR C18); PE detector: Sciex API III triple quadrupole [Q1, Q2, Q3] mass spectrometer with an APcI heated nebulizer† sample ionization system. MS data were collected and analyzed using the program MacQuan 1.1.2 (Sciex) with automatic determination of peak area.

MS entailed operating the nebulizer at 500° in a positive-ion APcI mode, with multiple reaction monitoring (MRM) using argon as collision gas and the parent/daughter transitions m/z 613.1 to 211.4 for UK-X methyl ester and 597.1 to 211.4 for UK-Y methyl ester (i.s.); 100 msec dwell time per transition. Data were acquired for 1 min; both compounds eluted at 0.5 min. The resolution for Q1 at m/z 613.1 and 597.1 was 2.2 a.m.u. peak width at half height, and Q3 resolution at m/z 211.4 was 2.0 a.m.u. peak width at half height.

Determining the derivatization time course entailed HPLC with a Merck-Hitachi AS2000 autosampler (Merck-BDH), a Varian (Walton-on-Thames) 9010 ternary gradient pump and a Coulochem EC detector (ESA, St. Ives, Cambs.; analytical electrode set at 700 mV). The column was from Hichrom (Reading): Spherisorb S5C6 (250 × 4.6 mm). The eluent (1 ml/min) was pH 2.25 20 mM sodium phosphate buffer containing 2.5 mM octanesulphonic acid and 35% (v/v) acetonitrile. Data were collected and integrated using the VG Multichrom chromatography data management system.

†inapplicable to the no-derivatization approach (below)

Sample collection and work-up,- Plasma samples from clinical trials were kept at -20° until analyzed. Samples for HPLC-MS(APcI) analysis were defrosted at room temperature for 1 h and briefly vortex-mixed. The calibration-line samples, QC's (1, 5 and 15 ng/ml; duplicates) and unknowns were extracted, derivatized, re-extracted and quantified as follows. From each, 1 ml was put into tapered polypropylene 10 ml tubes and, with vortex-mixing, 20 µl aqueous i.s. (20 ng UK-Y) added along with 1 ml pH 7 phosphate buffer. After loading (light vacuum) onto Bond-Elut Certify II cartridges pre-conditioned with 1 ml methanol and then 1 ml pH 7 buffer, and washing with 1 ml pH 7 buffer and then 1 ml 15% methanol/0.1% TFA, the cartridges were dried by application of vacuum and eluted into 10 ml polypropylene tapered tubes with 1 ml 80% methanol/0.1% TFA.

The tubes were briefly centrifuged to bring all the eluate to the bottom and then taken to dryness under a stream of N_2. When completely dry, 1 ml BF_3/methanol was added by Eppendorf dispenser with a 12.5 ml syringe. The tubes, tightly capped, were incubated at 65° for 30 min. After uncapping, 2 ml water was added and the volume reduced to ~1 ml at 65° (N_2 stream). The strongly acidic solution was then put onto pre-conditioned (1 ml methanol, then 1 ml water) C-18 cartridges using light vacuum. The tubes were washed out with 1 ml water which was added onto the cartridges. These were then washed with water (3 × 1 ml) and eluted into MTBE-rinsed polypropylene tapered tubes with methanol (2 × 1 ml). Brief centrifugation and taking to dryness at 37° (N_2 stream) was followed by addition, with vortexing, of 100 µl HPLC injection solution: 50% (v/v) aqueous methanol containing 20 mM AmAc and 5 mM TEA. After centrifugation and transfer to 200 µl crimp-top autosampler vials, these were capped, centrifuged at 10,000 rpm (for 5 min) and put into the autosampler; 80 µl was injected onto the column as above.

Method validation was done on three separate occasions by analyzing 7 replicates of 3 UK-X concentrations (0.5, 5.0 and 20.0 ng/ml) and, from the same stock solution, calibration-line spiked samples. The inaccuracy and imprecision cut-off levels for defining the LLOQ were 20%.

Immunoassay (IA)

Our initial strategy for raising antisera to UK-X was to use sheep and couple the t-butyl-protected hapten UK-Z (Fig. 2) to BSA and, *via* the lysine-like ε-amino group, couple to keyhole limpet haemocyanin. The protected conjugate was then unblocked using 90% (v/v) TFA, dialyzed extensively and injected as an emulsion in complete Freund's adjuvant into four sheep. Boosts with the same conjugate were done every 8 weeks with, 1 week after each, trial bleeds which were tested for binding of ^{3}H-UK-X using dextran-coated charcoal to separate free from bound label at various dilutions.

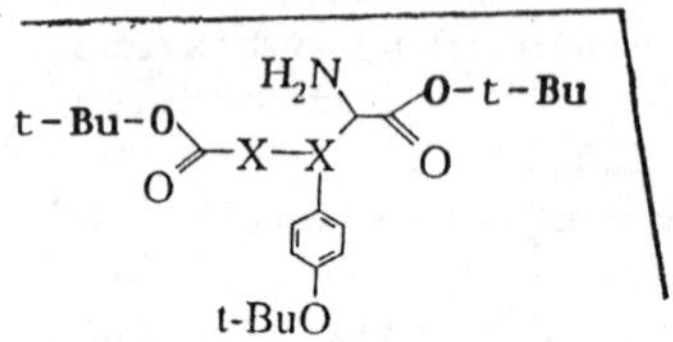

Fig. 2. Structure of UK-Z (initial hapten),

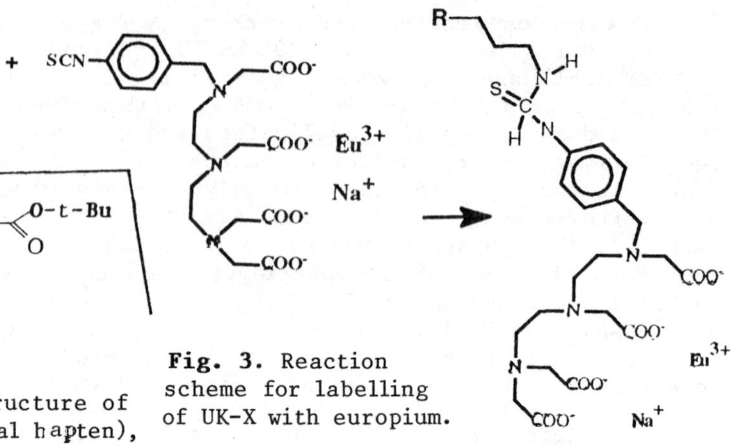

Fig. 3. Reaction scheme for labelling of UK-X with europium.

In rabbits anti-UK-X Ab's were raised by coupling UK-X to PPD similarly, *via* carbodiimide chemistry (as follows), and using this conjugate to prime and boost the animals.

Conjugation.- To 10 mg UK-X and 20 mg PPD or BSA in 0.5 ml dist. water, 0.5 ml (10 mg) 1-ethyl-3-[3-dimethylaminopropyl]-diimide solution was added dropwise with mixing. After adding 50 µl 0.1 M HCl the reaction was allowed to proceed in the dark (room temp.) for 18 h. The resulting conjugate was dialyzed against 0.09% NaCl (2 × 5 L) and the solution stored frozen in 1 ml aliquots containing 2 mg PPD or BSA conjugate.

Immunization of rabbits (6).- Each received, by i.m. and s.c. injections at 6 sites, 1 ml of an emulsion: 0.5 ml (1 mg) PPD conjugate and 0.5 ml Freund's complete (incomplete for the boosts) adjuvant. Bleeds were obtained 10 days after each monthly boost and terminal bleeds after a final boost 21 months after initial immunization. Ab titre and specificity were assessed using either plates coated with UK-X-BSA conjugate and Eu-labelled protein G to determine binding of rabbit-anti-UK-X Ab's, or goat-anti-rabbit coated plates to bind the anti-UK-X Ab's which then bound Eu-labelled UK-X.

UK-X was labelled with a Eu chelate by Dr Heikki Mikola (Wallac Oy, Turku) by reaction with the isothiocyanate group (Fig. 3), and purified by HPLC; collected fractions were stored at 4° and have been stable for >2 years under these conditions.

Assay method.- The following solutions were used: A, 50 mM TRIS-HCl buffer pH 7.75; B, same but with 75 mg/ml heavy-metal-free BSA; and (DELFIA) C: assay buffer and D: wash concentrate. Stock UK-X solution (~200 µg/ml) was kept at ~4° and was stable for 3 months; working solutions (10 µg/ml) were prepared daily.

The stock solution of anti-UK-X antiserum (1:100, rabbit L, bled 29.3.93) was prepared in A containing 0.9% NaCl and 0.02% sodium azide, and stored at ~4°. Working Ab solution (1:10,000) was prepared daily in C. Samples were assayed in duplicate. Analyte calibrants were made by serial dilution. For assays, the goat anti-rabbit coated plates were washed with 2 x 300 µl D (10-sec soak after each well fill). To each well, 50 µl of sample, standard or QC was added, then 100 µl Eu-labelled UK-X solution and 100 µl antiserum (Ab finally 1:25,000). The plates were kept for 2 h on the plate-shaker (low speed; room temp.). After 6 washes each well received 200 µl 'Enhance' solution. After shaking (5 min, room temp.) the plates were counted using the DELFIA fluorimeter, 1 sec per well at 1000 Hz.

The data were processed using the program 'MultiCalc'. Curve fitting was iterative using a 5-parameter logistic fit. Manual dispensing was used for plasma or urine and all other additions; but to further increase throughput we introduced a Packard 104DT 4-way robotic sampler with disposable tips, those used for plasma being pre-wetted and changed after each sample. To get the curve (as above) for the urine assay, water was used to dilute the calibrants; all urine samples were diluted ×10.

Method validation was carried out in three separate batches by analyzing 7 replicates at 3 concentrations (tabulated below) diluted from the same working solution as that for the calibration line. The accuracy and precision criteria for the LLOQ were again ±20%. The MultiCalc program generated an LOD for each batch which was always below the lowest calibration point.

RESULTS AND DISCUSSION

Early results from preliminary clinical studies indicated that the bioassay based on rat renal endopeptidase inhibition is too insensitive in man, even though it had served well for animal pharmacokinetic studies on UK-X*. Therefore we explored a number of analytical techniques for UK-X in human plasma, initially investigating SPE to separate analyte from the plasma matrix: the phases tried ranged from C-2 to C-18 and Certify II - the latter giving quantitative recovery as judged by the use of ^{3}H-UK-X and clean extracts as judged by HPLC. Then these extracts were analyzed either by HPLC-EC or by column-switching HPLC with fluorescence detection following pre-column derivatization with FMOC-Cl. Either method could intrinsically detect ~100 pg or less (data not shown), but neither was sufficiently robust to enable validation at 0.5-1 ng/ml. Our access to the mass spectrometer which had served well in our Department for several analytes including abanoquil at 10 pg/ml blood [1] led to exploitation of the mass-discriminatory power and sensitivity of HPLC-MS-MS(APcI) to facilitate rapid development of an assay for UK-X at sub-ng/ml levels. Preliminary work in the APcI mode

* Footnote θ on title p. applies

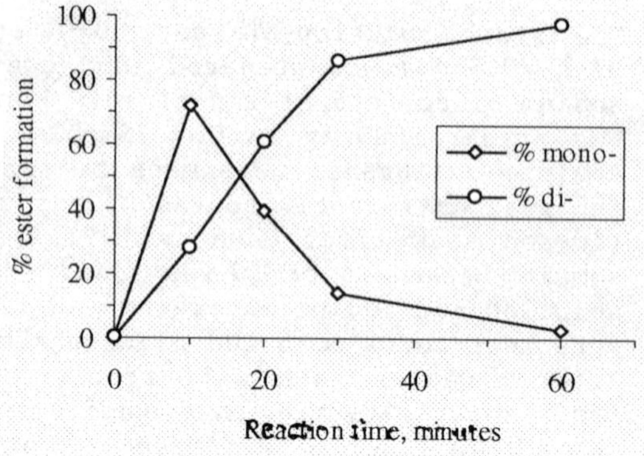

Fig. 4. Time course of methylation of UK-X (20 ng) for up to 60 min at 65° using BF_3/methanol. After SPE of the derivatives, the dry extracts dissolved in mobile phase were injected directly into the HPLC-EC system without switching. For details, see text.

suggested that the free-acid form of UK-X could not be detected with adequate sensitivity. For poorly understood reasons, such compounds with carboxylic acid moieties in their structure do not ionize well at low concentrations even in the negative-ion mode. This may be due to ion suppression by matrix components or lack of volatility or thermal stability. Carboxylic acid esters are known to ionize better [3] as was confirmed for UK-X by investigating the diester analogues and the dimethyl ester derivative - which could be produced quantitatively from UK-X in analytical samples by using BF_3/methanol as a methylating agent [4]. Diazomethane was another possible reagent but we considered this to be too hazardous and cumbersome for routine analysis. The LOD for dimethyl UK-X was ~100 pg.

MS-MS(API) assay

Derivatization with BF_3/methanol and subsequent dilution with water leads to a very strongly acidic solution. This acid had to be removed, and we found that SPE on C-18 BondElut cartridges was ideal for this purpose. The derivatized UK-X could be easily eluted using methanol, which was easy to evaporate to dryness with a N_2 stream. So we were able to develop rapidly a method for analysis of UK-X in human plasma, using SPE based on our earlier experience with Certify II cartridges, followed by methylation of the dry extracts, and by a further SPE and chromatography, with MS detection using MRM. We used an i.s., UK-Y, that was a close analogue and quantified UK-X using peak area ratios (analyte/i.s.). Using HPLC-EC for the parent and its mono- and di-methyl esters, the reaction was found to be >90% complete after 30 min at 65° (Fig. 4).

The mass spectrum obtained at Q1 for the derivatized UK-X showed a single ion at m/z 613 due to the protonated molecular ion of the dimethyl ester and m/z 597 for dimethyl-UK-Y. By selecting these parents and scanning for daughter ions in Q3 after collision with argon in Q2, dimethyl-UK-X and -UK-Y gave the same daughter ion of m/z 211 suitable for quantification. Fig. 5 shows the mass spectrum at Q3 for dimethyl-UK-X. Accordingly we selected the parent ions of m/z 613 and 597 and the m/z 211 daughter ions to quantify UK-X and UK-Y respectively.

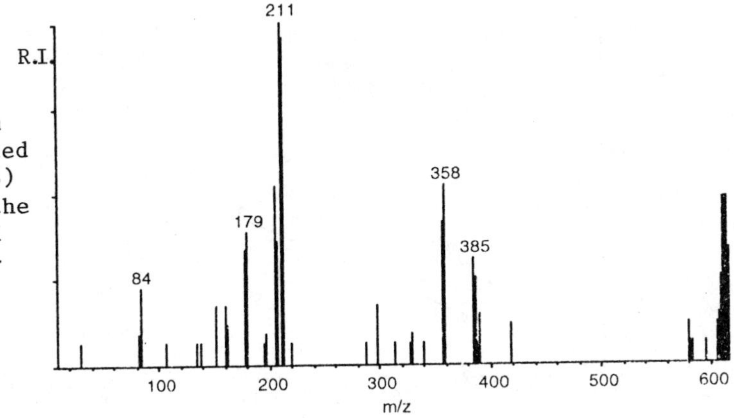

Fig. 5. Daughter ion mass spectrum of dimethylated UK-X (MH+ 613) obtained on the Sciex API III mass spectrometer.

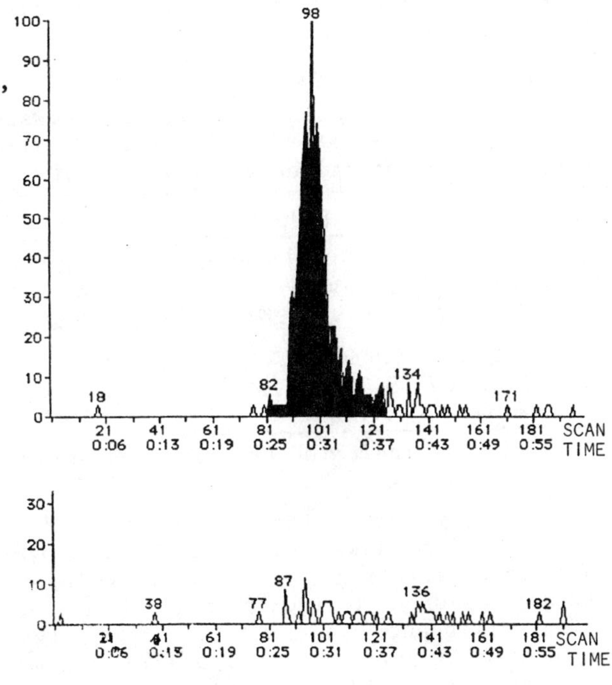

Fig. 6. HPLC-MS-MS (API; MRM) of extracted and BF_3/methanol-derivatized plasma, spiked with UK-X dimethyl ester, or *(below)* blank. Scan no. signified by numeral above the peak. 'TIME' is R_T (brief run times!).

Fig. 6 shows typical chromatograms for derivatized extracts of plasma, blank or spiked with UK-X. UK-X/UK-Y peak-area ratios were used to generate calibration lines (Fig. 7; 0.5-20 ng/ml) as established, for validation (Table 1A) on three occasions when **accuracy and precision** were found to be within 10%.

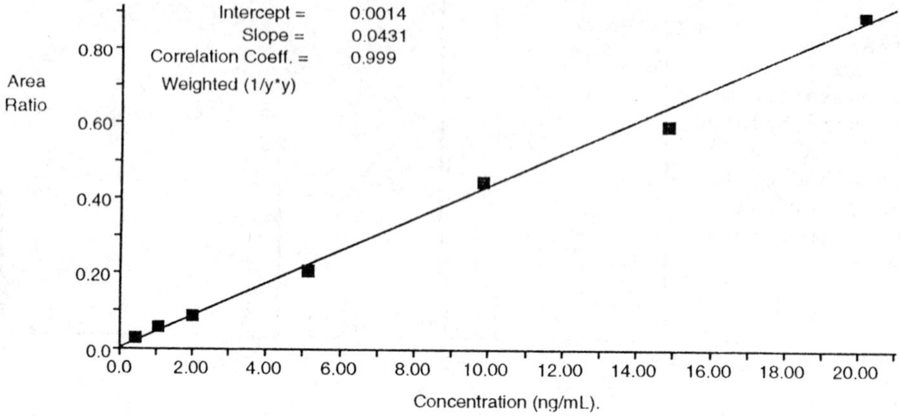

Fig. 7. Representative calibration line for UK-X in plasma after extraction, derivatization and analysis by the HPLC-MS-MS(API) method.

Table 1. Inaccuracy (%) and imprecision (%) in assay of UK-X: (A) HPLC-MS-MS(API), on 3 separate occasions distinguished by strokes //; (B) ISP MS-MS(API), on 2 occasions distinguished by a stroke. Concentrations are ng/ml, ±S.D. Generally n = 7.

	Spiked	Mean concentration measured	Inaccuracy	Imprecision
A	0.5	0.54±0.03/ 0.49±0.04/ 0.52±0.05	8.25/-2.12/ 4.22	6.32/7.98/ 9.30
	5	4.69±0.31/ 5.20±0.34/ 4.78±0.48	-6.15/ 4.02/-4.31	6.53/6.46/ 0.96
	20	20.48±1.43/20.12±0.88/20.32±0.36	2.38/ 0.61/ 1.60	7.01/4.39/ 1.78
B	0.5	0.51±0.05/ 0.57±0.08†	1.4 /14.7	10.5/13.0
	7.5	7.28±0.31†/ 7.22±0.30	-2.9/-3.8	4.2/ 4.1
	20	19.46±0.51/ 18.66±0.77	-2.7/-6.7	2.6/ 4.1

†n = 6 for the indicated occasion and spike level

This assay served for the early clinical studies with UK-X but suffered from low throughput and the need to extract twice. Physical constraints of the apparatus together with the length of the procedure limited throughput to 36 samples (including QC's and calibrants) per batch. Since two days were required to complete the assay, the speed of analysis available on the mass spectrometers was not fully utilized. For these reasons we undertook further assay development using the Certify II extraction method with an ISP interface for the API III mass spectrometer and obtained effective ionization of UK-X. Using a flow rate of only 200 µl/min and a mobile phase of methanol/water with 2 mM AmAc and 0.5 mM TEA, we validated this method with no i.s., obtaining the data in Table 1B. Accuracy and precision were within our accepted criteria for both batches.

At the lowest level (0.5 ng/ml) precision was less good than for the derivatization assay, but still <15%; much improvement would be obtainable by use of deuterated UK-X as i.s. Throughput of 70-80 samples per day was achievable, more than double that of the derivatization assay. However, we did not use the method for analysis of clinical samples since the IA described here, with much higher throughput, was developed concomitantly.

Immunoassay (IA)

Our initial strategy for raising antisera to UK-X in sheep by coupling the t-butyl-protected hapten UK-100,637 and keyhole limpet haemocyanin failed to generate any antisera as judged by the binding of ^{3}H-UK-X using the dextran-coated charcoal method of separation at various dilutions. We could find no binding of drug at all. Further immunizations were carried out using native UK-X as the starting material for conjugation; but still no binding was observed. We considered this surprising, since we expected the compound to be immunogenic. As a result, we also tested these bleeds with a DELFIA† system in a configuration such that UK-X-BSA conjugate was bound onto 96-well assay plates. Any Ab binding to the immobilized UK-X was detected using Eu-labelled Protein G. We then found that there were anti-UK-X Ab's present in these sheep bleeds.

Subsequent experiments showed that the binding was specific for UK-X and could be displaced by increasing concentrations of free UK-X in the assay. Fig. 8 shows typical data obtained for a sheep antiserum. It remains unclear why we found no binding of ^{3}H-UK-X to sheep antisera using the dextran-coated charcoal assay, but we assume that the affinity of the Ab-UK-X interaction was very low and that the charcoal stripped the bound UK-X from the Ab's before centrifugal separation occurred.

†defined on 2nd p. of text

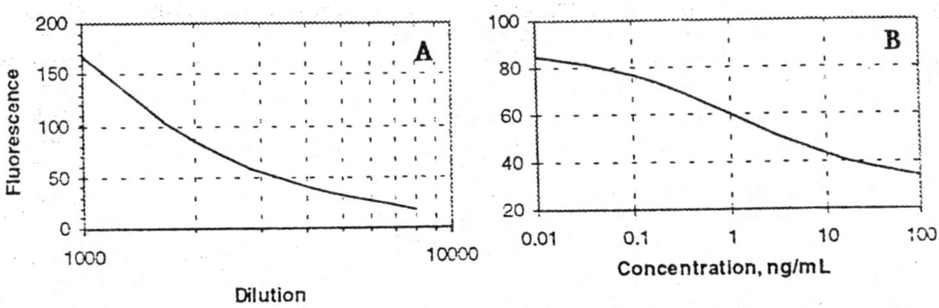

Fig. 8. Ab titre and concentration dependence for sheep antiserum (Sheep #1914). UK-X-BSA conjugate was coated onto 96-well plates, and the antiserum was diluted appropriately (**A**), or at 1:2000 final dilution increasing concentrations of UK-X (50 µl in buffer) were added (**B**). After 2 h (room temp.) Eu-labelled Protein G (1:200 dilution) was added. After a further 1 h the plate was washed, enhanced with 200 µl 'Enhance' and counted.

Table 2. Back-calculated calibration curve concentrations for DELFIA performed on 3 different occasions using a 5-parameter logistic unweighted algorithm.

	Concentration, ng/mL									
Theoretical	100	50	25	12.5	6.25	3.13	1.56	0.78	0.39	0.20
Found 1	98.7	51.4	26.4	12.7	6.24	2.95	1.60	0.81	0.38	0.20
Difference, %	-1.26	2.7	5.74	1.57	-0.20	-5.66	2.65	3.09	-3.39	0.95
Found 2	86.8	56.4	27.1	12.1	6.26	3.15	1.47	0.82	0.41	0.19
Difference, %	-13.2	12.9	8.4	-3.4	0.1	0.8	-5.6	4.2	3.9	-4.6
Found 3	103.9	49.0	24.7	12.5	6.34	3.13	1.53	0.81	0.39	0.20
Difference, %	3.9	-2.0	-1.0	-0.2	1.4	0.1	-2.4	3.0	-1.4	0.0

DELFIA on plasma

It would have been possible to develop and validate an IA for UK-X using the sheep antisera as described above with a Eu-labelled Protein-G end-point. This however was not as attractive a proposition as tailoring the assay to use of Eu-labelled UK-X. This was mainly because of the low titres of the sheep antisera, the commercial availability of quality-controlled plates coated with goat anti-rabbit antisera (thus we did not have to coat our own plates with UK-X-BSA conjugate), and the increased sensitivity available. Once high-titre rabbit antisera and the Eu-labelled UK-X became available, the assay as described was very quickly validated down to LLOQ 0.2 ng/ml. Later work showed that using a disequilibrium format for the assay (in which Ab was added 1 h prior to addition of label), the assay could achieve LLOQ 0.05 ng/ml (data not shown).

Fig. 9 shows a typical standard curve for the method. Back-calibration data appear in Table 2. Fluorescence counts typically ranged from ~80,000 at B_0 to 5,000 at 100 ng/ml and duplicates differed by <5% over the entire calibration range.

Manual method: accuracy and precision were assessed by analyzing 7 replicates at each of 3 concentrations (prepared from the same stock standard as that used for calibration-curve preparation) in duplicate as described above. A fourth validation was done using 8-way multi-channel pipettes (Table 3A).

Automated method: accuracy and precision were assessed whilst using for the assay a Packard automatic sample handling system (Multiprobe 104DT) - a 4-way instrument using disposable tips to dispense samples; we used a new tip after each addition to minimize the cross-contamination risk. As Table 3B shows, the automated method does not further benefit accuracy or precision. Its notable advantage was to throughput, 8 rather than 3 plates per day, thus freeing the analyst for other duties.

Table 3. Inaccuracy (%) and imprecision (%) for assaying UK-X in human plasma by DELFIA, manual (A) or automatic (B). Strokes // distinguish different occasions, as in Table 1; for the 4th A occasion (underlined values) an 8-way pipette was used to add label and Ab. Concentrations are ng/ml ±S.D.; n = 7.

Spiked		Mean concentration measured	Inaccuracy	Imprecision
A	0.195	0.195±.01/ 0.21±0.02/ 0.22±0.03/ 0.16±0.02	0.2/ 8.1/16.1/-17.6	6.7/10.0/14.1/12.9
	3.125	3.09±0.11/ 3.21±0.09/ 3.08/0.15/ 2.93±0.12	-1.0/ 2.8/-1.4/-6.3	3.7/ 2.7/ 4.8/ 4.2
	50	48.31±0.77/49.92±0.75/53.15±1.75/50.58±2.83	-3.4/-0.2/ 6.3/ 1.2	1.6/ 1.5/ 3.3/ 5.6
B	0.195	0.20±0.02/ 0.18±0.02/ 0.18±0.02	1/4/-6.5/-6.3	8.1/13.0/11.8
	3.125	3.20±0.09/ 3.08±0.04/ 3.02±0.09	2.4/-1.5/-3.5	2.9/ 1.4/ 3.0
	25	23.1±0.30/ 24.9±0.42/ 25.4±0.91	-7.5/ 0.4/ 1.6	1.3/ 1.7/ 3.6

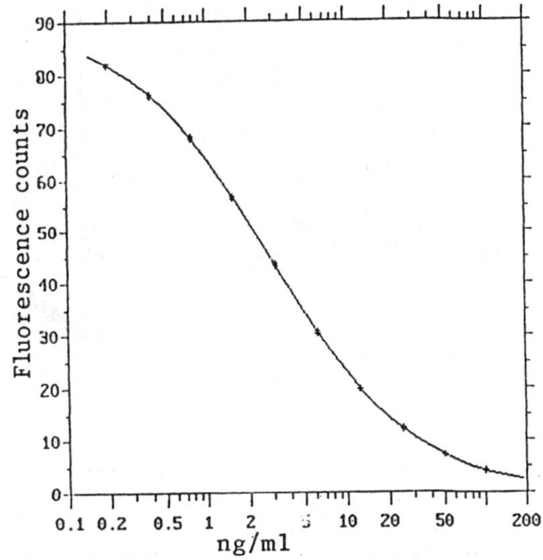

Fig. 9. A typical calibration curve for UK-X over the range 0.195-100 ng/ml from the program MultiCalc.

Sensitivity.- We defined the LLOQ as the lowest calibration point with acceptable precision - 0.195 ng/ml in human plasma. We set the ceiling at 50 ng/ml initially, but later experience led us to reduce this to 25 ng/ml since the greater range was not needed. Validation at 100 ng/ml was not attempted, but this top calibrant was used to assist calculation of the standard curve. We found with the program MultiCalc the 5-parameter logistic fit algorithm, unweighted, normally was best for calculating the standard curve as judged by the back-calculated values obtained from the standard curve.

Matrix effects.- A selection of 16 different control plasma samples was analyzed to check for interference from endogenous compounds. None was seen. QC's and blanks made up in Blood Transfusion Service plasma showed interference in the

Table 4. DELFIA values (ng/ml) for three different samples from a clinical trial to assess parallelism of the method. The indicated dilutions were done with blank plasma.

	Sample A			Sample B			Sample C		
Dilution	found	calc.	%diff.	found	calc.	%diff.	found	calc.	% diff.
neat	7.52			12.14			17.75		
2×	3.80	3.76	+1.1	5.96	6.07	-1.8	7.48	8.88	-15.8
4×	1.89	1.88	+0.5	2.64	3.04	-13.2	4.05	4.44	-8.8
8×	0.80	0.94	-14.9	1.33	1.52	-12.5	2.01	2.22	-9.5

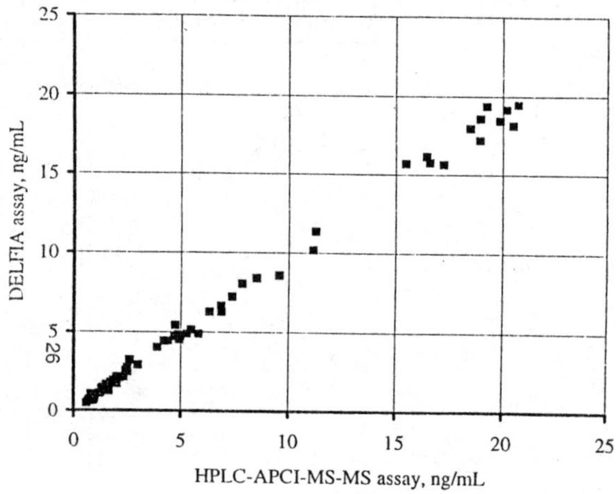

Fig. 10. Cross-validation of the DELFIA UK-X method with the HPLC-MS-MS(APcI) method. Samples from volunteers were analyzed by the two methods on the same day (see text). The individual concentrations obtained by each method were plotted against each other.

form of spuriously high concentrations of UK-X being reported. This is probably due to the presence of EDTA in the plasma which would strip the Eu^{3+} from the labelled UK-X and thus reduce the plate-bound counts, producing false high values.

Cross-reactivity cannot be checked with metabolites, as none are known for UK-X. However, as is manifested by cross-validation data (**Fig.** 10; see above) the selective MS method and DELFIA correlated excellently, which suggests that DELFIA is as selective as the physical method. No interferences indicative of cross-reactivity were found with a number of possibly co-administered drugs, e.g. captopril, enalopril, nifedipine, propranolol and paracetamol.

Parallelism,- Clinical samples containing high levels of UK-X (~10-20 ng/ml) were serially diluted down to ×8, as shown in Table 4, with blank human plasma to establish the parallelism of the assay. Evidently found and expected values agreed within acceptable limits, further confirming the assay's suitability for analyzing samples from volunteer in clinical trials.

Cross-validation between the DELFIA and HPLC-MS-MS methods was done by applying both methods to 6 sets of plasma samples from volunteers on the same day. Within each profile the mean differences were <20%. Fig. 10 shows a scatter diagram of the data; the correlation was 0.995 and the line's slope was 0.95.

Stability of UK-X in plasma, as evaluated with the HPLC-MS assay, was satisfactory for at least 6 months in frozen Transfusion-Service plasma and 1 month in clinical samples, and for 24 h in plasma kept at 37°. Frozen QC's were also stable for 6 months in Li-heparin plasma as by the DELFIA method.

DELFIA applied to urine

We also validated the DELFIA for analysis of UK-X in human urine (3 different batches). Some urines gave spuriously high blank concentrations for UK-X when the assay was done exactly as for plasma (50 µl sample). Hence each sample had to be diluted ×10 with water; likewise for the standard curve. From accuracy and precision data (Table 5) the LLOQ was set at 5 ng/ml.

Table 5. Inaccuracy (%) and imprecision (%) for UK-X, measured by automated DELFIA, in human urine: 3 batches, distinguished by strokes //. Concentrations are ng/ml, ±S.D.; n = 7; dilution ×10 or, for the 2000 ng/ml level, ×100.

Spiked	Mean concentration measured	Inaccuracy	Imprecision
5.0	0.42±0.03/0.41±0.05/ 0.52±0.07	-16.2/-17.2/ 4.9	8.1/12.6/13.6
31.3	3.24±0.11/2.97±0.25/ 3.45±0.21	3.5/ -5.0.10.1	3.3/ 8.3/ 8.7
250	22.7±0.61 /21.3±0.94/25/3±2.30	-9.0/-14.6/ 1.1	2.7/ 4.4/ 9.1
2000	18.0±0.49 /18.4±2.43/19.1±0.92	-9.8/ -7.8/-4.3	8.3/13.2/ 4.8

CONCLUSIONS

We have developed a fast, reliable high-throughput IA for UK-X in human plasma and urine that has been validated and is suitable for analyses on samples generated during the clinical development program. The method uses a Eu-labelled UK-X analogue with time-resolved fluorescence as an end-point which, in competitive IA's, has the major advantage that the samples are read very quickly (1 sec/sample *vs.* several min/sample for RIA). This benefits throughput and allows rapid feedback during assay development, with minimal disposal problems since no radioisotope is used. All samples (~600) from a small clinical trial can be analyzed within one week, and samples (>2000) from a large, multiple-dose cross-over study within one month.

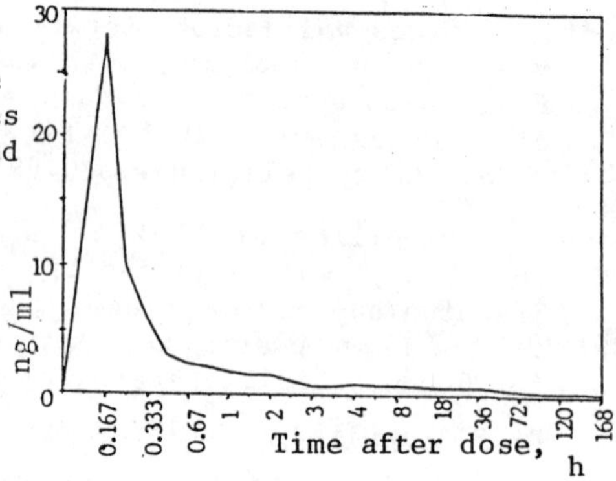

Fig. 11. Concentrations of UK-X in a volunteer following a single i.v. dose (0.5 mg). Blood samples (2 ml) were taken into heparinized tubes and centrifuged. DELFIA was performed in duplicate (see text). Note that the ordinate is not a linear scale.

This DELFIA method has served for all clinical trials on the compound after the first two, wherein the HPLC-MS-MS assay was used. The assay has also been successfully established in a contract research laboratory. Fig. 11 shows a typical plasma time course for a dosed volunteer. The DELFIA throughput was such that in this dose-escalation i.v. study all samples up to 24 h from all subjects dosed in one period were analyzed and the results reported within 2 days of receipt of samples. This is notable compared with assays using HPLC and, in particular, with the above HPLC-MS-MS(API) assay which involved, besides derivatization, two SPE and three evaporation steps; it took 2 days to complete and, with the apparatus available, was limited to <30 clinical samples per batch and hence only 120 per week. It did, however, serve well; it had LLOQ 0.5 ng/ml, and was robust, reliable and selective: compared with this physical method the IA showed no positive or negative bias.

Acknowledgements

We thank Dr Heikki Mikola of Wallac Oy for expert labelling of UK-X with Eu chelate. David Muirhead and Chris Brearley of our in-house Early Clinical Research Group gave plasma samples.

References

1. Kaye, B., Clark, M.W.H., Cussans, N.J., Macrae, P.V. & Stopher, D.A. (1992) *Biol. Mass Spectrom. 21,* 585-589.
2. Hemmilä, I., Dakubu, S., Mukkala, V-M., Siithari, H. & Lövgren, T. (1984) *Anal. Biochem. 137,* 335-343.
3. Beresford, A.P., Macrae, P.V., Alker, D. & Kobylecki, R.J. (1989) *Arz.-Forsch./Drug Res. 39,* 201-209.
4. VuDuc, T. & Vernay, A. (1992) *J. Pharm. Biomed. Anal. 10,* 187-191.

#A-4

A SENSITIVE SANDWICH CHEMILUMINESCENT ENZYME IMMUNO-ASSAY FOR THE DETERMINATION OF ADMINISTERED SALMON CALCITONIN IN HUMAN PLASMA

S-H. Hsu, Julie Wangsa, J. Lee & Lisa Rich†

Department of Drug Metabolism and Pharmacokinetics,
Rhône-Poulenc Rorer Central Research,
Collegeville, PA 19426, U.S.A.

A highly sensitive and rapid non-competitive sandwich-type ELISA using chemiluminescent detection was developed for measuring administered salmon calcitonin (sCT) in human plasma. The method comprised SPE with CBA cartridges and highly specific monoclonal combined with polyclonal Ab's for sandwich ELISA. The LLOQ is 5 pg/ml with a working range up to 150 pg/ml using 1.0 ml of plasma. Standards and QC's are prepared in human plasma preserved with EDTA. Intra- and inter-assay variations were <6% and the accuracy was 94-108%. The method has been satisfactorily applied to preclinical and clinical studies.*

Salmon calcitonin is a small polypeptide having 32 amino acids arranged in a single chain. CT* is a specific and highly potent endogenous inhibitor of osteoclastic resorption. In the normal turnover process, calcium transfers from blood to bone to support accretion, concomitantly with the ongoing bone-resorption process that results in continual calcium outflow from bone to blood. CT administration inhibits this outflow. In adult mammals, CT is believed to have a role in protecting the skeleton from loss [1-5]. The physiological role of sCT is not completely understood, but appears to be connected with skeletal maintenance during growth or other periods of calcium stress. sCT has been proposed for therapy of Paget's disease, postmenopausal osteoporosis and hypercalcaemia of malignancy.

Problems arise when measuring plasma sCT not only because of the low levels present in the systemic circulation [6, 7] but also due to heterogeneity [8-11]: thus there are high mol. wt. components, dimers and fragments. RIA determination of sCT

---

†address correspondence to Dr S-H. Hsu

**Abbreviations:* Ab, antibody; AKP, alkaline phosphatase; CBA, a weak cation-exchange bonded phase; ELISA, Enzyme-Linked Immunosorbent Assay; LLOQ, lower limit of quantification; QC, quality control; (s)CT, (salmon) calcitonin.

has been hampered by several problems such as LLOQ inadequacy and some lability of the ^{125}I-labelled sCT used as tracer [12]. These problems have led to the development of an ELISA. The sensitivity of ELISA's can be greatly improved by the application of an enzyme amplification system [13-15], of 'sandwich' type in the present method. In outline, CBA extraction was performed on plasma to concentrate the sCT and remove interfering proteins; then sCT was captured in microtitre wells coated with mouse anti-sCT monoclonal Ab. After washings and addition of rabbit anti-sCT polyclonal Ab, the amount of bound rabbit IgG was quantified by a goat anti-rabbit IgG-AKP conjugate. The bound conjugate's enzyme (AKP) activity, measured by luminescence after adding Lumi-Phos 530 as substrate, reflects the sample's sCT content. High sensitivity was achieved.

EXPERIMENTAL

Materials and reagents (analytical grade unless otherwise stated) were from the sources named.- HCl and methanol: Fisher (Pittsburgh, PA); BSA, Tween-20 and goat anti-rabbit IgG-AKP conjugate: Sigma (St. Louis, MO); sCT, Armour (Kankakee, IL); EDTA human plasma from anticoagulated whole blood: Biological Speciality (Lansdale, PA); Lumi-Phos 530: Lumigen (Detroit); Delbecco's PBS 10X: Mediatech (Washington, DC); Microlite 2, polystyrene base 96-well plates: Dynatec (Chantilly, VA); Bond-Elut cartridges: Varian (Sunnydale, CA). Mouse and rabbit anti-sCT Ab's came from the Immunochemistry and Drug Disposition Departments in-house. All solutions were prepared using deionized water. All plasma standards and QC's were stored at -20°.

Apparatus was procured thus: SpeedVac Concentrator #200: Savant Inst. (Farmingdale, NY); Luminometer #ML 1000: Dynatech; THERMOmax Microplate Reader/SOFTmax software: Molecular Device (Menlo Park, CA); Sorvall #RC 3B Refrigerated Centrifuge: DuPont Insts. (Newtown, CT); Vortex Genor 2: Fisher; Gilson #P200 & #P1000 Pipetmen: Rainin Insts. (Woburn, MA); Eppendorf Repeater Pipet: Brinkman Insts. (Westbury, NY); Speed-mate 30: Applied Separations (Bethlehem, PA); Balance #AE 100: Mettler Instr.(Hightstown, NY).

Procedures

Plasma sample extraction, with prior thawing of frozen standards and QC's.- Disposable CBA cartridges were conditioned with 1 column vol. of methanol and 2 of Milli-Q water; then 1 ml samples were passed through without vacuum, and washing done at low vacuum: 5 ml water, 5 ml 60% (v/v) aqueous methanol and 1 ml methanol. The eluate obtained with 2.5 ml methanolic HCl (0.833% w/v) was dried down in the centrifugal vacuum concentrator at room temperature; the residue was reconstituted in 220 µl buffer† with vigorous vortexing and centrifugation (3500 rpm; 20 min, 20°) and 200 µl placed into each plate well.

†phosphate-buffered saline (PBS) containing 0.1% BSA

Coating the plates was done by putting into each well 200 µl of the previously prepared monoclonal Ab solution and keeping overnight at 4° in a humid chamber. Unadsorbed Ab was removed by washing twice with the washing buffer, and any remaining binding sites on the well surface were blocked with 300 µl of blocking buffer, 1% BSA in PBS, in a humid chamber (1 h, room temp.). After 6 washes, samples were introduced.

Chemiluminescent ELISA procedures ('parking' was within a humid chamber).- Before use, anti-CT Ab's - mouse monoclonal (D10F4) and rabbit polyclonal - were purified on affinity columns, protein-G and sCT respectively. For ELISA, 200 µl of extract (standard, QC or sample) was introduced into each coated well. The plates were kept overnight at 4° and tben buffer-washed† (×6). With 200 µl of affinity-purified rabbit anti-sCT Ab working solution added to each well, the plates were shaken for 2 h (room temp.) and washed (×6). Bound sCT was determined by adding 200 µl goat anti-rabbit IgG Ab-AKP conjugate and keeping for 2 h at room temperature or overnight at 4°. After washing (×6), enzyme activity was measured by adding 200 µl Lumi-Phos solution and, after 5 min (room temp.), measuring light emission with the plate reader every 5 min.

Validation and quantification.- Human plasma standards (assayed in triplicate) and, from separate stock solutions, QC's (5 replicates) were prepared at concentrations tabulated below; for each value plasma (4 ml) was frozen in a 75 × 12 mm polystyrene test tube. Validation data as in Tables 1 & 2 were obtained. A 4-parameter logistic program (SOFTmax software program from Molecular Devices) was used to construct external standard curves: $y = [(A - D)/(1 + (x/C)B)] + D$. All curves were generated from the data collected at the 8th cycle of luminometer reading. Standard curves were plotted as RLE (relative light emitted) *vs.* log concentration of sCT. QC concentrations were interpolated from the standard curve. EDTA plasma (male) furnished non-specific binding (nsb) for background subtraction.

Test samples were collected from healthy volunteers given 1200 I.U. sCT. Blood collected in Sodium EDTA tubes was centrifuged at 4° (2000 rpm, 10 min). Plasma was stored at -20°.

RESULTS AND DISCUSSION

Fig. 1 shows typical calibration curves (r values at least 0.995). The method was satisfactory in respect of reproducibility and specificity. Precision and accuracy are manifest in Table 1 for spiked plasma and QC's. For standards the mean C.V.'s ranged from 0.38 to 5.33%, intra- and inter-assay, and the respective ranges for QC's were 3.98-7.46% and 3.75-5.24%.

For determining sCT in plasma the ELISA surpassed RIA in sensitivity and convenience. A typical RIA requires 4-5 days to complete an assay and is limited by the short half-life of

†PBS containing 0.05% (v/v) Tween 20

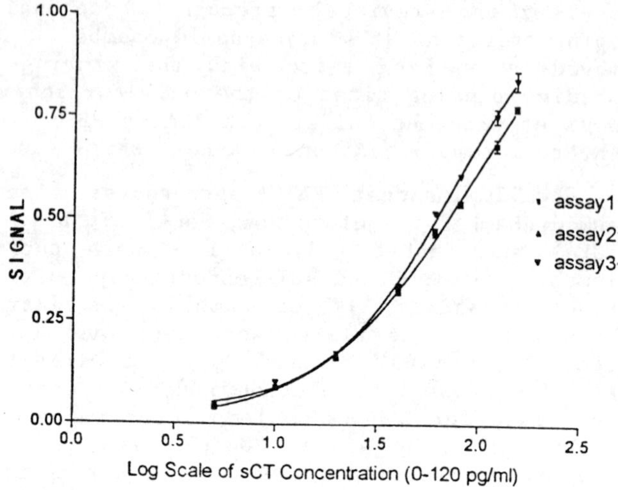

Fig. 1. Typical standard curves for plasma sCT: 3 assays (n = 3 for each); range 5-150 pg/ml.

Table 1. ELISA for sCT in human plasma: accuracy and precision

Set pg/ml:	5	10	20	40	60	80	100	150
Standards, intra-assay *(n = 3)*								
Mean found	4.82	10.84	18.92	39.61	61.11	78.47	123.43	147.97
C.V., %	1.11	1.58	0.84	4.38	3.81	0.38	1.33	0.81
Accuracy, %	96.31	108.40	94.58	99.03	101.84	98.09	102.86	99.98
Standards, inter-assay *(n = 9)*								
Mean found	4.83	10.77	18.83	39.22	62.90	79.12	120.41	151.39
C.V., %	5.33	4.58	1.78	3.48	4.13	2.74	4.80	2.35
Accuracy, %	96.60	107.69	94.16	98.04	104.84	98.90	100.34	100.93

QC's:	***intra-assay*** *(n = 5)*			***inter-assay*** *(n = 3)*		
Set pg/ml:	15	50	100	15	50	100
Mean found	14.73	50.18	100.6	14.81	49.72	104.8
C.V., %	3.98	7.46	4.74	3.75	4.23	5.24
Accuracy, %	98.20	100.4	100.6	98.73	99.44	104.8

^{125}I, whereas the ELISA requires only 2 days. The non-isotopic label and chemiluminescent end-point were conducive to ease and safety of operation and, through manipulating the time for the final signal to develop, to control of the dynamic range. The enhanced sensitivity is attributable to the specificity of the monoclonal Ab as well as the chemiluminescent detection system, with a 'sandwich' approach for enzyme amplification and, at the outset, SPE (CBA) extraction as a rugged and reproducible means of concentrating the sCT and removing interferants.

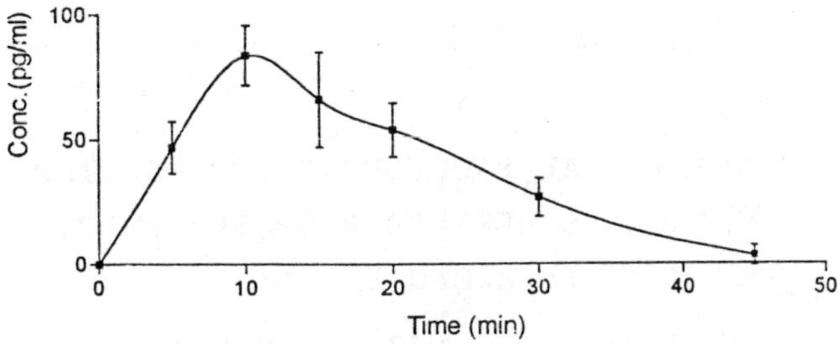

Fig. 2. Plasma levels of sCT, in relation to time, in four volunteers who got a dose of sCT (1200 I.U.). Bars indicate S.D.'s.

Fig. 2 shows a CT time course in volunteers. The method's range is 150 down to 5 pg/ml (LLOQ). The method is rugged and apt for measuring sCT in human plasma.

References

1. Milhaud, G., Perault. A.M. & Moukhtar, M.S. (1969) *C.R. Acad. Sci. Paris T 261,* 813-816.
2. Aliapoulios, M.A., Goldhaber, P. & Munson, P.L. (1966) *Science 151,* 330-331.
3. Friedman, J. & Raisz, L.G. (1965) *Science 150,* 1465-1467.
4. Martin, T.J., Robinson, C.J. & MacIntyre, I. (1966) *Lancet I,* 900-902.
5. Kohler, H. (1968) *Schweiz. Med. Wochen. 19,* 728-729.
6. Gudmunsson, T.V., Galantte, L. & *6 co-authors* (1969) *Lancet I,* 443-446.
7. Body, J.J. & Heath, H., III (1983) *J. Clin. Endocrin. Metab. 57,* 897-903.
8. Singer, F.R. & Habner, J.F. (1974) *Biochem. Biophys. Res. Comm. 61,* 710-716.
9. Deflos, L.J., Roos, B.A., Bronaert, D. & Parthemore, J.G. (1975) *J. Am. Med. Assoc. 227,* 403-406.
10. Sizemore, G.W., Heath, H., III & Larson, J.M. (1975) *J. Clin. Invest. 55,* 1111-1118.
11. Babe, M., Esbrit, P. & Rapado, A. (1985) *Clin. Chem. 31,* 786.
12. Heath, H., III, Body, J.J. & Fox, J. (1994) *Biomedicine & Pharmacotherapy 38,* 241-245.
13. Johansson, A., Stanley, C.J. & Self, C.H. (1985) *Clin. Chim. Acta 148,* 119-124.
14. Seth, R., Motte, P. & *6 co-authors* (1988) *J. Endocrin. 119,* 351-357.
15. Lynch, C., Seth, R., Bates, D.L. & Self, C.H. (1988) *J. Immunoassay 9,* 179-192.

#A-5

ANALYTICAL PROCEDURES FOR MONITORING HIGH MOL. WT. FRAGMENTS DERIVED FROM THE rRICIN IMMUNOTOXIN ZD 0490 IN HUMAN PLASMA

S.J. Bacon, C.J. Bailey† & R.J. Ruane

Safety of Medicines Department, Zeneca Pharmaceuticals, Alderley Park, Macclesfield SK10 4TG, U.K.

Zeneca ZD0490 is an Ab linked chemically at various sites to 1-4 molecules of recombinant ricin A and intended for treatment of specific antigen-bearing tumours. This article outlines the structure, pharmacology and analytical methods for ZD0490 and describes the assays adopted for determining, in human plasma, ZD0490-derived high mol. wt. fragments, free r-ricin A and free C242 Ab, focusing on method development for free C242 Ab.*

The procedure eventually adopted for free C242 Ab entailed SDS-PAGE/Western blotting followed by immunodetection and quantitation by scanning image densitometry. Complex problems were encountered with this methodology when applied to human plasma, notably the cross-reactivity of endogenous plasma material with the detection Ab. For the developed technique we ascertained the LLOQ and precision, both moderately satisfactory. Altogether the wide array of techniques used during this programme manifests the vast resources needed for regulatory analysis of such compounds.

ZD0490 is a tumour-selective Ab (C242Ab, mol. wt. ~150 kDa) linked at various sites *via* a hindered -S-S- bridge to between 1 and 4 molecules of r-ricin A (mol. wt. ~30 kDa). The intact immunoconjugate (ZD0490) has a mol. wt. in the range ~180 kDa 1:1 to ~270 kDa 4:1, and is a potential treatment for colorectal cancer. ZD0490 is shown diagrammatically in Fig. 1A.

The mode of action of this 'magic bullet' therapy can be broken down into three phases. The first (Fig. 1B) is localization. The C242Ab, specific for the antigen C242 expressed on the majority of tumour cell membranes, effectively localizes the toxin r-ricin A at the site of action, i.e. the tumour cell.

---

†addressee for any correspondence

**Abbreviations*.- Ab, antibody; BSA, bovine serum albumin; ELISA, enzyme-linked immunosorbent assay; HRP, horseradish peroxidase; LLOQ, lower limit of quantification; LOD, limit of detection; OPD, *o*-phenylenediamine dihydrochloride; PAGE, polyacrylamide gel electrophoresis; PBS, phosphate-buffered saline; PVDF, polyvinylidene difluoride; r-ricin, recombinant ricin A.

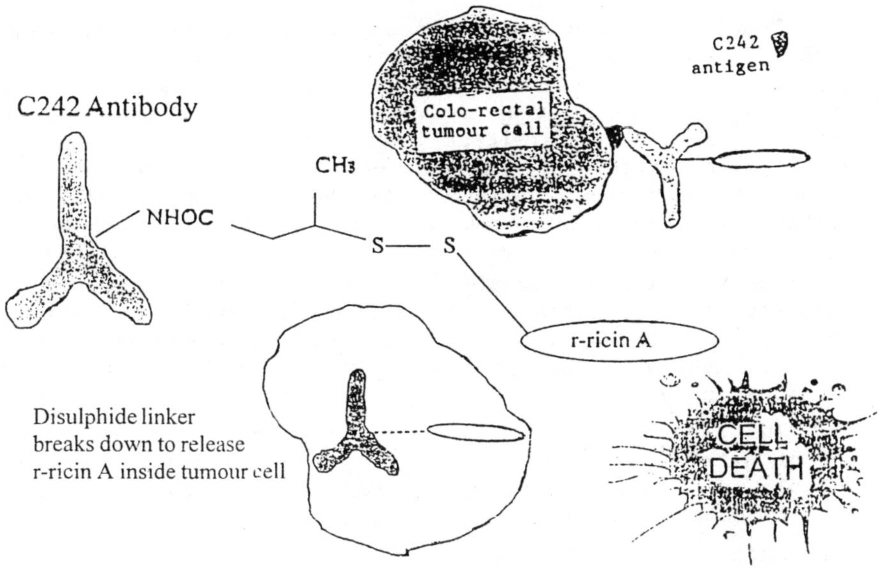

Fig. 1. **A**: Diagrammatic structure of ZD0490. **B**: Phase 1 localization; **C**: internalization; **D**, cell death.

The localized immunoconjugate (Fig. 1B) is internalized, and finally the -S-S- bridge is broken down and r-ricin A is released inside the cell, causing cell death (Fig.1, C & D); the killing ratio is believed to be 1:1, i.e. one cell death per molecule of ZD0490.

For the determination of ZD0490 plasma concentrations in man, analytical procedures were required for intact conjugate (ZD0490) and the possible high mol. wt. fragments, free ricin A and free C242Ab derived from it. The assays for determining these two fragments in plasma are described here, with particular attention to problems encountered during the development of the assay for free C242Ab.

EXPERIMENTAL (Incubations performed at room temp.)

ZD0490 was assayed by a non-competitive ELISA [1, 2] that was specific for intact conjugate.

Free r-ricin A

The assay involves a pre-filtration step which removes material of >100 kDa mol. wt., followed by a ricin-specific non-competitive double-Ab sandwich ELISA, utilizing an avidin/biotin system [3] to improve the LOD.

Centrifugal filtration.- PBS dilutions of standards and/or samples are aliquotted into non-sterile Ultrafree-MC filter units (100,000 NMWL; Millipore, Watford) and then centrifuged.

ELISA.- A 96-well flat-bottomed microtitre plate (Nunc, Roskilde, Denmark) is first sensitized for ricin using an appropriate dilution of rabbit anti-ricinus communus (Sigma). Following a blocking step using a BSA/Tween 20 solution, the antigen (in the ultrafiltrate) incubation is done, then the primary incubation of rat anti-ricin (Biotechnology Dept., Zeneca Pharmaceuticals) followed by the secondary incubation of biotinylated rabbit anti-rat IgG (Vector Labs., Peterborough). An avidin/HRP (enzyme)-labelled biotin complex (Vector Labs.), previously prepared, is then incubated on the plate prior to addition of the substrate, OPD (Sigma)/urea-adducted hydrogen peroxide (Merck/BDH, Poole). When adequate colour development has been achieved, the reaction is stopped with a concentrated citric acid solution.

The absorbance (optical density) response is measured at 450 nm with a suitable plate reader. Concentrations are then calculated from a 5-parameter (5PL) fitted curve of response *vs.* log concentration of the standards.

Free C242 Ab: ELISA method

This is a non-competitive ELISA whose purpose is to determine the total C242 Ab (C242Ab) present in human plasma. The difference between the results of this assay and those of the ZD0490 assay gives the plasma concentration of free C242Ab. This ELISA is therefore specific for both types of C242Ab.

A 96-well flat-bottomed microtitre plate is coated with polylysine (Sigma; 50 µg/ml in PBS). After antigen addition (colo-rectal membrane, from our Biotechnology Dept.; 250 µg/ml PBS) and incubation, 1% aqueous glutaraldehyde (Fluka Chemls., Gillingham) is briefly incubated on the plate, followed by a blocking step with BSA in PBS. Then standards/samples (maybe diluted) are added to the plate and incubated with shaking.

Thereafter an appropriate dilution of the detecting Ab, HRP-labelled goat anti-mouse IgG (Sera-Lab, Crawley Down), is added; after incubation, an OPD-based substrate solution is added. Subsequent procedures (colour-reaction stoppage, colour measurement and calculation) are as for free r-ricin A.

Free C242 Ab using SDS-PAGE with immuno-blotting/-detection

The SDS-PAGE separation is based on a SDS discontinuous system (U. Larmmli [4]) since it gives better resolution than a continuous system. Image densitometry is finally performed.

A 7.5% resolving or separating gel is prepared using a Protein II Slab Gel vertical gel-electrophoresis apparatus (Bio-Rad Labs., Hemel Hempstead). After its polymerization a stacking gel solution is prepared and layered onto the resolving gel and allowed to polymerize in the presence of a comb having enough teeth to provide the appropriate number of wells for the standards/samples to be analyzed.

Plasma samples/standards are diluted in buffer containing an electrophoretically transferable pink dye, pyronin Y (Sigma) and denatured by heating (few min. in water-bath). Then each is loaded into its appropriate well and the gel electrophoresed at a constant current using a suitable power pack (e.g. Bio-Rad PowerPac 3000) until the dye markers are 1-2 cm from the bottom of the gel. The stacking gel is removed and the resolving gel placed on a Hybond-PVDF membrane (Amersham Internatl., Little Chalfont) previously 'wetted' in aqueous methanol. Buffer-soaked filter papers are placed above and below the gel/membrane 'sandwich' and the stack is placed between the electrodes of an electrophoretic transfer or blotting unit (e.g. the Millipore Milliblot graphite electroblotter system). With a between-electrodes constant current, semi-dry transfer/blotting occurs.

When transfer is complete (no pinkness evident on the gel) the immobilized proteins can be probed immunologically. After a blocking step using non-fat powdered milk (e.g. Marvel), the membrane is incubated with HRP-labelled goat anti-mouse IgG (Sera-Lab). Enhanced chemiluminescence reagent (ECL; Amersham) is then poured onto the PVDF membrane (protein side up) and left to incubate for 1 min. The membrane is then quickly dried with tissue paper, wrapped in cling-film and exposed to photographic film, e.g. Hyperfilm-ECL (Amersham). Various exposure times are used, and the films generated are developed using an auto-developer (e.g. an X2 automatic film processor; X-ograph, Malmesbury) for speed and optimization of the exposure times. The visualized protein bands are then scanned using an image densitometer, e.g. Imagemaster Desk Top Scanner (DTS), Pharmacia Biotech, St. Albans), and the images analyzed/quantified using Imagemaster software (Pharmacia).

RESULTS AND DISCUSSION

Performance of analytical methods

ZD0490.- The assay for intact ZD0490 was used successfully for analyzing both toxicological and clinical samples. It had LLOQ 30 ng/ml and, at this level, C.V. ~20%.

Free r-ricin A.- The ricin-specific ELISA necessitates a filtration step to remove material of mol. wt. >100 kDa (including conjugated r-ricin A). The assay's quantitative range is 17-217 ng/ml with C.V. ~20% over this range. The assay served to analyze clinical samples, and no free r-ricin was manifest.

Free C242 antibody.- Since ELISA methods had proved successful with the two previous elements of this project, a third immunoassay, to determine plasma concentrations of free C242Ab, was considered. This ELISA, it was hoped, would be specific for both types of C242Ab, i.e. it would provide total C242Ab concentration data and the results of the analyses for ZD0490 could be subtracted to give a free Ab concentration:

total C242Ab[C] - ZD0490[C] = free C242Ab[C]

Past experience with difference assays such as this suggested that they were unreliable; but this path was pursued since a quick solution to the problem was needed. Initial experiments to optimize steps such as dilutions and incubations were carried out using free C242Ab, and these went well. However, when it came to the most important experiment, viz. to determine cross-reactivity, the results proved disappointing: responses for equimolar amounts of ZD0490 were significantly lower than those for free C242. For a difference assay to work in the present case, there must be 100% cross-reactivity between free C242Ab and conjugated C242Ab (ZD0490), i.e. the assay must give the same response to equivalent concentrations of the analytes for the estimate of total Ab to be correct.

Two explanations for why this did not happen are that the two forms of C242 might have different interactions with the antigen (C242Ag), or that the detection Ab's used in the ELISA might have different affinities for the free C242 and ZD0490 or a combination of the two. In either case, these differences may have been due to steric hindrance at the binding site.

The SDS-PAGE approach

At this stage in the assay development, the only techniques investigated had been those routinely used in an environment where assays have to provide high throughput and robustness, specificity, sensitivity and reproducibility so that the data produced can be used to derive pharmacokinetic and tentative metabolism information, for submission to regulatory bodies. Analytical techniques used in the biochemistry and toxicology areas were available but, to our knowledge, had not been tried and tested in an environment of routine analysis with high throughput. We sought, therefore, to establish whether or not one methodology in particular could fulfil the above criteria such as robustness, specificity and sensitivity.

SDS-PAGE, Western blotting and immuno-detection, practised by molecular biologists and protein chemists for analyzing high mol. wt. polypeptides and proteins [4-6], are techniques which, in principle, offer a way of separating, detecting and possibly quantifying free C242Ab. Peptide charge density is the basis for separations without SDS, but mol. wt. is the sole basis in the presence of SDS, which negates any residual charges that the peptides may possess. Once separated, the bands can be transferred, usually electrophoretically, to a more stable or manageable matrix such as nitrocellulose, nylon or PVDF - the 'Western blotting' process. The immobilized polypeptides can then be probed immunologically with enzyme-labelled, specific detection Ab's which in the presence of a suitable substrate, allow the bands to be visualized and measured.

The assay initially developed for C242Ab (Fig. 2) uses ECL as the measurable response because sensitivity is better than for either colorimetric or conventional CL methods. The light

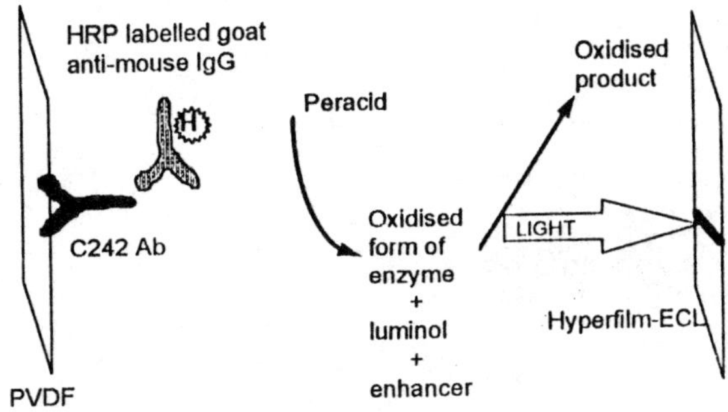

Fig. 2. Immuno-detection of free C242Ab following SDS-PAGE and Western blotting.

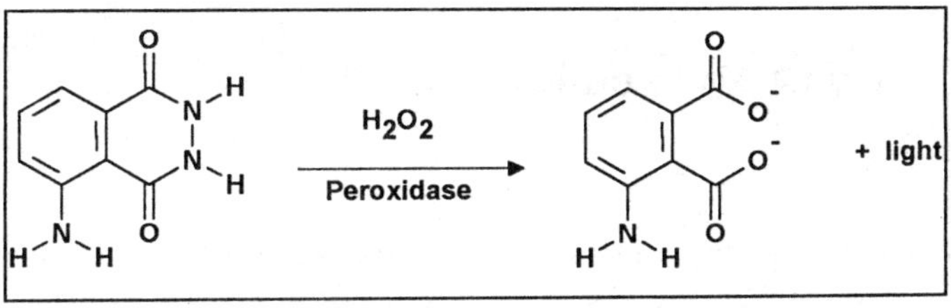

Fig. 3. The enzyme-mediated, light-producing oxidation of luminol.

is produced by the oxidation of luminol (Fig. 3). The phenolic enhancer added to the reaction mixture prolongs light emission, allowing band visualization, exposure to special photographic film and densitometric quantitation of the protein bands. This assay succeeds using spiked buffer samples, giving good LOD's and relatively 'clean' gel lanes and clear protein bands; but with spiked human plasma the detection Ab (HRP-labelled goat anti-mouse IgG) cross-reacts with plasma proteins and 'blacks out' the gels completely (Fig. 4).

The screening of different batches of HRP-labelled anti-mouse IgG Ab's, and of alternative dilutions, showed only one batch that gave satisfactory results. This was an HRP-labelled anti-mouse IgG Ab raised in a goat, produced by Sera-Lab. An example of the various calibration ranges investigated using this Ab is shown in Fig. 5. When the optical densities of the bands corresponding to free C242Ab are plotted against concentration, a graph such as appears in Fig. 6 is obtained. This 6-point curve (1.25-40 µg/ml) was used to generate precision data (Table 1), using concentrations expected to fit on the linear portion (~5-40 µg/ml; Fig. 6) of the sigmoidal calibration line (4-parameter logistic).

Table 1. Summary of precision data (C.V.'s, %; based on 3 replicates in each of 4 runs) in the linear range (see text).

µg/ml: Nominal	Mean found	C.V.: Intra-batch	Inter-batch	Total
10	10.2	14.7	8.0	16.7
20	20.0	9.9	12.3	15.9
30	29.5	13.2	5.8	14.4

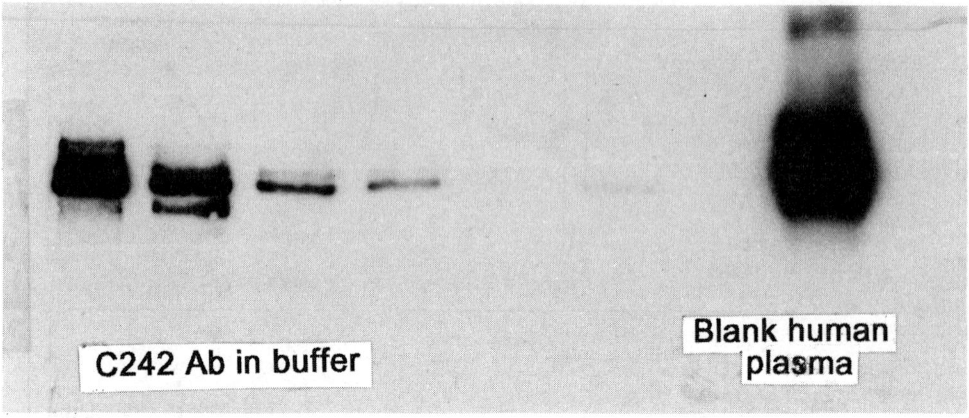

Fig. 4. Assay cross-reactivity with plasma proteins.

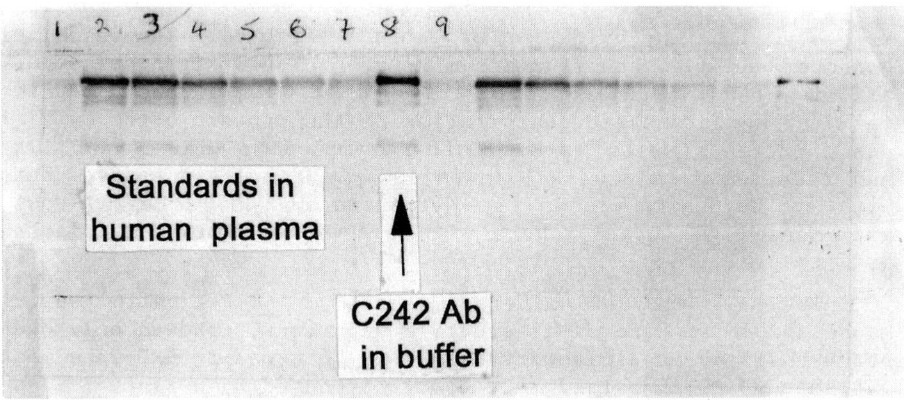

Fig. 5. Free C242Ab in human plasma: gel bands for a calibration range, 1.25-40 µg/ml (blank human plasma in lanes 1 & 9; lanes 10-16 are repeats of 2-8).

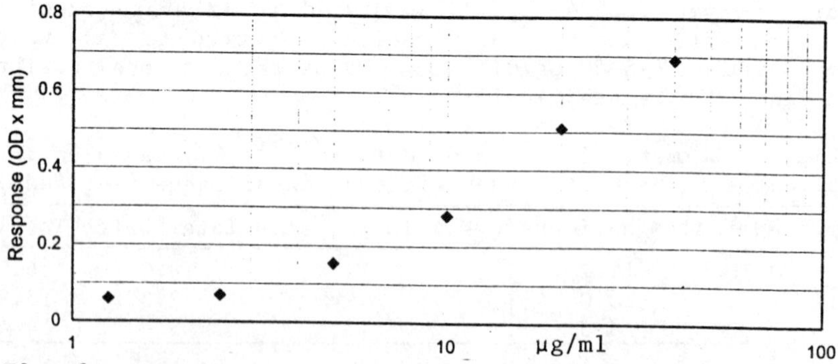

Fig. 6. Concentration *vs.* response plot for free 242Ab in human plasma in the range 1.25-40 µg/ml.

CONCLUSIONS

Validated quantitative assays for intact immunoconjugate, ZD0490 and free r-ricin A are in place. For analysis of free C242Ab over the linear range, 5-40 µg/ml, SDS-PAGE with Western blotting and immuno-detection appears to work well; the assay is specific, reproducible (C.V. ~15%) and adequately sensitive.

The results obtained during these investigations suggest that the SDS-PAGE separatory approach represents an alternative to conventional techniques that may be of use for both qualitative and quantitative work with future immunoconjugates that may come into development. However, further work is needed, which will probably include lowering the calibration range, direct quantitation, and methods to improve sample throughput.

Acknowledgement

Thanks are expressed to Sue Hall and Rick Wright (Research Department, Zeneca Pharmaceuticals).

References

1. Goldstein, L., Klinman, N.R. & Manson, L.A. (1973) *J. Natl. Cancer Inst. 51,* 1713-1715.
2. Engvall, E. & Perlmann, P. (1971) *Immunochemistry 8,* 871-879.
3. Kendall, C., Ionescu-Matiu, I. & Dressman, G.R. (1983) *J. Immunol. Meth. 56,* 329-339.
4. Larmmli, U. (1970) *Nature 227,* 680-685.
5. Towbin, H. & Gordon, J. (1984) *J. Immunol. Meth. 72,* 313-340.
6. Hames, B.D. & Rickwood, D., eds. (1990) *Gel Electrophoresis of Proteins. A Practical Approach.* 2nd Edn., IRL Press/Oxford University Press, Oxford. See particularly pp. 1-139 (Hames).

#A-6

SAMPLE PREPARATION AND SUBSEQUENT STEPS: STRATEGIES FOR PEPTIDES AND PEPTIDOMIMETICS

H.M. Hill†, J.N. Robson, T.A.G. Noctor, K. Tennant, D. Thomas, S.D. Clarke & J.E.C. Burnett

Corning Hazleton Europe, Dept. of Biopharmaceutical Analysis, Harrogate, N. Yorks. HG3 1PY, U.K.

Selected endogenous proteins and peptides have been used in replacement therapy for many years, e.g. insulin. These and a wider range of proteins are now produced largely through recombinant technology, e.g. GH, GCSF and EPO. Many of these are quantitated in biological matrices with minimal sample clean-up followed by IA or in some cases enzymic assay or bioassay. Such compounds usually have mol. wt. >2 kDa.*

Low-mol. wt. peptides have been modified in order to extend their biological half-lives and increase their potencies. The resulting physico-chemical characteristics are complex, making sample clean-up a largely empirical procedure. Techniques used include various protein precipitation procedures such as addition of methanol, MeCN or $ZnSO_4$ (with sulphuric acid as well as along with methanol); polar extracting solvents such as ethyl acetate and DCM may also be used. Varying the ratio of organic solvent (extractant or precipitant) may give some selectivity. Further refinements are possible using SPE. Yet the presence of related or similar compounds in the extract may call for further differentiation by separatory techniques including HPLC and CE. Whilst CE has limited use in assaying biofluids for peptide-related analytes, their assay at low concentrations has been revolutionized by linking HPLC on-line to MS(-MS).

Peptides used therapeutically are a diverse group containing synthetic and/or naturally occurring amino acids and at least one peptide bond. The examples cited here range in mol. wt. from 140 Da to 2 kDa, equivalent to 2-15 amino acid residues.

---

†addressee for any correspondence

**Abbreviations include:* AVP, Arg-vasopressin; GH, growth hormone; GCSF, granulocyte colony stimulating factor; EPO, erythropoietin. APcI, atmospheric pressure chemical ionization; CE, capillary electrophoresis; EC, electrochemical (detection); MALDI, matrix assisted laser desorption ionization. DCM, dichloromethane; MeCN, acetonitrile; PCA, perchloric acid; PSA, pentanesulphonic acid; TBP, tributyl phosphine; TCA/TFA, trichloro/fluoro-acetic acid. IA, immunoassay; ELISA, enzyme-linked immunosorbent assay IRMA, immunoradiometric assay.

Peptides can be synthetically modified in order to increase absorption (through the gut, nasal membrane or lung), minimize degradation, or improve action at the receptor. In consequence the physico-chemical properties of peptidomimetics ('pseudo-peptides') can differ significantly from those of 'naturally' occurring peptides.

The range of analytical methodologies available for quantitating such peptides/peptidomimetics is now briefly described in respect of relative advantages and disadvantages together with examples from differing molecular types. Based on these experiences, some basic strategies are proposed for the quantitation of peptides/peptidomimetics. The analyte range reviewed includes hormones (both native and modified), metalloproteinase and protease inhibitors, and immunosuppressants.

QUANTITATIVE METHODS

IA's are widely used for quantitation of proteins and peptides. While there are few alternatives where the mol. wt. is >3 kDa, smaller molecules can be chromatographically separated and quantitated by various detection systems.

IA's (e.g. RIA, ELISA, IRMA) [1] have the advantage of high throughput and sensitivity coupled with minimal sample preparation. However, there are a number of major disadvantages which prevent them being the first line of attack. Smaller molecules in particular can suffer matrix and other interferant effects. Thus non-specific binding can interfere with the quantitation process, as can interference by other matrix-related components such as lipids. Some such effects can be minimized by dilution or Ig addition prior to assay, or by some form of sample preparation, e.g. SPE, or solvent extraction to remove interferants. Such problems associated with protein/peptide IA are mentioned in other arts. in Sects. #A and #B and in a review [1].

Closely related to IA's are assays dependent on 'biological' effects, such as receptor binding, enzyme activity changes and others with biological/physical endpoints, e.g. reductions in viscosity.* In general these assays require no sample preparation and, while not being specific for the analyte, they do reflect its biological activity.

Application of HPLC with quantitative endpoints such as UV, EC, fluorescence or MS(-MS) detection is more apt for low-mol. wt. peptides/peptidomimetics. Although concentrations of naturally occurring peptides in biomatrices are low, typically sub-ng/ml, many peptidomimetics are present, after dosing, at ng/ml levels. Even if higher, UV detection is of little use at

---

*'*Edited out*': the term 'bioassays', best confined to assays where culture growth is measured. Examples of radio-receptor assays (RRA's): Vol. 16 [Plenum; ed. Reid *et al.*] (p. 45 and, for β-endorphin, p. 70); Vol. 20 (p. 289). In Indexes see 'Enzyme'.

therapeutic concentrations, although it may serve for *in vitro* stability studies both in biomatrices and in the presence of cell-free extracts, as well as for method development. Most peptides/peptidomimetics have little native fluorescence; but using fluorescent reagents, notably *o*-phthaldialdehyde, assays can achieve sensitivities in the sub-ng/ml range - as can assays with EC detection and MS(-MS) systems although in either case the inherent properties of the molecule significantly influence assay sensitivity. However, the use of MS(-MS) is of especially wide applicability, and so is becoming the technique of choice.

CASE STUDIES

Ideally, minimizing sample preparation should result in a high throughput along with improved precision and accuracy (due to minimizing the number of manipulative steps). In practice, some sample preparation may be needed, usually to improve assay sensitivity by concentrating the analyte in a smaller volume prior to chromatography. Further clean-up may be necessary, depending on the nature of the detection system. EC systems are relatively sensitive and specific, but are prone to contamination from dirty samples.

The examples reviewed span quantification of drugs in biofluids and stability studies both *in vitro* in biological matrices and under varied physico-chemical conditions (especially pH).

Metalloproteinase inhibitor BB-94, considered more deeply by Moore *et al.* in art. #B-4.- In brief, the method uses automated SPE. Following protein precipitation with methanol, the sample is applied to the cartridge and washed with buffer and then methanol/water; the cartridge is dried and washed with hexane, prior to elution with hexane/ethyl acetate (or more appropriately methanol, as proposed by Moore). Analysis is by gradient RP-HPLC, with MS-MS(APcI) detection.

Protease inhibitors.- A wide range of analytes has been evaluated, all of which are in an early stage of development and subject to secrecy agreements. In general they bear many similarities to one another including the presence of at least one peptide bond. However, depending on the substituents in the molecule they vary in polarity, and hence water-solubility, from high to low. Therapeutic concentrations range from the low ng/ml to µg/ml. Accordingly sample preparation ranges from simple protein precipitation, with methanol or MeCN, through extraction with ethyl acetate (and mixtures with hexane) to neat hexane. Although extractability from water/buffers may be good (>50%), recoveries are drastically reduced when extracting from plasma, perhaps ameliorated by initial protein precipitation. Using the most elementary sample-preparation procedures, and using methanol/ammonium acetate (or formate) mobile phases, sensitivities in the low ng/ml range are attainable, using LC-MS(ESP). Increased sensitivity can be achieved by further concentration of the extract and/or use of LC-MS-MS(APcI).

In general, recoveries from rat plasma may be significantly lower than from human plasma, using the same extraction technique. Recoveries can be improved by denaturing or solubilizing the protein by addition of agents such as ethylene glycol. Even with low recoveries, e.g. 18% for a protease inhibitor using ethylene glycol/hexane (only 1-2 % with neat hexane), the assay sufficed for toxicokinetic studies, the intra-day values being:

nominal concn. 3 ng/ml:	accuracy 96.7%;	precision 17.2%;
15	103.3	5.6
40	104.5	5.4

Immunosuppressants typically have mol. wt. ~1 kDa. Cyclosporin, a cyclic peptide having 11 amino acid residues, mol. wt. 1.2 kDa, is typical of this group. Generally extraction with 'ethers' and/or C-18 SPE cartridges provides a good yield with sufficient cleanliness to run on LC-MS-MS(ESP). LC-MS-MS(APcI) gives added sensitivity. Isocratic or gradient mobile phases based on ammonium acetate/MeCN (range of ratios ~1:1 to 1:2) are used. Temperatures of 60°-70° using RP columns (including CN) result in acceptable chomatography.

Acetyl neurotensin [2], Ac-Arg-Arg-Pro-Tyr-Ile-Leu, was studied in respect of proteolysis by intestinal enterocytes, the major products being Ac-Arg-Arg-Pro-Tyr and Ile-Leu. The conditions were as follows: Beckman C-18 column; for gradient: A, 0.05% (v/v) TFA and B, 95% MeCN/5% TFA; sample prepared by stopping enzymic hydrolysis with 0.1 M HCl; 210 nm detection.

Bradykinin analogue, RMP-7 [3], Arg-Pro-Hyp-Gly-Thr-Ser-Pro-4-Me-TyrΨ(CH_2NH_2) Arg: *in vitro* hydrolysis was shown to lead to four major degradation products, using a Vydac C-18 column with a wide range of gradients - typically A, 0.1% TFA and B, MeCN (detection at 210 and 280 nm).

Dynorphin A 1-13 [4], Tyr-Gly-Gly-Phe-Leu-Arg-Arg-Ile-Arg-Pro-Lys-Leu-Lys: 'metabolism' in blood or plasma was found to give major (4) and minor (>10) identifiable transformation products. HPLC was done with Spherisorb ODS 2 and Waters C-18 (10 µm) columns using a gradient: 0.03% aqueous TFA 30 mM PSA containing MeCN, 10% (A) to 90% (B). For sample preparation, 1 vol. of precipitating agent (5% aqueous $ZnSO_4$/MeCN/methanol, 5:3:2 by vol.) is added to plasma; the supernatant from centrifugation at 10,000 **g** is injected. MALDI-MS was used off-line to quantitate the 'metabolites'.

Ergopeptines yield, on hydrolysis, lysergic acid, ammonia, pyruvate, Pro and another amino acid - Phe, Leu, Ile or Val. Typical ergopeptines (and doses) are: ergotamine (2 mg), dihydroergotamine (30 mg) and bromcriptine (25 mg). These compounds are highly metabolized, which, together with the relatively low doses involved necessitates, for defining kinetic profiles, highly sensitive assays (2-50 pg/ml) that demand good clean-up and concentration. This was achieved using a disc-based, mixed mode SPE system (SPEC MP1): following application to the cartridge and washing with methanol and acetic acid, the analyte was eluted with a mixture of DCM, propan-2-ol and ammonia. Quantitation was carried out by LC-MS-MS(APcI).

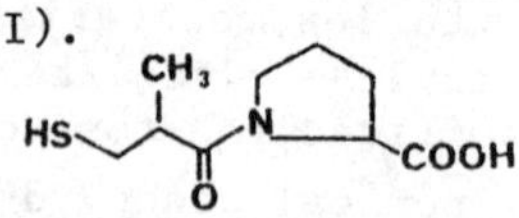

Captopril, 1-[2-(S)-3-mercapto-2-methyl-1-oxopropyl]-L-proline, is related to a naturally occurring nonapeptide teprotide, Pyro Glu-Trp-Pro-Arg-Pro-Gln-Ile-Pro, that inhibits peptidyl dipeptidase, which catalyzes the destruction of bradykinin as well as the synthesis of angiotensin II. As a thiol, captopril reacts with free thiols (including itself to form a dimer) and with plasma proteins to form adducts. [These aspects, and possible assay methods, are surveyed by A. Rakhit in Vol. 18 (Plenum; ed. E. Reid *et al.*).- *Ed.*] Free captopril must therefore be 'protected' or captured as soon as the sample has been taken, in order to prevent the oxidative formation of disulphides. This can be carried out by immediately reacting with methylmaleimide. The free thiol can then be released with 1% TBP and further reacted with a fluorescent maleimide, e.g. **N**-(1-pyrenyl)-maleimide; the product may be extracted with 1-chlorobutane. The process is entirely dependent upon the reactivity of TBP, which degrades rapidly.

The derivatized captopril is subjected to RP-HPLC using a mobile phase of 1% acetic acid/MeCN/methanol (50:40:10 v/v); the fluorescence parameters are ex: 273 nm, em: 389 nm.

Model hexapeptides [5], e.g. Val-Tyr-Pro-Asp-Gly-Ala (all L).- The amino acid before and after the Asp residue was modified so as to evaluate the Asp residue's susceptibility to hydrolysis at different pH's. A range of HPLC systems was used to evaluate the hydrolysis products: typically a C-18 column (Hypersil 5 µm or Spherisorb 3 µm); elution by 12% MeCN in 0.1% TFA, with 214 nm detection (also characterization by FAB-MS).

Metkephamid [6], Tyr-D-Ala-Gly-Phe-N-Me-Met-NH_2.CH_3CO_2H, was run on a Lichrospher 100RP-8 (5 µm) column; a step gradient was used (A: pH 6.9 phosphate buffer; B: methanol). Detection was by fluorescence, the residue from drying down the original extract (ethyl acetate pH 9, ×4) having been reacted with OPdiA.

Other case studies featured in Vol. 16 of this series [7], **including** the HPLC-EC measurement of neuropeptides, e.g. AVP, oxytocin, Met- and Leu-enkephalin, angiotensin II and neurotensin (G.W. Bennett & co-authors). Sample preparation entailed homogenization and sonication in 1 M acetic acid, followed by centrifugation at 4000 **g** for 20 min; a Waters C-18 column (10 µm) was used, with 0.15 M pH 5.8 NaH_2PO_4/methanol (60:40).

Also in Vol. 16 is an elegant article by R.F. Venn and co-authors [cf. #B-3 in present vol.] on assay of β-endorphin and related peptides. Losses due to adsorption onto glass are emphasized; the glassware used for sample preparation had to be silanized. Sample clean-up was done with C-18 SPE cartridges, pre-washed with methanol/0.1% TFA. After loading and washing, peptides were eluted with 1.5 ml methanol/0.1% TFA (80:20). It was vital to add mercaptoethanol to protect S-containing peptides such as Met-enkephalin [formula on p. 61 - *Ed.*].

The articles in Vol. 16 (1986; [7]), generally pre-dating the LC-MS-MS(APcI) era, focus on achievement of sensitivity but do provide useful guidance on sample preparation. Guidance is also given, in the article by J.A. Smith & M.J. O'Hare on assay of protein growth factors, concerning choice of LC conditions, with attention to size-exclusion as well as RP packings. Until fairly recently, IA has provided the most sensitive detection system, albeit in conjunction with LC.

STRATEGY FOR SAMPLE PREPARATION AND SUBSEQUENT STEPS

In general, the physico-chemical nature of peptides and peptidomimetics is such that they tend to be readily soluble in water, as must be considered in the sample-preparation process - which should also take account of the separation/quantitation steps. The first line of attack in our laboratory is LC-MS-MS. Use of single-stage MS can be a solution where plasma levels are high (μg/ml). At lower concentrations (ng/ml) the presence of a myriad of naturally occurring peptides influences both the sensitivity and the selectivity of LC-MS methodology, entailing long separation times and sophisticated chromatography.

Use of LC-MS-MS significantly improves the inherent specificity of the assay. Until the advent of Turbo Ionspray, the general philosophy was to firstly use Ionspray (pneumatically assisted electrospray). If the required sensitivity could not be achieved, APcI often provided it - but sometimes necessitated modification of the sample preparation procedure. Little or no modification is needed now that Turbo Spray is available and, in contrast with Ionspray, gives sufficiently improved sensitivity. The relative merits of these interfaces are discussed in art. #E-1 (D. Thomas) and other articles in this volume.

Columns used by us for LC-MS-MS are generally RP, C-18 or C-8 (50 × 4.6 mm), usually base-deactivated. However, LC-MS generally needs longer columns (100-150 mm) to improve specificity, or may need gradient elution to separate low mol. wt. endogenous peptides not removed by sample preparation. Even the short columns used for LC-MS-MS may be run with a gradient if, despite the added specificity of MS-MS, improved chromatographic separation is needed.

Mobile phases used for assaying peptides/peptidomimetics with UV, fluorescence or EC detection appear, from perusal of the above case studies, to be firstly compatible with LC-MS and secondly very similar to one another. Most use methanol and/or MeCN as the organic component along with ammonium acetate (or formate) buffers modified with TFA (0.05-1.0%). In many cases aqueous solutions of TFA with organic modifier suffice. Use of high concentrations of organic modifier (>60%) can minimize molecular aggregation of analytes which results in a reduced response at relatively high analyte concentrations.

Sample preparation [Vol. 23 has an elementary survey, #I-4]

The overall strategy in developing a sample preparation protocol is to begin simply, and increase complexity only if required. Wherever possible, the simplest technique compatible with LC-MS(-MS) is used. A confounding phenomenon which seems to conflict with the fact that peptides/peptidomimetics are typically water-soluble is the low recovery in sample preparation, even if not by solvent extraction. The probable reason is that many of these agents have companion binding proteins; they may have an inhibitory or stimulatory influence, affecting both metabolism and pharmacokinetics as well as pharmacological manifestations (art. #A-1; e.g. [8]). Thus, binding proteins for GH modulate its pharmacological response and also $t_{\frac{1}{2}}$.

Similarly, naturally occurring protease inhibitors exist in many cases as an irreversible complex between the protease and its inhibitor [8]. This may have significant implications in devising assays for synthetic protease inhibitors that have been developed as a therapy for control of HIV. On this basis it would not be a surprise to find that many peptides/peptidomimetics have been designed to modify the activity of endogenous proteins (e.g. enzymes or receptors) that are likely to be closely associated with them, such that it may be difficult to recover them from the biological matrix.

Protein precipitation.- The range of precipitants used is extensive, including MeCN, methanol, PCA, tungstic acid, ammonium sulphate and various combinations thereof (cf. a sample-preparation review that ante-dated the HPLC era [9] - *Ed.*). While the main function of protein precipitants is to remove plasma proteins, it is obvious that other factors are operative. Varying degrees of selectivity can be achieved by use of these agents.

Table 1 summarizes a report on the protein-precipitation effectiveness of different agent using various volumes.

Table 1. Relative efficacy of various protein precipitants, reported by J. Blanchard [10], *courtesy of author and Elsevier.*

	Vol. of precipitant added per vol. of plasma			
Precipitant	0.2	0.8	2.0	4.0
10% (w/v) TCA	99.5	99.7	99.8	99.8
Tungstate-H_2SO_4	3.3	99.7	99.8	100.0
$ZnSO_4$-NaOH	41.1	92.7	99.3	99.8
Acetonitrile (MeCN)	13.4	88.1	99.7	99.8

Evidently 'complete' removal of proteins by MeCN needs a 2:1 ratio; with 1:1, ~10% remains in the supernatant, and can physically interfere with the chromatography and/or the MS interface - especially with APcI, where blockages that reduce assay robustness may occur. If the solvent mix resulting from deproteinization is incompatible with the mobile phase, which would be significantly 'weaker' than the injection medium, it may be necessary to evaporate under N_2 at an elevated temperature or, preferably, by vacuum centrifuge. The residue can then be taken up in mobile phase or a compatible solvent.

Other procedures for precipitating proteins may be more efficient on a volume-for-volume basis; but most entail radical changes in pH or addition of ions which may significantly complicate further sample processing or subsequent HPLC. Hence use of miscible organic solvents should be the first choice. Use of MeCN or methanol results in a homogeneous supernatant of solvent and plasma water. However, if NaCl or other 'salting out' agents are added, this can afford a partition between the aqueous portion and MeCN, the drug usually partitioning into the organic fraction. This may not always happen, and it may be a fallacious belief that analyte recoveries will be high.

Unreliability of recovery is confirmed by unpublished work (H.M. Stevens & V.W. Bunker) which D.B. Campbell cited early in this series [11]. The recoveries of 36 basic analytes from whole blood were shown to be highly variable. The procedures tried were also fairly harsh - tungstic acid, aluminium sulphate or chloride, 5 M HCl at 80°, TCA and PCA - and it was therefore advocated that aluminium chloride be used as "being the safest". Poor recovery following protein precipitation is usually due to occlusion of the analyte into the precipitate, rendering it inaccessible to the solvent.

Occlusive analyte loss in protein precipitation techniques can be minimized, but not necessarily overcome, by solubilization of the protein using ethylene glycol and/or pH adjustment. Other protein precipitants [10] can be used to afford subtle sample clean-up prior to processing on cartridges; but usually some protein escapes precipitation and so further processing by SPE is requisite (see art. #D-4n). Such subtle separations can result in the removal of endogenous peptides that co-elute with the analyte of interest and have the same MS-MS characteristics. However, it is not possible to derive empirical rules to effect such separation; hence the procedure is largely trial-and-error.

Procedural variations including liquid-liquid extraction.- Where sample concentration is necessary to achieve the desired sensitivity, extraction by organic solvent or by SPE is needed instead of, or as well as, protein precipitation. In general, because of the polar nature of peptides, ethyl acetate is the most favoured extraction solvent, and pH values in the region of 4-7.

For modified peptides/peptidomimetics, solvents less polar than ethyl acetate may be used. Increasing the complexity of the work-up procedure has a further advantage which D. Thomas mentions in art. #E-1: matrix effects can be minimized.

CONCLUDING COMMENTS

In summary, LC-MS(-MS) provides a sensitive and relatively specific detection system which, in our hands, has served for all peptides/peptidomimetics investigated. Moreover, sample preparation can be minimized. Provided that pitfalls of such an approach are recognized, this should be the first approach.

References

1. Ferraiolo, B.L. & Mohler, M.A. (1992) in *Protein Pharmacokinetics and Metabolism* (Ferriaolo, B.L., Mohler, M.A. & Gloff, C.A., eds.) [*Pharmaceutical Biotechnology*, Vol. 1], Plenum, New York, pp. 1-33. [Cf. art. #A-1, this vol.]
2. Bai, J.P.F. & Chang, L-L. (1995) *Pharm. Res. 12,* 164-167.
3. Straub, A.J., Akiyama, A., Parul, P. & Musso, G.F. (1995) *Pharm. Res. 12,* 305-308.
4. Müller, S. & Hochhaus, G. (1995) *Pharm. Res. 12,* 1165-1175.
5. Oliyai, C. & Borchardt, R.T. (1994) *Pharm. Res. 11,* 751-758.
6. Langguth, P., Merkle, H.P. & Amidon. G.L. (1994) *Pharm. Res. 11,* 528-535.
7. Reid, E., Scales, B. & Wilson, I.D., eds. (1986) *Bioactive Analytes, including CNS Drugs, Peptides and Enantiomers* [Vol. 16, this series], Plenum, New York, 422 pp.
8. Mohler, M.A., Cook, J.E. & Baumann, G. (1992) as for 1., pp. 35-71.
9. Reid, E. (1976) *Analyst 101,* 1-18.
10. Blanchard, J. (1981) *J. Chromatog. 226,* 455-460.
11. Reid, E., ed. (1976) *Assay of Drugs and other Trace Compounds in Biological Fluids* [Vol. 5, this series], North Holland, Amsterdam; see pp. 105-106 (D.B. Campbell); also pp. 107-120 (arts. on solvent extracion, including pH aspect).

#A-7n

A Note on

LC-MS PROFILING OF PEPTIDOMIMETICS AND PEPTIDES FROM BIOLOGICAL SOURCES

Gerald C. DiDonato,† Anne Arroyo, Mark Hail, Susan Taylor, Bethanne Warrack & Mike Lee

Analytical Research & Development,
Bristol-Myers Squibb Pharmaceutical Research Institute,
P.O. Box 4500, Princeton, NJ 08543, U.S.A.

Peptidomimetic drugs are of increasing interest to the drug industry due to their similarity to many pharmacologically active compounds. Concomitantly, much interest is aimed at the characterization of peptides and proteins as targets for pharmacological intervention. Characterization both of peptidomimetic drugs in biological fluids and of peptides and proteins is often rate-limiting in the discovery and development of novel pharmaceutical agents. To accelerate the process we have developed rapid protocols, based on LC-MS, for the profiling of metabolites [1], for the determination of drug levels in rat plasma, and for the mol. wt. characterization of proteins.

The determination of HIV protease inhibitors in rat plasma utilizes a water/acetonitrile mobile phase (0.05% TFA) with a small Keystone Partisil C-18 column (100 × 2 mm). Preparation of rat plasma samples following drug administration involves extraction with ethanol, evaporation, and reconstitution in ethanol/water (1:1 by vol.). The use of targeted column-eluate analysis employing MS-MS with MRM* eliminates signals due to other components present in the sample and thus facilitates the use of an isocratic, fast method of analysis. Analysis times for various HIV protease inhibitors are 2-4 min per sample.

Universal LC-based methodologies applicable to a wide variety of compounds have been developed for the profiling of biotransformation products. Our present mode of operation involves the implementation of an LC method for each metabolism sample, often entailing use of several different columns.

---

†addressee for any correspondence
**Abbreviations*.- MRM, multiple reaction monitoring; MS-MS, tandem MS; ESI (synonymous with ESP), electrospray ionization. LC connotes HPLC.

Methods require ~1 h of analysis time. The development of a universal method for analyzing metabolites, which embraces peptidomimetic drugs in biological samples including bile, eliminates the need for lengthy method-development time and hardware conversion, thus reducing the overall analysis time.

The characterization of proteins has been accelerated by development and proliferation of MS with ESI (ESP). However, the quality of data is often compromized by the presence of salts and detergents. Infusion of such samples leads to the overlap of multiply charged peaks, making mass assignments difficult and decreasing the signal/noise ratio. Currently, e.g. for large hydrophobic peptides, we utilize on-line MS(ESI) methods using polymeric microbore HPLC columns for separation and polymeric protein-trap cartridges for sample concentration and clean-up.

Reference

1. DiDonato, G.C. & Gaskell, S.J. (1994) in *Biological Mass Spectrometry: Present Future* (Matsuo, T., ed.), Wiley, Chichester, pp. 541-557.

Comments on #A-1, M.A. Mohler: PROTEIN PHARMACEUTICALS
#A-2, D. Gygax: 'BIOLOGICAL' ASSAY APPROACHES
#A-3, R.F. Venn: NEUROPEPTIDE STUDIES
#A-6, H.M. Hill: SAMPLE PREPRATION & THEREAFTER

Concerning binding of a protein agent (or Ab) to its binding protein, **D.J. Anderson asked Mohler** whether in general this would be predicted to result in decreased clearance, rather than increased clearance such as occurs with haptoglobin when heamoglobin binds to it. **Reply.**- There is no guideline for predicting how a protein complex will clear compared with the free protein. Each case has to be investigated individually. **I.D. Wilson asked** whether iodination of the protein affects its metabolic fate. **Reply**: no marked influence; but one should check whether any properties of the protein are altered.

Comment by H. Birnböck to Mohler, concerning Ab influences on antigen determinations by IA, and conversely.- Those working in the field must be aware of such interdependencies so as to interpret results correctly and detect artifacts. Effects are strongly dependent on the kind of Ab's used in the assay system (e.g. monoclonal or polyclonal, species, etc.).

Replies by Gygax to H. Birnböck: (1) Only one of the PK trials was done with FOBIA rather than ELISA. (2) With FOBIA, operation with fibres in parallel might indeed increase sample throughput; but we have not tried this. **Answer to R. Venn,** concerning the finding that the sandwich FOBIA gave a sigmoid rather than linear curve, with a dynamic range of only 2 orders of magnitude.- This seems to be typical, but the results were limited by the number of observations made with the fibre.

Editor's annotation.- Assay (SPE, RP-HPLC-UV) of an analog of hirudin's C-terminal region in rat/monkey plasma (LLOQ 0.1 µg/ml): Knadler, M.P. (1992) J. Chr **579**, 371-375.

Venn, replying to J. Stobaugh.- The locus for hydrolysis is the fluorophor backbone, not the coupling functional group.

Replies by Hill.- To R. Laplanche.- Indeed one must employ an i.s. for LC-MS work. **To J. Moore** concerning the lability of many peptide analytes in rat plasma:- use of acetonitrile-treated whole blood for assay provides a remedy. **Answer to H. de Bree**: chlorobutane gives a good extraction recovery and has the advantages of a low b.p. and low water-solubility.

P. Heizmann raised a general point, namely whether it is advantageous to use, as a protein precipitant, $ZnSO_4$ rather than acetonitrile or methanol. He wondered if it breaks drug-protein binding more efficiently. **Hill's reply.**- Our use of $ZnSO_4$ in a peptide assay served not to increase recovery but to remove an interferant. Different approaches have subtle specificities. Often one has to try a whole range (useful guidance has been given by Blanchard - cited in #A-6).

Citations contributed by Senior Editor For journals, Chr denotes Chromatography

PARTICULAR PEPTIDES (besides entries prefixed ° in Sect. #ABC)

Neuropeptide 'bioanalytical case histories'

Prior to C-18 HPLC-fluor of rat-plasma hexapeptides having **neurotensin** activity, RP-SPE was applied: Machida, R. et al. (1993) Biol. Pharm. Bull. **16**, 43-47.

Neuropeptide GF antagonists in mouse samples (including 1 M acetic acid homogenates of tissues) were isolated by SPE and run on a C-18 column with EC detection.- Cummings, J. et al. (1994) J. Chr **B 653**, 195-203. A similar method with HPLC-UV, adaptable (e.g. HPLC-RIA) to various neuropeptides, was used for synthetic opioids/anti-opioids in plasma, e.g. enkephalins: Partilla, J.S. et al. (1995) J. Chr **B 667**, 49-56.

On facing p.: formulae for some ***enkephalins*** *(including synthetic)*

Another approach for serum enkephalins is immunoaffinity isolation, derivatization and HPLC-fluor: Hendrickson, T.L. & Wilson, G. (1994) J. Chr **B 653**, 147-154. With a methionine-enkephalin RIA kit, opioid precursors in bovine pituitary (pH 8 homogenates; + enzyme inhibitors) were determined after gel permeation and trypsin/CNBr treatment of fractions.- Fridland, G.H. & Desiderio, D.M. (1994) Anal. Lett. **27**, 2117-2125.

'Bioanalytical case histories' for other peptides

Salmon calcitonin (salcatonin; cf. S-H. Hsu's art., #A-4) in 5 µl samples was assayed by acetonitrile/ClO_4^- precipitation, derivatization (precautions against adsorption losses) and gradient C-18 (TSK gel) HPLC-fluor.- Fukuda, T. et al. (1995) Biomed. Chr **9**, 52-55 [cf. Lee, H.S. et al. (1995) J. Chr **B 673**, 136-141].

SPE (carefully designed) and HPLC (column unstated) followed by ELISA served for an immunomodulatory peptide, **thymosin-α_1**: Naylor, P.H. et al. (1994) J. Liq. Chr **17**, 3541-3551.

'Determination of **Peptide S20** in Human Plasma Using Post-column Photolysis...' [a pentapeptide; Genentech]: an HPLC-EC approach, usable for others that lack aromatic amino acids.- Chen, L. et al. (1993) J. Pharm. Biomed. Anal. **11**, 999-1007.

Studies on a **renin inhibitor**: Ziegler, K. & Seeberger, A. (1993) Biochem. Pharmacol. **45**, 909-916 [cf. below, Vol. 22].

Albumin-aurothiomalate complex separated from other human serum components using an N-methylpyridinium polymer column: Nishimura, K. et al. (1992) J. Liq. Chr **15**, 791-803.

In the PRESENT SERIES, peptide material appears in Vols. 16 and (IL assay) 18 [publ. Plenum], and 22 (p. 133; enzyme-inhibition assay of a small renin-inhibitory peptide) and 23 (p. 296). In each Analyte Index (Cumulative in Vol. 22) peptides are in #IIa.

[Reminder: the SERIES TITLE prior to Vol. 23 was *Methodological Surveys in Biochemistry and Analysis*; all ed. by E. Reid *et al.*]

The renin inhibitor study in the Vol. 22 citation (Burton & Simmonds) focused on screening candidate drugs by adapting a kit method (^{125}I) for renin, based on its catalysis of the conversion of angiotensin to ANG I. It was, fortunately, possible to selectively inactivate endogenous renin by brief heating.

Another Vol. 22 art. (Niessen & co-authors; p. 255) amplifies Mohler's allusion, in art. #A-1, to the effectiveness of LC-MS in characterizing bioactive proteins.

J. Chamberlain, in the 2nd edition of his pioneer guide 'The Analysis of Drugs in Biological Fluids' (1995; CRC Press), cites bioanalytical methods for endogenous proteins and peptides that can be used therapeutically. For natriuretic factors of atrial (**ANF**) and related type, specific immunoassay poses problems, as discussed in a survey of ways to investigate output and fate by J.A. Clerico & co-authors (1995; Int. J. Clin. Pharm. Res. **15**, 65-86).

Since **opioid/enkephalin** agents, still a 'growth area', get several mentions in the present book, some formulae are given.-

Chemical structures for opioid peptides. a) Leu⁵-enkephalin; b) Met⁵-enkephalin; c) DAGO; d) [D-Cys², D-Cys⁵] enkephalinamide

Enkephalins (some synthetic, e.g. C). [From an art. in 'IBL', Oct. 1985]

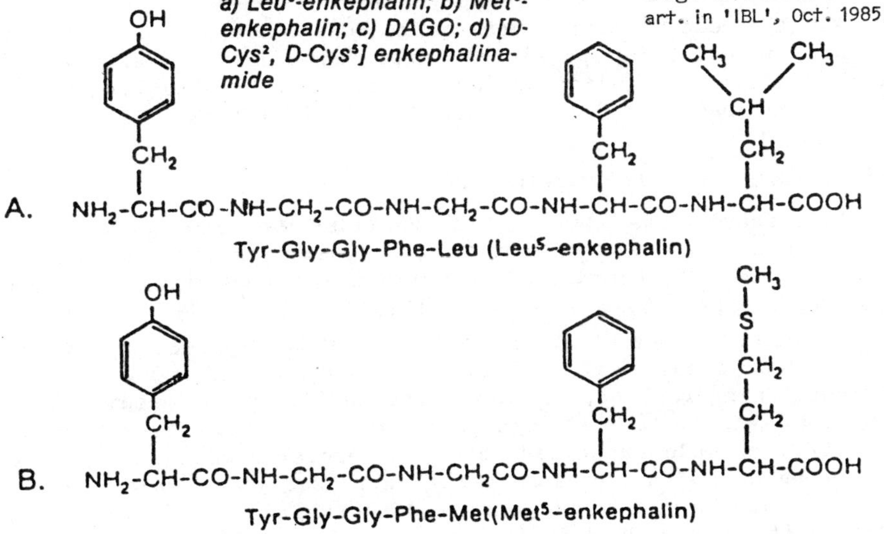

C. Tyr-D-Ala-Gly-MePhe-NH$(CH_2)_2$-OH(DAGO, μ specific agonist)

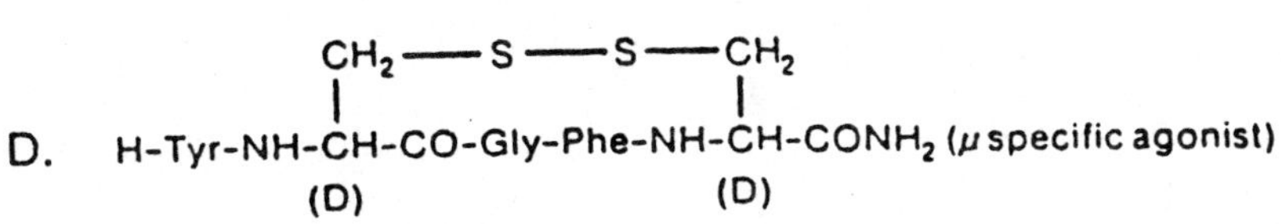

SOME GENERAL APPROACHES (not all bioanalytical)

Proteins, peptides and polynucleotides: HPLC surveys occupy two issues (1992) of J. Chr **B 599** (1-274).

'Separation and Analysis of Peptides and Proteins': a review, according to separation technique, by C. Schöneich & 7 co-authors: Anal. Chem. **65**, 67R-84R (1993).

'Sample Preparation [etc.]...Peptides': a survey nominally on substance P (unadecapeptide; fragments too) but bears on other peptides; embraces HPLC, detection and RIA including labelled antigen isolation.- Rissler, K. (1995) J. Chr **B 665**, 233-270.

Literature centred on isolation and separation

For SPE a range of bonded phases has been studied with small peptides, synthetic and casein-derived, differing in polarity.- Herraiz, T. & Casal, V. (1995) J. Chr **A 708**, 209-221.

Compared with porous poly(styrene-DVB), polyalkylene HPLC stationary phases (hydrophobic; wide pH-stability) were very apt for isolating peptides, e.g. substance P, GnRH diastereo-isomers and fibrinogen peptides.- Krause, E. & Bienert, M. (1992) J. Liq. Chr **15**, 1773-1784 (see also 1797-1806, Zou, H. et al.: RP-HPLC retention behaviour of various dipeptides).

Alkyl-bonded alumina, preferably long-chain and narrow-pore (fused-microplatelet particles surpassed spherical particles), was effective for octapeptides and protein tryptic fragments: Haky, J.E. et al. (1992) J. Liq. Chr **15**, 1831-1852.

Literature centered on detection

'Selective Determination of Peptides Containing Specific Amino-acid Residues... .': review, embracing selective derivatization and CE, besides HPLC-EC (if a tyrosine ring), HPLC-UV and HPLC-fluor.- Cui, H. et al. (1995) J. Chr **A 704**, 27-36.

Coulometric detection of di-/tri-/tetra-peptides separated by RP-HPLC can be effected by post-column conversion, with a Cu-containing reactor, to oxidizable Cu(II)-peptide complexes: Kubo, H. et al. (1994) Anal. Sci. **10**, 579-582. From a study (by CZE) of model peptides devoid of aromatic amino acids, it was concluded that UV absorbance increases with chain length; hence proteins are most readily detectable.- Becklin, R.R. & Desiderio, D.M. (1995) Anal. Lett. **28**, 2175-2190.

Section #B

AGENTS WITH PEPTIDE-TYPE TARGETS

The suffix '**n**' in #B-2n signifies a 'NOTE' rather than a main article. COMMENTS (Forum discussion) are on p. 64.

OVERLEAF: 'protease' nomenclature and types

Proteolytic enzymes: class names and types

The following recommendations have not been enforced in the editing of the present book:-

Enzymes that cleave peptide bonds

The International Union of Biochemistry recommends the term **'peptidase'** to signify a peptide (bond) hydrolase, but does not debar the older terms given in parentheses ():-

Peptidases (Proteases):
- Endopeptidases (Proteinases)
- Exopeptidases

A fuller classification was included in an art. (J.K. McDonald; 55 refs.) on connective tissue degradation/remodelling in this series, Vol. 17 (1987; ed. E. Reid *et al.*; Plenum). Its focus on tumour spread relates to arts. #B-3 and #B-4 in this Section. (Vol. 17 title: *Cell Membranes and Disease, including Renal.*)

#B-1

DEVELOPMENT AND CROSS-VALIDATION OF RIA AND LC-MS-MS METHODS FOR DETERMINING THE FIBRINOGEN-RECEPTOR ANTAGONIST L-734,217 IN PLASMA

J.D. Gilbert†, E.L. Hand, D.A. McLoughlin, J.D. Ellis, T.V. Olah, A.S. Yuan & C. Fernández-Metzler

Department of Drug Metabolism I,
Merck Research Laboratories, West Point, PA 19486, U.S.A.

L-734,217 is an orally active, non-peptide fibrinogen receptor antagonist. For its measurement in plasma, two sensitive and specific assays have been developed. (1) After SPE isolation the drug is converted to its N-*TFA* derivative prior to analysis by HPLC with APcI-MS-MS detection, using a homologue as i.s.; the LLOQ is 0.4 ng/ml (C.V. 9.3%). (2) Direct RIA is performed on 50 µl plasma samples, the immunogen used to raise the Ab having been prepared by coupling to BSA* via *a dinitrophenylene bridge and the radioligand by reaction of the drug with* ^{125}I*-labelled Bolton-Hunter reagent. The RIA is sensitive to 0.2 ng/ml (C.V. 6.2%) and shows no matrix interference. Its specificity and accuracy were established by parallelism and plasma-fractionation experiments, and also by cross-validation with the LC-MS-MS procedure.*

Platelet aggregation and activation play a major part in thrombosis, and anti-thrombotic therapy can be attained through pharmacological antagonism of platelet aggregation [1-4], whose final step involves binding of plasma fibrinogen to an activated platelet membrane-bound glycoprotein complex termed GpIIb/IIIa [5-7]. A minimal amino acid sequence necessary for this reaction has been identified as Arg-Gly-Asp in fibrinogen [8], and it is possible to design synthetic analogues that can block the interaction competitively and thus inhibit platelet aggregation.

L-734,217 is a specific, orally active fibrinogen receptor antagonist [9]. Sensitive and specific methods for its quantitation in biofluids were required to support the pre-clinical and clinical studies needed to develop this drug candidate. We had previously developed RIA's for analysis of the i.v.-active fibrinogen receptor antagonists MK-852 [10] and MK-383 [11];

†addressee (location WP 26-354) for any correspondence

**Abbreviations* [others in Preface; some introduced or varied editorially].- APcI, atmospheric pressure chemical ionization, generally +ve ion mode; DFDNB, 1,5-difluoro-2,4-dinitrobenzene; DMF, dimethylformamide; DNFB, dinitrofluorobenzene; MeCN, acetonitrile; MRM, multiple reaction monitoring; MS-MS, tandem mass spectrometry; TFA, trifluoro-acetic acid/-acetyl.

RIA was chosen because sensitivity had to be high (0.2-1 ng/ml) and the lack of good chromophores and limited scope for derivatization handicapped the development of HPLC-UV assays. Both RIA's had proved sensitive and, as predicted, specific: after i.v. administration, little if any metabolism took place and the drugs were excreted essentially unchanged.

L-734,217, being given orally, might be more prone to biotransformation. Whilst RIA's are generally sensitive and very convenient where large numbers of samples are to be assayed, their potential lack of specificity is always a concern, and one that can only be fully addressed after the assay is established and plasma/urine specimens from test animals or man have become available. As it typically takes 4-6 months to develop an RIA, much development time is lost if the assay ultimately proves non-specific and alternative methods then need to be sought.

LC-MS-MS has lately come into use for sensitive and very specific measurement of drugs and their metabolites in biofluids at low concentrations [e.g. 12] (cf. several arts. in this book and Vol. 23 - *Ed.*). The technique is particularly useful for substances which lack good chromophores; being rugged and highly specific, it needs only the simplest of extractions and enables use of fast LC with conventional columns and mobile phases. It has served for one of the two methods - the other being direct RIA - now developed for L-734,217 in animal and human plasma, meeting strict criteria even at <1 ng/ml concentrations.

EXPERIMENTAL

Materials were as follows.- Ammonium acetate, dimethylaminopyridine, sodium heparin, BSA (#A7638): Sigma (St. Louis). BSA to prepare immunogen, and rabbit gamma globulin: Calbiochem (La Jolla, CA). Sheep anti-rabbit gamma globulin serum: Arnel Products (New York). DFDNB and TFA: Aldrich (Milwaukee, WI). TFA anhydride and diisopropylethylamine: Pierce (Rockford, IL). Other chemicals/solvents: Fisher Scientific (Fair Lawn, NJ). ^{125}I-labelled Bolton-Hunter reagent, 5700 µCi/µg, in anhydrous benzene: NEN ((Wilmington, DE). Bakerbond C-18 cartridges, 3 ml × 200 mg: J.T. Baker (Phillipsburg, NJ). Hydrocarbon-free air, and other gases, all 99.999%: Matheson (Morris Plains, NJ).

Standards.- Stock solutions of L-734,217 and its i.s. contained 1 mg/ml in deionized water. Working standards (Table 1) were prepared in assay buffer for RIA or, water-diluted, added to control plasma (50 µl to 0.5 ml) for the LC-MS-MS assay.

RIA

Immunogen.- For coupling L-734,217 to BSA *via* a dinitrophenylene bridge (Fig. 1), using ^{3}H-drug to determine the extent of incorporation into BSA, the drug (17.7 mg, 50.2 µmol, 2.5 ml) containing a trace of ^{3}H-drug (1.6 pmol, 3.4 µl, 22.4 Ci/mmol; in ethanol) was reacted with 10.2 mg (50 µmol) DFDNB in ethanol (1 ml) with stirring in the dark (room temperature). Reaction was complete within 30 min as determined by TLC.

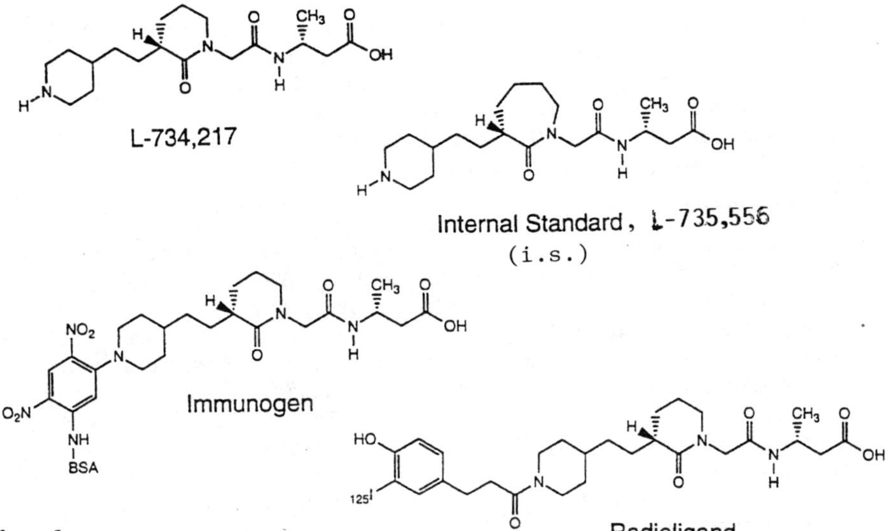

Fig. 1.
Structures of L-734,217, L-735,556 (i.s. for the LC-MS-MS assay) and the immunogen and radioligand used in developing the RIA.

Immunization.- Each of 4 female New Zealand white rabbits was given ~1 mg of immunogen, emulsified with 1 vol. of Freund's Complete Adjuvant (Sigma), by intradermal injections at multiple sites with s.c. and i.m. injections also. Rabbits were given booster doses i.m. and s.c. in Freund's Incomplete Adjuvant at 1, 3 and 6 months. Antisera were collected just prior to the 3- and 6-month boosts and were stored at -70° prior to use.

Radioligand.- Benzene was removed from a vial of ^{125}I-labelled Bolton-Hunter reagent by a dry N_2 stream just before use. To the residue (1 mCi, 0.45 nmol) was added 10 µg L-734,217 (0.028 µmol) in 50 µl ethanol containing 6 µM diisopropylethylamine. The mixture was stirred vigorously for 1 min and then kept for 5 h (room temp.) with stirring. The reaction mixture (~45 µl) was fractionated by HPLC using a 3.0 × 30 cm microBondapak C-18 column (10 µm; Waters, Milford, MA). The mobile phase (1 ml/min) comprised MeCN (35% v/v) in 0.1% aqueous TFA. Counts were monitored by a flow-through radioisotope detector, #170 (Beckman, Irvine, CA). No unreacted Bolton-Hunter reagent was found; the product radioligand (structure: Fig. 1) eluted at 11.3 min, well resolved from unreacted drug (~2 min). Fractions (4 × 0.5 ml) were collected (#203 Gilson fraction collector) as the tracer came off the column. The two with highest radioactivity were tested for immunoreactivity and then pooled and stored at -20°; the radioligand thus stored had ~8 weeks' useful life.

Fractionation of plasma samples was done to verify assay specificity. Plasma was collected from rats (at 1 and 24 h) and dogs (1 and 4 h) given L-734,217 orally [dose: see later]. For

RIA specificity checking, 200 µl MeCN was added to 100 µl plasma to precipitate protein, and 50 µl aliquots of the supernatant were put onto a Spherisorb C-18 column (5 µm; 250 × 4.6 mm). A linear gradient of 5 to 40% MeCN was applied at 1 ml/min, and fractions collected at 0.5 min intervals were analyzed using RIA. L-734,217 eluted after 15 min.

RIA procedure

The buffer was pH 7.5 0.05 M phosphate containing 0.05 M EDTA, 0.1% BSA and 0.04% sodium azide. The assay was generally configured to use 50 µl test plasma or, for standards, control plasma from the species concerned; if animal plasma with high drug concentrations, test samples and all standards and QC's were pre-diluted 1:10 with buffer. For QC's, drug-free plasma was spiked at 1, 10 and 100 ng/ml. The rabbit antiserum ('1st Ab') was diluted 1:30,000 in assay buffer containing 50 µg/ml rabbit gamma globulin to provide bulk for the double-Ab precipitation. The radioligand (200,000 cpm/ml) was, for pipetting convenience, pre-mixed with buffer containing sheep anti-rabbit globulin ('2nd Ab') serum at 1:50 dilution. In our experience the binding at equilibrium is similar whether the 1st and 2nd Ab's are added simultaneously or separately.

Assay steps.- With an automatic pipetting station (APS; Micromedic, Horsham, PA) reagents were added to 75 × 12 mm glass culture tubes:- 50 µl standard solution/test plasma/QC plasma; then 50 µl control plasma/buffer/buffer respectively (standards as in Table 1; 0.01-10 ng/assay tube). To all tubes were added 100 µl 1st Ab/gamma globulin reagent, 100 µl 2nd Ab/radioligand (20,000 cpm/tube) and 700 µl assay buffer. Non-specific binding (nsb) was determined in tubes lacking 1st Ab and analyte but containing the 'carrier' gamma globulin. The tubes were kept overnight (~18 h, room temp.), centrifuged (45 min, 800 **g**; supernatants decanted and discarded), and inverted to drain on paper towels. The consolidated pellets were counted for 3 min in a Micromedic #Apex 10/600 multi-detector gamma counter. All samples were assayed in triplicate.

Calculation.- Binding (B, cpm) at each concentration was expressed as % of the observed binding (B_o) without analyte:

$$\%B = 100 \times (B - nsb)/(B_o - nsb)$$

Calibration curves were constructed by plotting $\log_{10}$ of the plasma concentration in standards against Logit B where:

$$\text{Logit B} = \text{Log}_{10}[\%B/(100 - \%B)]$$

A third-order polynomial curve fit, rather than the conventional linear regression, is routinely used in our laboratories since this enhances accuracy at the concentration range's extremities. Appropriateness of fit was evaluated by reading standards as though unknowns. L-734,217 concentrations in test samples were determined by interpolation from the standard curve. Plasma containing >200 ng/ml was diluted with control plasma and 50 µl aliquots taken for re-analysis.

LC-MS-MS assay

Separation was effected using Hewlett-Packard (Wilmington, DE) modules - #1050 solvent delivery system and #1050 autoinjector having a 100 µl loop; 20 µl injected - and an Ultracarb C-18 250 × 4.6 mm column (5 µm; Phenomenex, Torrance, CA). Elution, at 1 ml/min, was with MeCN/methanol/10 mM ammonium acetate (54:4:42 by vol.) containing 0.1% v/v TFA. Processing (80-100 per batch) was controlled by a Macintosh IIFX Computer running Sciex (Thornhill, Ont.) RAD (routine acquisition & display) software.

MS-MS was done with a Sciex #API III triple quadrupole mass spectrometer and an interfaced nebulizer probe set at 500°. The nebulizing gas (air) was at 800 psi; the auxiliary (make-up) gas setting was 1 L/min. APcI was effected by a corona discharge needle (+5 µA). Positive ions were sampled into the analyzer (*via* a 0.0045-inch pinhole; curtain gas N_2 at 0.7 L/min), programmed to admit the protonated molecules $[M+H]^+$ at m/z 450 and 464 for the N-TFA derivatives of the drug and i.s. respectively, *via* the first quadrupole mass filter (Q1) with collision-induced fragmentation in Q2 (collision gas argon, 370 × 10^{12} atoms/cm^2); *via* Q3, product ions were monitored at m/z 347 (drug) and 361 (i.s.). The orifice potential, electron multiplier and interface heater settings were 60 V,-3.5 kV and 50° respectively; the dwell time was 200 msec. Peak area ratios arising from MRM of the derivatives, analyte (m/z 450 → 347)/i.s. (464 → 361), were computed using Sciex's McQuan software. Calibration curve construction employed a weighted (reciprocal of concentration) linear least-squares regression; interpolations gave the drug values.

Extraction prior to LC-MS-MS.- Into a 75 × 12 mm glass culture tube were placed 0.5 ml plasma, 50 µl (10 ng) i.s. and 1.0 ml 0.1 M K_2HPO_4 (pH 10.6). After brief vortexing the mixture was transferred to a Bakerbond C-18 cartridge preconditioned by washing with methanol and water (each 2 × 3 ml). The samples were loaded initially by gravity and then with vacuum (~1 ml/min). After washing with 3 ml 0.1 M K_2HPO_4 and 3 ml water (washings discarded), the drug and i.s. were eluted with 3 ml of methanol, which was then removed at 40° in a TurboVap TV Evaporator (Zymark, Hopkinton, MA). Traces of water were removed by reconstituting in an azeotropic mixture (400 µl) of toluene/ethanol (1:1) and re-evaporating. N-TFA derivatives were prepared by adding to the dried residue 400 µl 10 mM dimethylaminopyridine in MeCN and 40 µl TFA anhydride. After 30 min at 40° acylation was essentially complete. The reagents were removed under a N_2 stream and the derivatized extracts reconstituted in 200 µl of mobile phase prior to analysis by LC-MS-MS. QC spiking levels were 0.4, 1, 4 and 40 ng/ml. The absolute recoveries of L-734,217 and its i.s. were respectively 74.6% and 74.4%.

Statistical analysis of data

Data from cross-validation experiments were compared using the paired t-test and regression analysis programs in a statistical analysis package (Excel, Version 4.0; Microsoft, Redmond, VA). Fixed and proportional errors were determined from the 95% confidence intervals around the linear regression's slope and intercept functions; random error was assessed from the C.V. of the individual ratios for (test data)/(reference data). Predicdicted reproducibility (R_C for plasma concn. C) between different analytical methods (here n = 2) was calculated by a modification of a relationship proposed by Caulcutt & Boddy [13]:

$$R_C = t\sqrt{n}\,\sqrt{\sigma^2_{C1} + \sigma^2_{C2}}$$

t being the t-statistic ($P = 0.05$) at the combined D.F. used to compute the precision of the 2 individual assays, whose variances at concn. C are σ_{C1} and σ_{C2}. If no significant differences exist between methods (i.e. reproducibility of results from the two methods is defined only by the precision of the individual assays) then 95% of test-method results should be within the range C ± R_C obtained by the reference method.

RESULTS

The aim was to develop a reliable, fast assay for L-734,217 in plasma down to ~0.5 ng/ml. Very satisfactory RIA's had been developed for two drugs in the same class; but RIA development using polyclonal antisera typically takes 3-6 months, and an interim assay was needed to support preclinical development. LC-MS-MS was selected because of its ruggedness, ease of automation and high sample throughput. Accuracy and specificity were obligatory, with intra-assay C.V.'s at the LLOQ not exceeding 10%.

LC-MS-MS

Attempted RP-HPLC on underivatized L-734,217 with no ion-pairing agent was not fully satisfactory, and sensitivity was poor in both -ve and +ve ionization modes. Hence the drug was analyzed after conversion to its N-TFA derivative which considerably reduced its polarity and enabled use of MS-compatible LC conditions. Fig. 2 shows +ve product mass spectra, and Fig. 3 MRM results for extracts of plasma from a dosed dog; the chosen packing (see above) yielded symmetrical peaks that had convenient retention times.

Assay parameters.- Calibration curves (Fig. 4) showed good linearity and reproducibility using a weighted (1/C) regression. Peak-area ratios were virtually zero for drug-free plasma. No interferences from endogenous plasma components were seen with pre-dose samples from rats (6), dogs, monkeys and volunteers. Using QC's prepared independently of the standards, values for precision and accuracy (mean % found/actual) were acceptable over the range 0.4-200 ng/ml (Table 1).

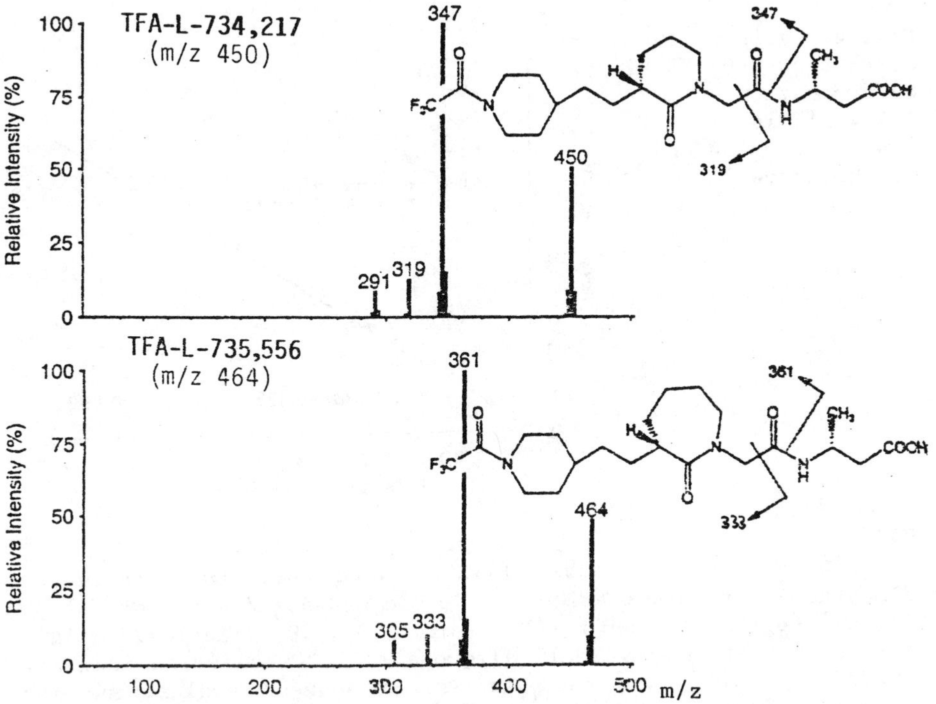

Fig. 2. Positive product ion mass spectra (background subtracted) of the protonated molecular ions of drug and i.s. TFA derivatives.

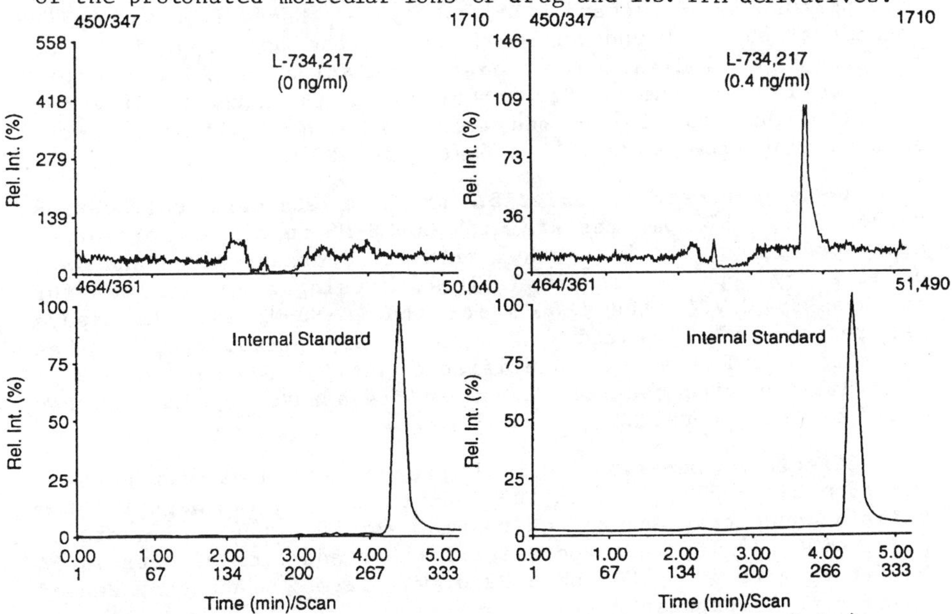

Fig. 3. Chromatograms obtained by MRM of extracts of dog plasma containing i.s. and also *(right)* 0.4 ng/ml L-734,217.

Fig. 4. Calibration curve for LC-MS-MS determination of L-734,217 in plasma. (INSET: the low end of the curve.)

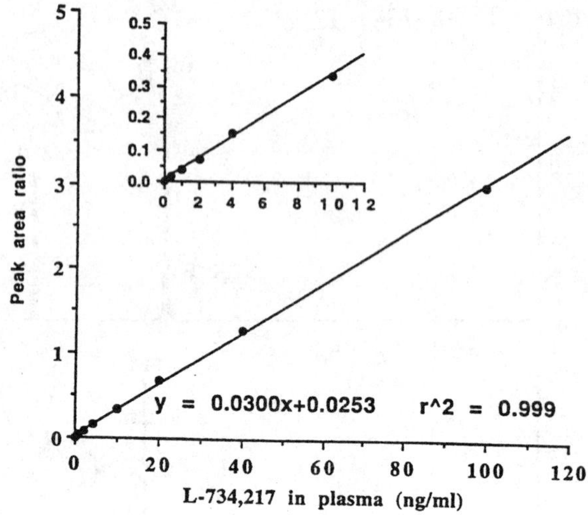

RIA

Antisera.- One rabbit (#1330) gave antiserum, collected 3 months after immunization, that yielded 50% binding of radioligand (B_o/total counts = 0.5) at 1:30,000 dilution. This antiserum, stored as small aliquots at 1:300 dilution in assay buffer at -70°, was used in all subsequent evaluations and assays. Binding was essentially unaffected by pH in the 6-8.5 range, and the assay was run routinely at pH 7.5. The I_{50} was ~300 pg/tube. Scatchard [14] analysis (bound/free *vs.* molar concentration of bound antigen) showed the anticipated curved relationship, indicating a mixed population of Ab's in this polyclonal antiserum. Deconvolution of the curves using the method of Hunston [15] showed the Ab's to have affinity constants ranging from 1.1×10^8 to 6.76×10^9 M^{-1}.

Assay parameters.- Fig. 5A shows a standard calibration curve. Fig. 5B compares RIA with LC-MS-MS in respect of intra-assay precision, which together with inter-assay values and data on accuracy appears also in Table 1. Using a 10% maximum for intra-assay C.V.'s the LLOQ's for the LC-MS-MS and RIA assays are respectively 0.4 and 0.2 ng/ml. Any plasma samples that exceeded the RIA upper quantifiable limit, 200 ng/ml (B/B_o), were diluted with control plasma and re-assayed. The nsb was 0.82% of total counts.

Effect of heparin.- Normal drug-free plasma was harvested from blood collected in tubes containing sodium heparin; the heparin concentration in such plasma was ~15 u./ml. Aliquots of plasma (50 µl) were assayed after further addition of heparin in buffer to give 50, 500 and 5000 u./ml plasma. Accuracy seemed unaffected by heparin.

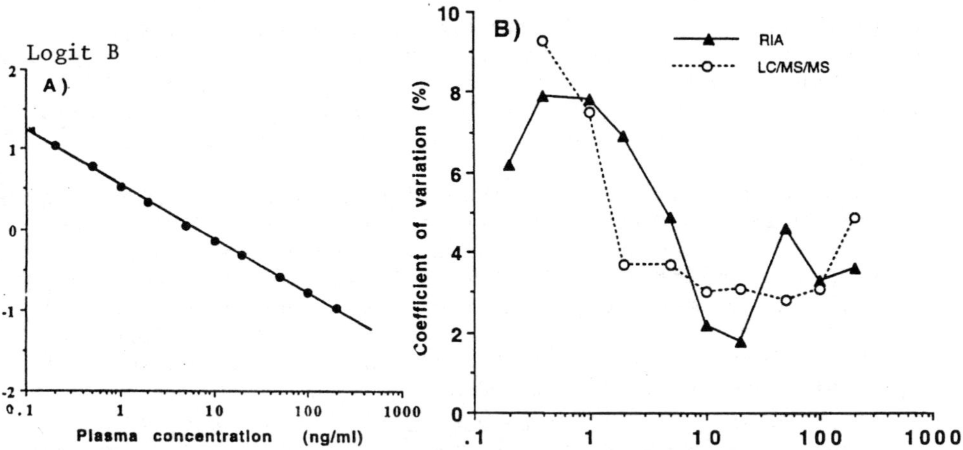

Fig. 5.
A) Calibration curve for the RIA.
LogitB= 0.5396 - 0.70026LogC + 0.02799(LogC)^2 - 0.00375(LogC)^3.
B) Intra-assay precision profiles for the two assay procedures.

Table 1. Intra-assay and *inter-assay* recovery (% found/added; accuracy) and C.V. (RSD, %; precision) by LC-MS-MS [n = 5] and RIA [n = 7 or (*inter-assay*) 10].

LC-MS-MS			RIA		
ng/ml	Recovery	C.V.	ng/ml	Recovery	C.V.
			0.2	110	6.2
0.4	93, *89.7*	9.3, *14.8*	0.5	96	7.9
1.0	104, *103.9*	7.4, *6.6*	1.0	94, *94.0*	7.8, *7.5*
2.0	104	3.7	1.0	95	6.8
4.0	98, *110.5*	3.7, *3.7*	5.0	98	4.9
10.0	100	3.0	10.0	98, *100.5*	2.2, *3.2*
20.0	97	3.1	20.0	100	1.8
40.0	97, *106.3*	2.8, *3.1*	50.0	101	4.6
100.0	102,	3.1,	100.0	99, *98.0*	3.3, *4.4*
200.0	101	4.9	200.0	95	3.9

Parallelism and recovery.- Parallelism experiments done on RIA's may disclose unknown drug-related substances that might compete with the drug for binding sites on the Ab's; they are generally metabolites, and frequently show cross-reactivities which are not constant but increase with increasing dilution. After giving L-734,217 orally, samples were taken: rat plasma, suitable for direct analysis, and dog plasma which was bulk-diluted with buffer to fall within the calibration curve. Both stock plasmas were then serially diluted with control plasma, and 50 µl aliquots taken for assay. Control plasma spiked with L-734,217 at 1950 ng/ml was diluted and assayed concurrently.

For plotting in Fig. 6, the measured concentrations of the drug were multiplied by the applicable dilution factors. The curves demonstrate reproducibility of the values irrespective of the dilution, with no divergences at concentration extremes and thus no indication of the presence of cross-reacting metabolites.

RIA accuracy was also examined by addition of standards using plasma samples obtained at different times after one oral dose (60 mg/kg) to both rats and dogs, with pre-dose controls. After assaying the samples, further drug in fixed amount was added and the ratio (%) of measured total drug to expected value (original measured value + the supplement) ascertained. For the different original values (ng/ml) as underlined, the following ratios that represent '% recovery' were found.-

DOG (supplement 50 ng/ml), pre-dose, nil: 103.0, 98.4, 97.6%; 0.5 h post-dose, 66.9: 99.1; 1 h, 96.0: 99.2; 6 h, 267.5: 102.8%

RAT (supplement 45 ng/ml), pre-dose, nil: 98.9, 95.1%; 2 h post-dose, 166.8: 98.7; 4 h, 120.3: 93.5; 8 h, 33.9: 95.4%

Evidently assay accuracy is not compromised by the test matrices.

RIA specificity.- After RIA on aliquots, post-dose samples furnished HPLC fractions for RIA: ng/ml values were as follows.-
DOG, 1 h post-dose: pre-HPLC, 1214; main peak, 1162 (95.7% rec.); 4 h post-dose: pre-HPLC, 135; main peak, 135 (100.0% rec.).
RAT, 1 h post-dose: pre-HPLC, 615; main peak, 658 (106.9% rec.); 24 h post-dose: pre-HPLC, 210; main peak, 204 (97.1% rec.).

Thus in both species the immunoreactivity in fractions accorded well with the pre-HPLC measurement; it was confined to the drug position and no cross-reacting substances were evident.

Cross-validation of the RIA and LC-MS-MS methods

Analyses on plasma from animals given 30 mg/kg L-734,217 orally were compared, firstly on 20 samples from rats. The mean ratio RIA/(LC-MS-MS) was 0.996 with C.V. 11.2%. The observed t-value (-0.795 with 19 D.F.) when compared with t-critical (2.10 for $P = 0.05$) showed no significant between-method difference. As shown in Fig. 7, linear regression analysis gave values for intercept and slope (confidence intervals respectively -34.4 to 36.5 and 0.959 to 1.059) that indicated neither fixed nor proportional biases.

For 35 dog samples (Fig. 7), the mean ratio RIA/(LC-MS-MS) was 0.910 (C.V. 5.5%). The observed t-value was 5.25 (34 D.F.; cf. t-critical, 2.03), indicating that RIA gives significantly lower concentrations. The regression analysis showed no fixed bias (confidence interval -71.2 to 77.4 for the intercept), but the 0.901 slope value (0.87 to 0.94) demonstrates a negative proportional bias of between 6 and 13%.

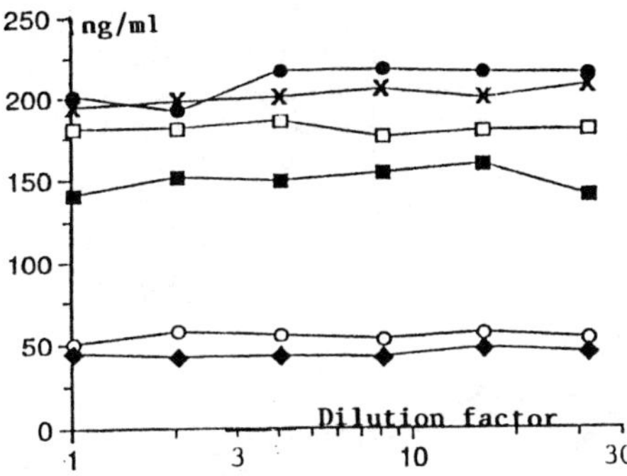

Fig. 6. Demonstration of satisfactory parallelism between rat and dog for RIA determination of L-734,217 in plasma.

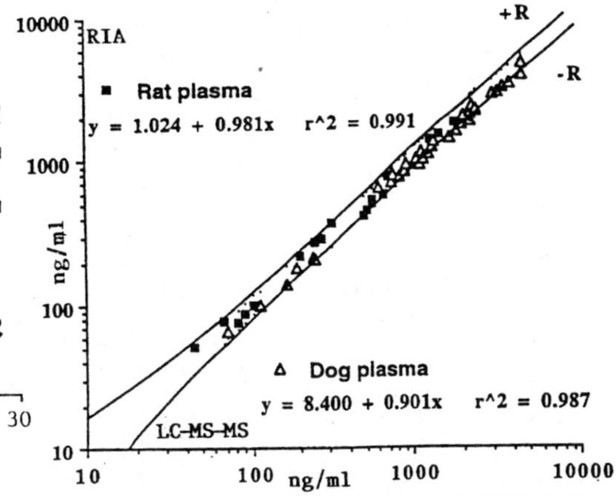

Fig. 7. Cross-validation of L-734,217 assay by RIA and LC-MS-MS: plasma concn. data comparing rats and dogs after oral dosing (30 mg/kg).

DISCUSSION

RIA, provided it is specific, is a preferred bioanalytical procedure in our laboratories due to its simplicity, sensitivity and ability to easily process samples in large batches. RIA had originally proved the method of choice for analyzing two related i.v.-administered fibrinogen receptor antagonists, MK-852 and MK-383. The present RIA for the orally active drug L-734,217 was sensitive to 0.2 ng/ml with acceptable accuracy and precision. Specificity was demonstrated by parallelism experiments on animal post-dose plasma samples and by HPLC of such samples with RIA examination of the fractions. The accuracy of the RIA was established both by spiking control plasma and and by adding standards to test plasma. The test plasma volume can be up to 50 µl with no adverse effect on binding. The assay has served for analyzing samples from pre-clinical and clinical studies. Fig. 8 shows representative plasma concentration-time curves for animals and volunteers receiving single doses of L-734,217.

The RIA, however, took >6 months to develop, and an interim method was needed to support the pre-clinical program. As there was no suitable chromophore, HPLC with MS-MS was selected, since APcI combined with triple quadrupole detection affords notable versatility for assaying trace levels of drugs in biofluids. Of particular importance are the high sensitivity and specificity of the approach, achievable using conventional HPLC columns with rapid assay development, and affording fast sample throughput.

Our assay acceptance criteria require that the intra-assay C.V. not exceed 10%. The LC-MS-MS method had, accordingly, an LLOQ of 0.4 ng/ml (C.V. 9.3%) and further served to evaluate the

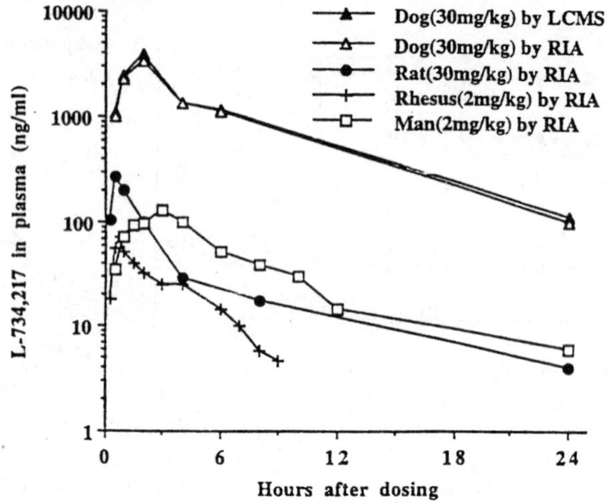

Fig. 8. L-734,217 concns. in the plasma of animals and volunteers receiving oral doses of the drug.

specificity of the RIA. Useful criteria for cross-validation of methods have recently been described, based on objective common-sense aided by a few simple statistical treatments, e.g. t-test and regression analysis [16, 17]. Strictly, the use of a more complex errors-in-variables regression is recommended for cross-validation experiments, since a simple linear regression assumes no errors in the reference method [18]. However, for comparing analytical data spread over several orders of magnitude, the two types of regression differ little in confidence intervals [17].

While the cross-validation of rat plasma data was unambiguously satisfactory, there was a small but significant negative proportional bias in the RIA when applied to dog plasma. Its cause remains unknown. Although this observation is of concern, it is notable that 30 out of the 35 dog plasmas assayed showed reproducibilities within the limits predicted from the individual variances of the assays. Altogether the results allowed us to conclude that the RIA and the LC-MS-MS assay were essentially equivalent and that the RIA was acceptably specific.

LC-MS-MS has lately evolved into an invaluable quantitative bioanalytical tool because of its sensitivity and unparalelled specificity. RIA, however, is very convenient and highly cost-effective. For example, U.S. contract laboratories charge ~$40 per sample for RIA but $150-200 for LC-MS-MS assays. In our own laboratories we find the credibility of RIA's is much enhanced when they are successfully cross-validated against equally sensitive but intrinsically specific methods based on LC-MS-MS.

Acknowledgements.- The authors thank Ms. Maureen Hetzel and Ms. Nancy Simpson for their help in preparing this manuscript.

References

1. Stein, B., Fuster, V., Israel, D.H., Cohen, M., Badimon, J.J. & Chesebro, J.H. (1989) *J. Am. Coll. Cardiol. 14,* 813-836.
2. Fitzgerald, D.J., Roy, L., Catella, F. & Fitzgerald, G.A. (1986) *New Engl. J. Med. 315,* 983-989.
3. Fitzgerald, D.J., Catella, F., Roy, L. & Fitzgerald, G.A. (1988) *Circulation 77,* 142-150.
4. Hamm, C.W., Lorenz, R.L., Bleifeld, W., Kupper, W., Wober, W. & Weber, P.C. (1987) as for 1., *10,* 998-1004.
5. Marguerie, G.A., Plow, E.F. & Edgington, T.S. (1979) *J. Biol. Chem. 254,* 5357-5363.
6. Hawiger, J., Parkinson, S. & Timmons, S. (1980) *Nature 283,* 195-197.
7. Peershke, E.I., Zucker, M.B., Grant, R.A., Egan, J.J. & Johnson, M.M. (1980) *Blood 55,* 841-847.
8. Ruoslahti, E. & Pierschbacher, M.D. (1987) *Science 238,* 491-497.
9. Duggan, M.E., Naylor-Olsen, A.M., Perkins, J.J., Anderson, P.S., Chang, C.T-C., Cook, J.J., Gould, R.J., Ihle, N.C., Hartman, G.D., Lynch, J.J., Lynch, R.J., Manno, P.D., Schaffer, L.W. & Smith, R.L. (1995) *J. Med. Chem. 38,* 3332-3341.
10. Yuan, A.S., Hand, E.L., Hichens, M., Olah, T.V., Barrish, A., Fernández-Metzler, C. & Gilbert, J.D. (1993) *J. Pharm. Biomed. Anal. 11,* 427-434.
11. Hand, E.L., Gilbert, J.D., Yuan, A.S., Olah, T.V. & Hichens, M. (1994) *J. Pharm. Biomed. Anal. 12,* 1047-1053.
12. Gilbert, J.D., Olah, T.V., Barrish, A. & Greber, T.F. (1992) *Biol. Mass Spectrom. 21,* 341-346 (see also 63-68).
13. Caulcutt, R. & Boddy, R. (1993) *Statistics for Analytical Chemists,* Chapman & Hall, London; see pp. 126-147.
14. Scatchard, G. (1949) *Ann. N. Y. Acad. Sci. 51,* 660-672.
15. Hunston, D.L. (1975) *Anal. Biochem. 63,* 99-109.
16. Gilbert, J.D., Greber, T.F., Ellis, J.D., Barrish, A., Olah, T.V., Fernández-Metzler, C., Yuan, A.S. & Burke, C.J. (1995) *J. Pharm. Biomed. Anal. 13,* 937-950.
17. Gilbert, M.T., Barinov-Colligon, I. & Miksic, J.R. (1995) *J. Pharm. Biomed. Anal. 13,* 385-394.
18. Roy, T. (1994) *J. Pharm. Biomed. Anal. 12,* 1265-1269.

#B-2n

A Note on

RETENTION OF FIBRINOGEN IN ION-EXCHANGE HPLC: EVIDENCE FOR A DOMAIN BINDING MODEL

Yansheng Liu & David J. Anderson†

Department of Chemistry, Cleveland State University, Cleveland, OH 44115, U.S.A.

An ion-exchange HPLC method for fibrinogen has now been developed (below). Literature on its HPLC analysis is scant, since its strong adsorptive properties lead to irreversible retention. Its non-specific adsorption has now been minimized by using a polymethacrylate HPLC support [1]. Injection of a pure fibrinogen standard furnishes multiple peaks, seemingly due to different conformations that are differently oriented in their binding to the ion-exchange packing material; evidence is now presented.

EXPERIMENTAL

HPLC.- The system comprised two Solvent Delivery Module pumps with a System Controller (#112, #421; Beckman, Berkeley, CA), a Variable Wavelength Detector (Dionex, Sunnyvale, CA) set to 280 nm, an SP 4270 Integrator (Thermo Separation Products, San Jose, CA) and a 386SX computer. Using an HPLC slurry packer (Alltech, Deerfield, IL) at 500 psi and a pH 7.4 buffer (0.5 M NH_4Cl, 25 mM Tris/HCl), a 50 × 4.1 mm column was packed with anion-exchanger, diethylaminoethyl-functionalized polymethacrylate (Millipore, Milford, MA; 15 µm; 100 nm pore size).

The loading buffer ('A') was pH 7.4 25 mM Tris/HCl which, where so stated, contained 3 M urea. The elution buffer ('B') was 25 mM Tris/HCl, 2 M NH_4Cl, pH 7.2. Three linear-gradient speeds were employed: **fastest**, 0 to 20% B in 40 min, 0.6 ml/min **(Δ%B/Δmin = 0.5; Δ%B/Δml = 0.83)**; **intermediate**, 0 to 18% B in **72 min, 0.3 ml/min** (0.25; 0.83); **slow**, 0 to 25% B in 150 min, **0.15 ml/min (0.17; 1.11)**. Gradient speed connotes the change in mobile phase strength with respect to time (Δ%B/Δmin), not to volume (Δ%B/Δml; gradient steepness). Gradient speed in this work was varied while gradient steepness was kept near-constant.

Samples.- Fibrinogen was from Calbiochem (human plasma, plasminogen-free, clottable proteins >95%; #341578, Lot 229491; La Jolla, CA) or ICN (human plasma, >90%; #151123, Lot 54304; Cleveland, OH). Fragments D and E were provided by Dr Shainoff, Section on Thrombosis Research, Cleveland Clinic Foundation. All protein solutions were prepared with buffer containing 25 mM Tris/HCl, 0.15 M NaCl. pH 7.4; generally 20 µl was injected.

---

†addressee for any correspondence

RESULTS

Multiple peaks.- Fig. 1 shows that fibrinogen chromatographs as multiple peaks depending on the gradient speed: their number decreases as the speed decreases, the slowest giving predominantly one peak. The phenomenon of multiple peaks is not an infrequent occurrence in the HPLC of proteins, particularly RP-HPLC. It is usually explained by either heterogeneity (structural or conformational) or intermediate denaturation kinetics compared to the chromatography time (producing different conformational forms that are chromatographed on the column) or self-association effects. Self-association of fibrinogen was ruled out as an explanation, by experiments which showed no change in peak pattern when different concentrations of fibrinogen were run. Although structural heterogeneity of fibrinogen has been reported extensively [2], Fig. 1 indicates that this is not the overriding explanation for the multiple peaks, as essentially only one peak is produced by the slowest gradient. In addition, fractions collected from each peak of the multi-peak fibrinogen sample produced multiple peaks upon rechromatography, which indicates convertibility between the peaks. These results point to different conformational forms of fibrinogen being present.

Domain binding model.- Fibrinogen has 3 distinct domains: a central E domain between 2 identical terminal D domains, being connected to D through a coiled coil peptide structure. Extended proteolysis of fibrinogen by plasmin furnishes an intact E domain plus 2 small portions of 2 coiled regions (fragment E) and an intact D domain plus a small portion of the coiled coil region (fragment D).

Comparing the chromatography of fibrinogen with that of these fragments allows the study of domain influences on the chromatographic pattern. Fig. 2 shows that most of the fibrinogen peaks can be assigned to either a D-domain retention (the early-eluting peaks) or an E-domain retention (later-eluting peaks). The presence of urea in the mobile phase is seen to significantly improve resolution compared to Fig. 1.

Consequent explanation of the Fig. 1 results.- Injection of fibrinogen onto the ion-exchange column leads to orientation-dependent diversity in the binding of molecules to the packing. Most bind through the D-domain, because the E-domain in native fibrinogen is sterically unavailable for binding [3-5]. But some fibrinogen molecules are denatured (or partially proteolysed), exposing the E-domain, and bind to the packing material through the more strongly binding E-domain. Hence there is orientation-dependent peak diversity with faster gradients. But the lengthened elution process with slower gradients leads to much more denaturation, exposing the E-domain such that desorbed fibrinogen molecules will become re-bound *via* E rather than D as originally; eventually most of the molecules are denatured, so leading to the one E-domain binding peak seen in Fig. 1.

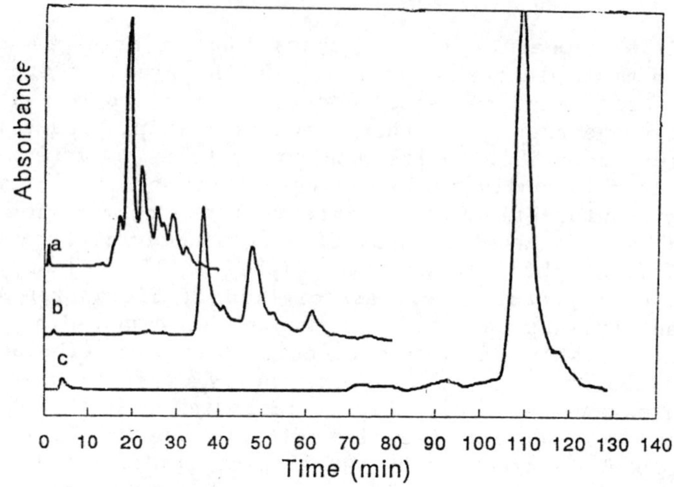

Fig. 1. Comparison of the chromatograms (no urea present) of fibrinogen (Calbiochem) obtained with different gradient speeds: **a**, fast; **b**, intermediate; **c**, slow. The fibrinogen amounts injected were respectively 200 µg, 120 µg (30 µl) and 100 µg.

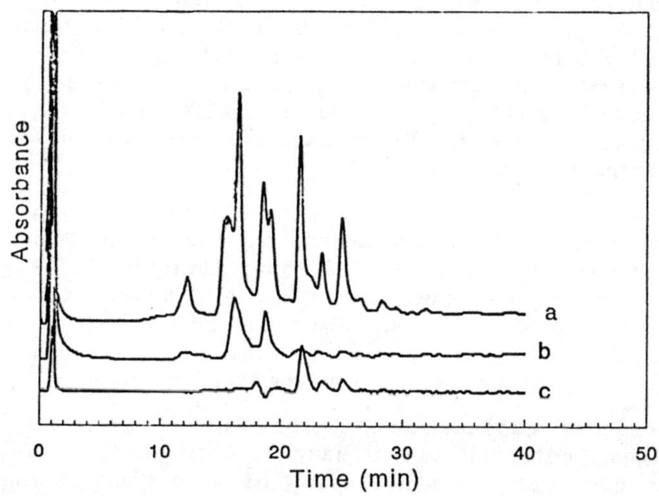

Fig. 2. Comparisons of chromatograms (with urea present and using a fast gradient): **a**, fibrinogen (ICN); **b**, fragment D; **c**, fragment E. The amounts injected were respectively 200 µg, ~35 µg and ~30 µg.

CONCLUDING COMMENTS

Chromatography of fibrinogen on an ion-exchange HPLC column gives multiple peaks resulting from the native and denatured forms binding to the column through different domains on the molecule; the native form binds through the D-domain and the denatured form binds through the E-domain. With a fast gradient multiple peaks are observed in the chromatography of fibrinogen. This is due to the elution of different conformational forms of fibrinogen, which bind through different domains. At slower gradient speeds most of the fibrinogen becomes denatured, and hence the E-domain binding peaks increase at the expense of the D-domain binding peaks. No such domain-dependent mechanism of retention in HPLC has hitherto been reported.

Acknowledgements

The authors thank Dr John Shainoff, Head of the Cleveland Clinic Foundation's Section on Thrombosis Research, for purified samples of fibrinogen and fragments D and E. This study was supported by a Grant-in-Aid from the American Heart Association and a Research and Creative Activities Grant from the Graduate College at Cleveland State University.

References

1. McConnell, J.P. & Anderson, D.J. (1993) *J. Chromatog. 615,* 67-75.
2. Shafer, J.A. & Higgins, D.L. (1988) *Crit. Rev. Clin. Lab. Sci. 26,* 1-41.
3. Weisel, J.W., Stauffacher, C.V., Bullitt, E. & Cohen, C. (1985) *Science 230,* 1388-1391.
4. Apap-Bologna, A., Raitt, F., Webster, A. & Kemp, G. (1987) in *Fibrinogen 2: Biochemistry, Physiology and Clinical Relevance* [Proc. Fibrinogen Workshop, Glasgow, 1987] (Lowe, G.D.O., *et al.,* eds.). Excerpta Medica, Amsterdam, pp. 31-34.
5. Erickson, H.P. & Fowler, W.E. (1983) *Ann. N.Y. Acad. Sci. 408,* 146-163.

#B-3

BIOANALYTICAL STRATEGIES TO SUPPORT THE DEVELOPMENT OF AN HIV PROTEASE INHIBITOR, INCLUDING GC-MS, HPLC AND CHIRAL ANALYSIS

C.A. James†, N.K. Burton, T.A. Enos, R.J. Simmonds, S.A. Wood, S.A. Rees, ᵒG. Padbury & ᵒF.J. Schwende
Pharmacia & Upjohn: Fleming Way, Crawley RH10 2LZ, U.K. and ᵒKalamazoo, MI 49007, U.S.A.

Requirement *Quickly developed assays for the drug, U-96988 (racemic) in rat and dog plasma, for formulation and toxicology studies, and in human plasma and urine, for Phase I clinical studies. Chiral analysis was later required.*

End-step *GC-MS for initial clinical studies; thereafter (faster) HPLC-UV. HPLC-UV served for animal biofluids and human urine and, with a cyclodextrin column, for enantiomers.*

Sample handling *Animal plasma: precipitate with MeCN* and put diluted supernatant onto the column. Human plasma: extract with solvent and derivatize (BSTFA) for GC-MS; or precipitate with MeCN and concentrate for HPLC. Urine: load direct.*

Comments *Variable recovery when, to render the HPLC method more sensitive, extraction of larger human-plasma samples (especially by SPE) was tried, probably because of high protein binding and very poor aqueous solubility (which also raised suspicion of loss with urine if frozen). GC-MS assay (LLOQ 2 ng/ml) was superseded by a simpler HPLC method after refining the precipitation procedure to achieve 0.02 µg/ml LLOQ. Chiral assay studies were deferrable.*

U-96988 is an HIV protease inhibitor [1] intended for the treatment of AIDS and HIV infection. The molecule, possessing two asymmetric carbon atoms, was available only as a racemic

†correspondent; Editor will forward

**Abbreviations:* BSTFA, *N,O*-bis-trimethylsilyltrifluoroacetamide; GLP, Good Laboratory Practice; i.s., internal standard; LLOQ, lower limit of quantification; MeCN, acetonitrile; TFA, trifluoroacetic acid.

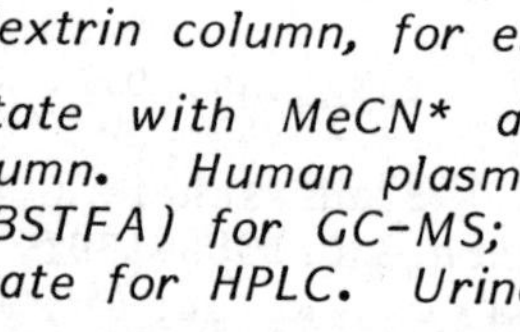

U-96988

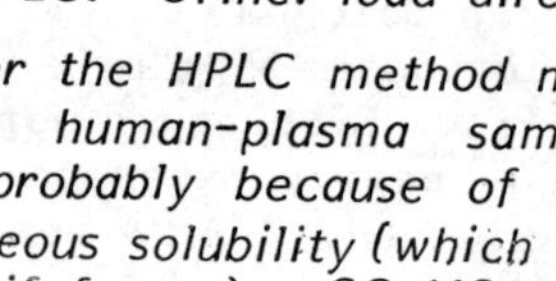

U-97960
(LC i.s.)

mixture containing all four possible stereoisomers. Whilst it was recognized that eventually a stereoselective analytical method would be required, initially the need was for suitable methods to support pre-clinical toxicokinetic, formulation and bioavailability studies, and initial tolerance studies in man.

Initial assays

Discovery studies were supported with a simple assay that involved protein preparation of plasma (typically 50 µl) with MeCN, followed by injection of a large volume of the diluted supernatant onto an HPLC column. With UV detection at 295 nm adequate sensitivity was achieved (range ~0.1-20 µg/ml). With minor modifications the assay was validated and proved suitable for rat and dog GLP toxicokinetic studies. A GC-MS assay with liquid-liquid extraction from acidified plasma was developed for the first clinical study. This assay was very sensitive with LLOQ 2 ng/ml, but was complex and time-consuming.

Following the first clinical trial it became apparent that the high sensitivity of the GC-MS method was not required and that the precision of the assay was rather poor in routine use. A simple routine HPLC assay for human plasma was therefore required, but with a significantly better LLOQ (0.02 µg/ml) than the existing assay for rat and dog plasma. Producing such an assay would normally be straightforward through accumulated experience with the compound. However, the development of a viable assay took many months, although there were delays due to the need to frequently provide urgent support for ongoing studies with the established assays.

METHODS FOR ASSAY OF U-96988

HPLC for rat and dog plasma.- To 50 µl plasma was added 100 µl of 10 µg/ml U-97960 in MeCN. After vortex-mixing and centrifugation (12,000 **g**, 2 min), the supernatant was diluted (50 µl water) and injected for HPLC. The column (Zorbax RP-8, 5 µm, 150 × 3.2 mm) and pre-column (Brownlee MPLC NewGuard RP-8, 15 × 3.2 mm) were maintained at 40° in a forced air column oven. The mobile phase was MeCN/water/TFA (55:45:0.1) at 1.5 ml/min flow rate. Detection was at 295 nm. The calibration range was typically 0.2-50 µM [~0.1-20 µg/ml], although an interfering peak was seen in certain individual rats that raised the LLOQ to 0.5 µM (Fig. 1).

GC-MS for human plasma.- To 0.5 ml was added 100 µl i.s. (U-95929, 1 µg/ml). After vortexing, 1 ml 10% (v/v) sulphuric acid and 4 ml chlorobutane were added. The capped tubes were shaken (20 min; 350 cycles/min). The solvent layer was dried down at 65° under N_2, 50 µl silylating reagent added (BSTFA 1% TMCS), and transferred to vials in an autosampler (Hewlett-Packard HP 7673). From the capped vials, heated for 5 min at 65°, 2 µl was injected into the GC capillary column - J&W DB1, 15 m × 0.25 mm (film thicknesses used: 0.1, 0.25 & 0.33 µm).

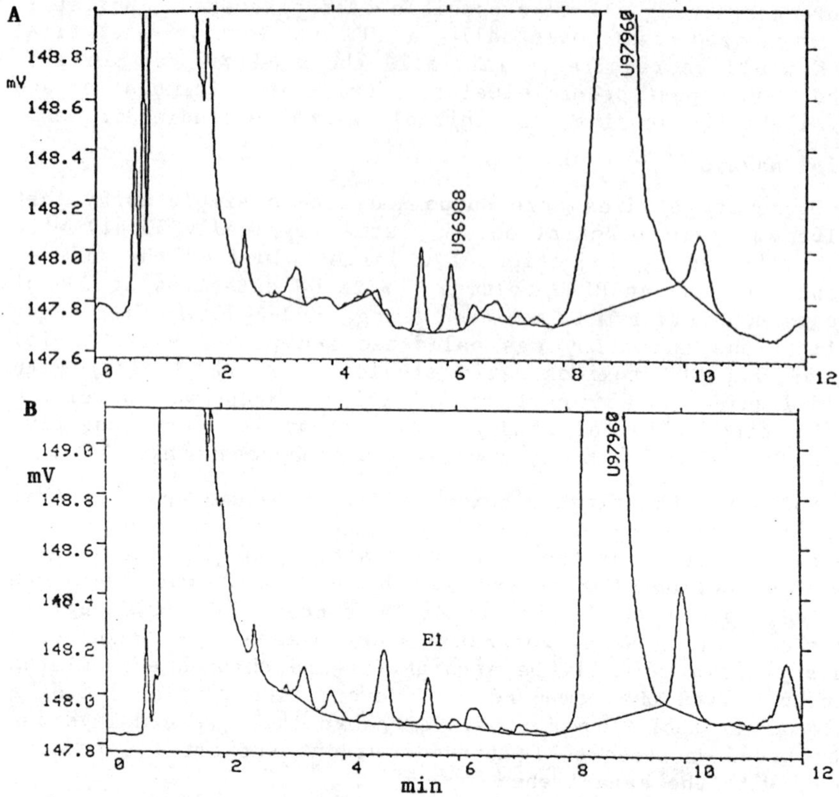

Fig. 1. HPLC of rat plasma: **A**, spiked with 0.1 µg/ml U-96988; **B**, a blank showing endogenous interference (EI) in the U-96988 peak position. U-97960 is the i.s. for HPLC.

The GC oven program ran from 150° to 320° with 30°/min increase (final temperature hold: 2 min). The carrier gas (0.5 ml/min flow rate) was helium; the injector and transfer line were at 300°. Analytes were quantified with a HP 5971 Mass Selective Detector (EI mode), monitoring (SIM) at m/z 377 and 405.

As U-96988 contains two asymmetric centres, two of the stereoisomer pairs might be separable in an achiral environment. Whilst a single peak was seen for U-96988 with RP-HPLC, an incompletely resolved peak was seen by GC (Fig. 2, B). Since complete resolution was not achievable and only racemic reference material was available, the sum of the areas of the two partially resolved peaks was used for quantification. The i.s. lacks a second asymmetric centre and is seen as a single peak (Fig. 2A). The typical GC calibration range was ~2-400 ng/ml. QC results were significantly more variable than for HPLC assay.

Fig. 2.
GC-MS of human plasma spiked with: **A**, U-95929 as i.s.
B, U-96988, 1.8 ng/ml.

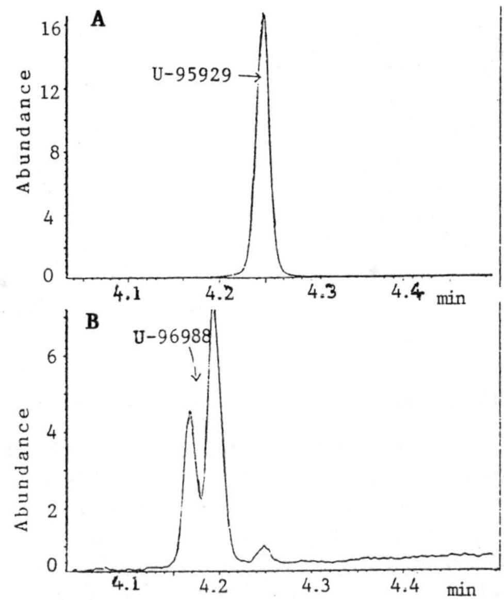

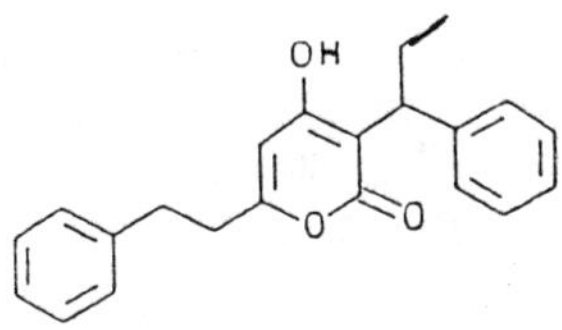

U-95929

Fig. 3.
HPLC of human plasma spiked with U-96988 (0.02 µg/ml) and U-97960 as i.s.

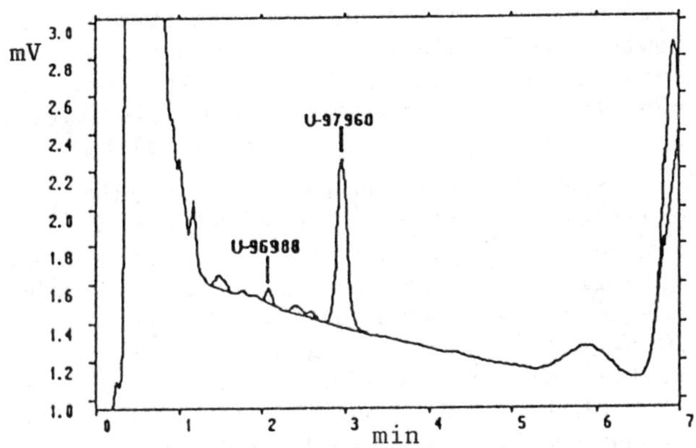

HPLC for human plasma.- To 100 µl plasma was added 200 µl i.s. (0.1 µg/ml in MeCN). After vortexing and centrifugation as above, the supernatant was transferred to 75 × 12 mm tubes and partially evaporated (40°, 5 min; TurboVac LV). The concentrate was diluted with 50 µl of mobile phase, and 80 µl run on an HPLC system at 40° as before except that the column was 80 mm long and the flow 1.7 ml/min. Fig. 3 shows a chromatogram. The calibration range was typically 0.02-8 µg/ml. Above this, non-linearity was apparent; but such samples could be analyzed by diluting with control plasma.

Chiral analysis.- The stereoisomers of U-96988 could be separated on a Cyclobond I 200 column, 250 × 4.6 mm, at 5° with MeCN/diethylamine/acetic acid (99.7:0.1:0.6) as mobile phase (1.0 ml/min; detection at 295 nm). Fig. 4 exemplifies the HPLC.

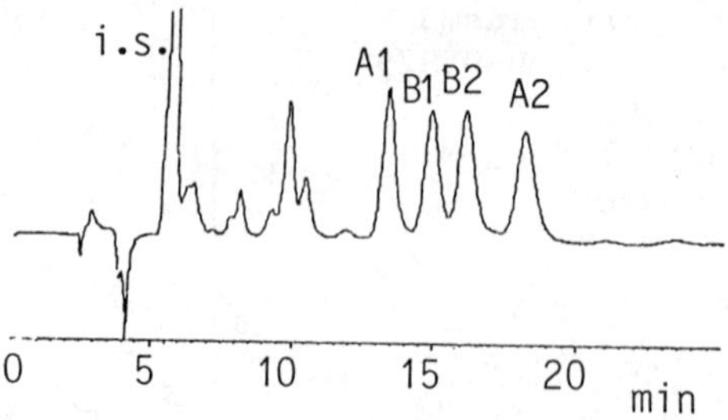

Fig. 4. HPLC with a chiral column, showing separation of stereoisomers in rat plasma spiked with U-96988.

DISCUSSION

Our success in developing the HPLC assay for dog and rat plasma, together with the GC-MS method, led us to anticipate that a routine HPLC method for human plasma, with improved sensitivity, could be rapidly developed by combining an extraction procedure, SPE or liquid/liquid, with the existing HPLC separation. Initial results indicated that following acidification good retention was achieved on non-polar SPE phases. However, reliable recovery could not be achieved; other options such as ion-exchange fared little better. Solvent extraction showed some promise, but variable recoveries and very poor precision were found in initial validation experiments.

Some explanation for these results comes from the physicochemical properties of U-96988 and its protein binding. The compound has pKa 4.9 and is poorly water-soluble (intrinsic solubility in water is 0.09 µg/ml; 28 µg/ml in buffer at pH 7.4 but only 0.09 µg/ml at pH 2.0). High solubility (well over 10 mg/ml) is seen in organic solvents such as MeCN and ethanol. Protein binding is >99%. At first, high protein binding was suspected as the cause of poor recoveries with SPE; but later it became apparent that protein binding was responsible for maintaining the drug in solution in plasma, far above the limits of solubility in an equivalent aqueous solution. Some of our confusing results were probably due to precipitation of drug from solution when, through protein precipitation or SPE, it was no longer in a plasma matrix.

Additional problems were found in redissolving U-96988 following an evaporation step, although a variety of organic solvent and solvent/aqueous mixtures were tried. Liquid/liquid extraction worked reasonably well prior to GC-MS analysis, but analyte recovery may have been improved by the direct addition of derivatizing agent to the tube after evaporating the chlorobutane. Moreover, the i.s. used for GC-MS was unsuitable for HPLC analysis, and the alternatives tried may not have compensated as well as U-95929 for variations in U-96988 recovery.

The remaining problems could have been overcome with time and systematic evaluation; but this was precluded because yet another set of clinical samples was due to arrive. Initial attempts to 'scale up' the protein precipitation extraction used for rat and dog assay had resulted in very 'dirty' samples after solvent evaporation. We re-examined this approach with the idea of always keeping the drug in an aqueous/MeCN environment. The sensitivity gain needed was achieved by extracting a larger sample volume by MeCN precipitation and concentrating the supernatant by partial evaporation. A more sensitive UV detector and a shorter column (80 mm), introduced to reduce run-time, also gave small but useful benefit to sensitivity.

Even with such a simple extraction scheme the poor solubility of U-96988 was apparent, as the assay was non-linear above 20 μM (~8 μg/ml) although the detector was linear with standard solutions to far higher concentrations which, if present in plasma, were coped with by initially diluting the sample with control plasma, so getting accurate results.

Urinary assay development centered especially on the relatively poor solubility and steep pH-solubility curve of U-96988, such that it could precipitate if samples were cooled or frozen or if pH changed during storage. As little excretion of unchanged drug into urine was anticipated, we had to demonstrate that a 'none-detected' result did imply no excretion of unchanged drug rather than 'loss' during sample storage. Urine samples could not be analyzed immediately and were to be stored on ice, aliquots being frozen when the collection ended. As an extra precaution we had the entire collections frozen from the first clinical study, with the idea that we could recover precipitated U-96988 by adding MeCN or adjusting pH if required.

U-96988 was assayed in urine by directly injecting diluted samples onto the same HPLC system as for dog and rat plasma. We spiked large collections of control urine (taking the highest and lowest pH samples obtained from a number of volunteers) and demonstrated that adequate recovery of U-96988 was obtained, even after freeze-thaw cycles; yet no drug was detectable in stored aliquots, or bulk collections, following the highest doses in the clinical study.

CONCLUDING COMMENTS

At first sight, most problems in bioanalysis of U-96988 might be expected to reside in its chirality. Whilst a chiral separation was developed and would have been needed eventually, achiral methods fully sufficed to progress the drug through various pre-clinical studies and to reach key decision points in early clinical trials. Whilst many drugs have poor aqueous solubility, this is normally a problem in the formulation area and not in bioanalysis where analyte amounts are miniscule.

For U-96988, concentrations found *in vivo* were far above the solubility limits in simple aqueous solution, reflecting high protein binding. These characteristics combined to frustrate development of a number of extraction schemes. Whilst these problems were circumventable (perhaps with a combination of ion-exchange SPE and very careful control of pH and of water/solvent concentrations during the entire extraction), a 'back-to-basics' approach was adopted, and careful refinement of our original, simple protein precipitation method generated the required assay.

In spite of these problems, methods were successfully developed for analyzing U-96988 in biofluids to support a very rapid program for progressing this drug into early clinical trials.

Acknowledgement

The GC-MS assay for U-96988 was developed by N.K. Brown shortly before leaving to work at SKB Pharmaceuticals (Welwyn).

Reference

1. Craig, J.C., Grief, C., Mills, J.S., Hockley, D., Duncan, I.B. & Roberts, N.A. (1991) *Antiviral Chem. Chemother.* *2,* 181-186.

#B-4

ANALYSIS OF PSEUDOPEPTIDYL MATRIX METALLOPROTEINASE INHIBITORS IN PLASMA

J. Moore†, M. Churchill, J. McGowan, C. Mason, S. Richards & W.A. Galloway

Department of Cellular Biochemistry and Bioanalytical, British Biotech Pharmaceuticals, Watlington Road, Cowley, Oxford OX4 5LY, U.K.

Matrix metalloproteinases comprise Zn-containing enzymes that degrade molecules such as collagen and underly processes such as tumour spread. Inhibitors of these enzymes, containing hydroxamic acid, may become therapeutically important. Bioanalysis of such MMPI's is achievable by LC-MS and LC-MS-MS methods. Their development entailed overcoming some problems as now considered for two compounds, batimastat and marimastat. A general approach is described. As exemplified by a comparison of APcI and ESI interfaces, some problems may be solved by changes in LC-MS hardware rather than in sample preparation.*

Matrix metalloproteinases are a family of Zn-containing enzymes that can degrade structurally important macromolecules such as collagen and proteoglycans. Constitutive expression in tumour cells is believed to facilitate tumour metastasis and invasion in cancer patients. Consequently, hydroxamic acid-containing MMPI's* have been developed and are currently under investigation as pharmaceutical products for the treatment of cancer and of arthritis and other tissue-degenerative diseases.

For the compound type we are investigating, the structural elements to exhibit inhibition of matrix metalloproteinases are possession of a peptide backbone, a non-polar side-chain and a hydroxamic acid group. Fig. 1 shows a generic MMPI structure.

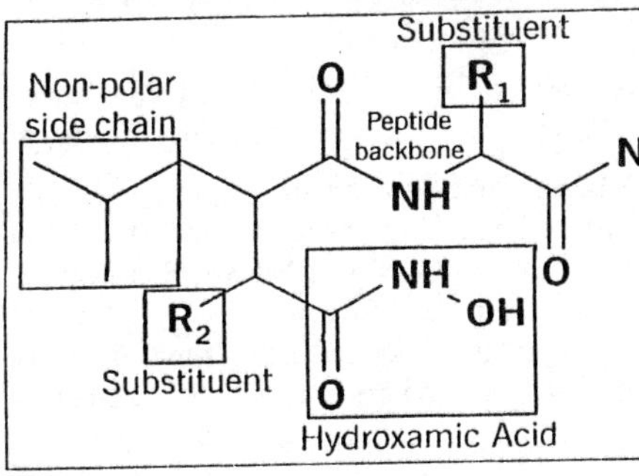

Fig. 1. Generic MMPI structure

†correspondent
**Abbreviations.*- MS interface: APcI, atmospheric pressure chemical ionization; ESI (or ESP): electrospray. MMPI, matrix metalloproteinase inhibitor; QC, quality control (sample); TFA, trifluoroacetic acid.

With multiple potential sites for substitution, compounds of this type can exhibit a very wide range of physico-chemical properties. Accordingly, LC-MS or LC-MS-MS was the detector of choice for the establishment of a new Bioanalytical Laboratory to support drug development for the MMPI's. The very high cost of the instruments was justified by the anticipated high sensitivity, selectivity and minimal method-development requirements.

Examples of methods for the analysis of MMPI's in a range of matrices will be used to demonstrate some of the problems encountered during method development and analysis by LC-MS and LC-MS-MS and the steps taken to overcome them.

GENERAL APPROACH (amplified in companion art., #C-7)

Several guiding principles for method development apply, besides simple HPLC with methanol or acetonitrile, and use of APcI.

- Most methods involved detection in the MS rather than MS-MS mode since these compounds did not produce sensitive daughter-ion spectra.
- Acid or solvent deproteinization as a means of sample pre-treatment was avoided since inherent sample dilution limits sensitivity and instrument contamination problems can occur.
- Sample pre-treatment was automated wherever feasible, giving data of higher quality than manually executed methods and resulting in more efficient use of laboratory staff.

BATIMASTAT (formula on opposite p.)

Batimastat is highly lipophilic, insoluble in water and extremely stable. Methods were required for its analysis in rat, dog and human plasma with LOD 5 ng/ml.

Extraction scheme.- Analysis of the compound was achieved by adding BB-2028 (a close structural analogue) as i.s. and extracting the analytes at pH 4 onto C-18 SPE cartridges. These were washed with 50 mM pH 4 phosphate buffer (3 ml), methanol/water (10:90 by vol.; 3 ml) and hexane/ethyl acetate (90:10; 3 ml). The extracts were dried down in a vacuum centrifuge and reconstituted in mobile phase.

HPLC was performed with a Novapak C-18 analytical column (150 × 3.9 mm) and a mobile phase of methanol/acetonitrile/water (60:14:26 by vol.) at 0.75 ml/min flow rate. The R_T (min) was 2.0 for batimastat and 2.3 for i.s.

MS-MS with an APcI interface was performed with a Finnigan TSQ 700 instrument, monitoring at m/z 267 and (i.s.) 281.

Problems encountered.- Samples prepared as above often had an oily precipitate when reconstituted. Endogenous interferences were less marked with LC-MS-MS rather than LC-MS; but injecting only a few samples into the system resulted in a rapid loss of chromatographic separation and dramatic suppression of ionization. Thus, despite rigorous clean-up, samples were too dirty

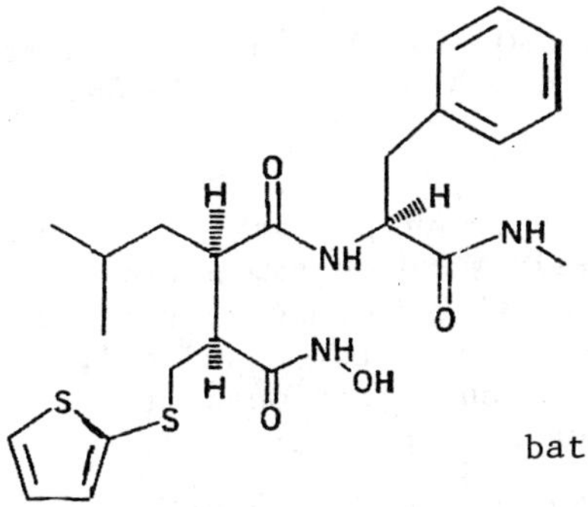

batimastat

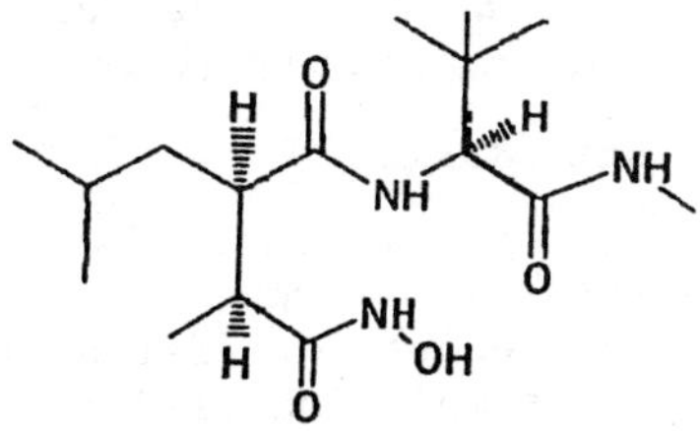

marimastat

for analysis. When the SPE cartridge was eluted with methanol, however, the interferant was held back. The modified method was successfully validated and later applied to the analysis of several thousand toxicology and clinical trial samples.

MARIMASTAT

Marimastat is readily soluble in water and stable, but has no chromophore. Methods were required for its analysis in rat, marmoset and human plasma and urine, with LOD 5 ng/ml.

Human and marmoset plasma

Extraction scheme.- During method development a range of approaches were investigated, including protein precipitation with acetonitrile or trifluoroacetic acid, SPE using C-2, C-8, phenyl, cyanopropyl and ion-exchange cartridges, and finally solvent extraction. Acid or solvent deproteinization approaches resulted in methods with low sensitivity. It was not possible to bind the compound strongly enough onto an SPE packing to allow inclusion of a wash step, which was shown to be necessary to prevent MS contamination and eventual blockage of the inlet capillary by endogenous plasma proteins.

A solvent extraction was developed using ethyl acetate at pH 7; but extracts still tended to block the MS inlet capillary after only a few injections. This was circumvented by adding ammonia solution (0.1 M; 100 µl) to the separated solvent layer and vortex-mixing the tube, to back-extract the analytes; after neutralization with HCl (0.1 M; 100 µl), any residual solvent was removed by vacuum centrifugation for 5 min. Since the plasma extracts thus produced were non-volatile, the extract did not have to be transferred to an autosampler vial with capping; injections were made directly from the tube used for the back-extraction. The presence of a small amount of NH_4Cl in sample extracts might reasonably have been expected to cause MS interface problems; but trials proved that its presence improved sensitivity.

HPLC was performed with a column as for batimastat and a mobile phase of methanol/water (55:45; 1 ml/min). The R_T (min) was 2.0 for marimastat and 2.3 for i.s.

MS was done as above (Finnigan SSQ 710) with an APcI interface (but see below), monitoring at m/z 332 and (i.s.) 366.

Rat plasma

During analysis of rat plasma samples prepared as above, the chromatograms became progressively more contaminated as an endogenous interferant built up in the MS interface. Increasing the sample run time gave only a temporary remedy, and only ~45 samples could be analyzed in a single run; this meant that the method was impracticable for the routine application.

The APcI interface was exchanged for an ESI interface, and an in-line flow splitter (split ratio 50:1) was installed after the analytical column. Thereby the flow into the ESI interface was reduced to only a few µl/min and hence less endogenous material was available to contaminate the MS; yet there was no reduction in sensitivity of analyte detection since the ESI interface operates optimally at very low flow rates.

With this simple modification up to 100 rat plasma samples could be injected with no recurrence of contamination problems.

Human urine

For human urine the strategy adopted was to add an i.s. (BB-1090) to the sample, mix, and inject directly into the MS with no further pre-treatment. It was anticipated that the absence of any lipid or protein in the sample would make urine the ideal matrix for analysis by LC-MS with good detectability and none of the blockage or contamination problems described above.

The technique was subjected to a full method validation, which gave excellent results with very low C.V.'s for both inter- and intra-day variation and good accuracy and precision data. However, when the assay was applied to the analysis of routine samples, QC sample data upon which the acceptance of an analytical batch is based was very variable. The inclusion of a mobile phase wash injection or a study sample immediately prior to QC sample dramatically affected the interpolated value, indicating that a co-eluting peak was affecting the response of marimastat. Investigations suggested that the co-eluting material was present in urine at varying concentrations between individual samples.

It was concluded that direct-injection analysis of urine by LC-MS is not recommended unless the analyte is present in such high quantities that a dilution of the sample to overcome matrix effects can be performed, and there is still a risk that erroneous results may be obtained. The solution adopted was to use a simple two-pump column switching system which pumped clean mobile phase into the MS instrument for most of the HPLC run and switched only the analyte and i.s. peaks into the MS for detection at the appropriate time, avoiding any solvent front material. Thus any early or late eluting endogenous peaks are switched to waste (Fig. 2).

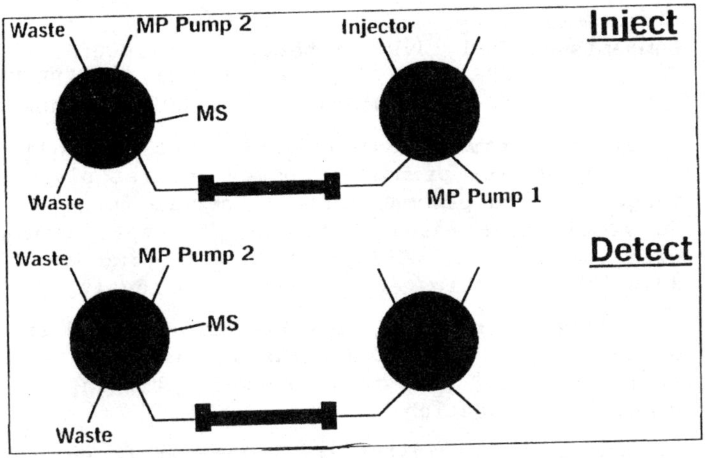

Fig. 2. The switching system used for human urine (see text).

Using this simple system, several hundred urine samples were successfully analyzed with no recurrence of the imprecision problems described above.

SENSITIVITY

During the development of the various methods for analysis of MMPI's, it became clear that the limit of quantification of the assays was consistently in the low-ng region. This was higher than anticipated since manufacturers claim that analytes can be routinely determined at pg levels. It was not known at the time whether this was due to inherent limits of sensitivity of the instrumentation or the nature of the analytes. Using the same method-development approaches and instrumentation, methods have been developed for various analytes and close structural analogues for which very widely differing LOD's were observed, suggesting that the achievable sensitivity for MS methods is largely a function of analyte structure. The expectation that highly sensitive methods can be developed for every analyte with minimal method-development effort is therefore unrealistic.

CONCLUSIONS

The wide range of physico-chemical properties exhibited by MMPI molecules makes LC-MS a good choice for their analysis since adequate sensitivity and detection selectivity can be achieved. The ultimate sensitivity may vary among a series of analogues, and significant method-development effort can be required to produce robust and reliable methods. Matrix effects do occur, and may present major problems which are best solved by the use of conventional sample clean-up techniques and simple column-switching systems. Many analytical difficulties are specific to a single species or matrix and, when developing methods, instrumental changes must be considered independently of changes to the sample clean-up scheme since both can be used successfully to eradicate problems with interferences.

Comments on #B-1, J.D. Gilbert: A FIBRINOGEN RECEPTOR BLOCKER
#B-2n, D.J. Anderson: HPLC OF FIBRINOGEN DOMAINS
#B-4, J. Moore: A METALLOPROTEINASE INHIBITOR

S. Wring remarked to Gilbert, who had found RIA equivalent to LC-MS-MS for primate, monkey and rat plasma but negatively biased for dog plasma, that it is rare for a 'non-specific' RIA to yield lower results than a reference technique. Agreeing with this remark, **Gilbert** said that time was lacking to track down the cause; to get dog results urgently, LC-MS-MS was used.

Anderson, replying to R.F. Venn.- Indeed the chromatography of fibrinogen D and E domains is temperature-dependent: going from 12° to 25° and 40° we showed a binding shift, reflecting D-domain degradation.

Annotation by Editor (cf. #B-2n): fibrinogen peptides were included in a study of polymer-type HPLC stationary phases by Krause, E. & Beinert, H. (1992) J. Liq. Chromatog. **15,** 1773-1784.

Moore, answering R.J. Simmonds who wondered why a liquid-liquid extraction had been adopted for marimastat despite the stated preference for SPE methods, confirmed his preference for SPE as allowing automated sample preparation, which saves staff time compared with conventional 'wet' methods. Both approaches were explored for marimastat; a single ethyl acetate extraction at pH 10 gave excellent sensitivity and specificity [art. #B-4], such that the analyst felt it unjustified to spend more time developing an SPE method.

Section #C

VARIOUS DRUG ANALYTES: PROBLEMS AND STRATEGIES INCLUDING MS

The suffix '**n**' as in #C-2n distinguishes 'NOTES' from main articles. COMMENTS (starting on p. 159) comprise Forum discussion remarks, and Editor's 'annotations' particularly on microdialysis and assay of retinoids.

Assay annotations on VARIOUS DRUG ANALYTES, grouped by therapeutic class, also feature in Sect. #ABC.

#C-1

BRAIN DISTRIBUTION OF A CNS-ACTIVE DRUG STUDIED BY MICRODIALYSIS COUPLED WITH MS-MS(TSP) IN FLOW INJECTION MODE

C. Nuez†, R. Laplanche, P. Guntz & M. Lemaire

Sandoz Pharma AG, Dept. DS/DMP, CH-4002 Basel, Switzerland

The distribution of SDZ ENA 713 (ENA) and its phenolic metabolite (MET) in respect of the brain was quantified using microdialysis in the frontal cortex of freely moving rats. Dialysate concentrations of each were corrected for recovery by retrodialysis using deutero-ENA and -MET as reference compounds. Simultaneously ENA, MET and their deuterated forms were determined by MS-MS(TSP) with 'FI' sample introduction from an autosampler after a simple liquid-liquid extraction of microdialysis samples. For plasma samples the analysis was by GC-MS.*

ENA was administered to four rats by infusing 4 µg/min i.v. for 4 h (total dose infused: 0.96 mg). The steady-state brain ECF concentrations varied between 32 and 44 ng/ml for ENA and between 9 and 14 ng/ml for MET. At steady-state, the brain ECF/plasma concentration ratio averaged 35 ±4% (S.D.) for ENA and 29 ±6% for MET. These results indicate that ENA significantly penetrates into brain.

SDZ ENA 713, 3-[(1-dimethylamino)ethyl]phenyl ethylmethylcarbamate, a new drug for symptomatic treatment of Alzheimer's disease, selectively inhibits the activity of brain acetylcholinesterase [1]. The mechanism of action involves the release of a phenolic metabolite, 3-[(1-dimethylamino)ethyl]-phenol. It was therefore of interest to study the pharmacokinetics of both compounds in the brain. Microdialysis associated with an MS assay seemed to be the most promising method to meet that aim. Developed initially for measuring neurotransmitters in brain [2, 3], microdialysis is also a powerful tool with which to study the distribution of exogenous compounds [4-6].

*Abbreviations.- ENA denotes the drug and MET the metabolite named in the text; for their D_6- forms hexa-D- [typographical advantage]. MS-MS signifies tandem operation; SRM = selected reaction monitoring, TSP = thermospray. ECF, extracellular fluid; FI, flow injection; i.s., internal standard.

†addressee for any correspondence

This moiety lacking in MET

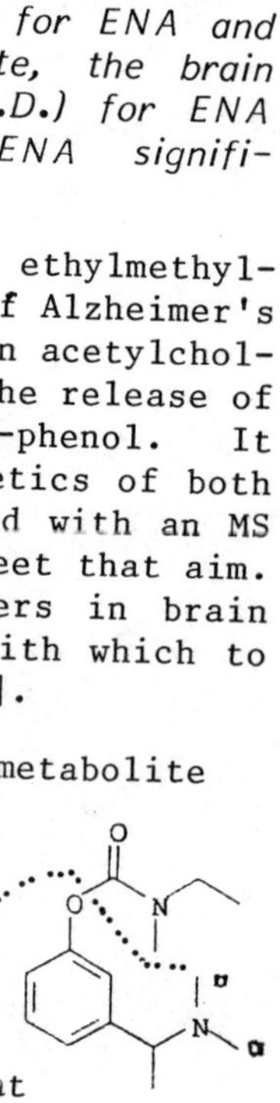

ENA

CD_3 label at ¤ positions

The technique [7] involves the introduction of a narrow dialysis probe into the tissue of interest and, through it, perfusion at a low flow rate with a fluid of physiological nature. Low mol. wt. compounds diffuse across the dialysis membrane into the perfusion medium, which is collected in fractions for analysis. The determination of the actual (free) concentration of a compound in the tissue implies the assessment of a diffusion recovery, which is dependent on the probe, the perfusion flow and the nature of the dialyzed compound. Because of the small sample volumes (typically <100 µl), a sensitive assay is needed. In past studies HPLC was used with UV [8] or EC [9, 10] detection; but such detection methods are too insensitive [11, 12] for some molecules. In 1990 Caprioli & Lin [13] introduced continuous detection of the microdialysis effluent by MS-MS with fast atom bombardment, whereas another group [14] reported off-line analysis of collected fractions by MS-MS with TSP ionization.

Here we describe an MS-MS assay with FI and TSP ionization for brain microdialysis fractions; stable isotope-labelled com-compounds serve, besides i.s. use in the MS-MS analysis, for determination of the recoveries of the dialyzed compounds by the retrodialysis method [15]. The analytes were first extracted from the dialysates in order to prevent a clogging of the TSP capillary with salts in the perfusion medium. The assay was applied to the determination of ENA and MET within a microdialysis experiment in rats. In addition, the assay of plasma samples by GC-MS aimed to estimate the passage of ENA and MET through the blood-brain barrier.

EXPERIMENTAL

Reagents.- ENA and MET were synthesized in-house, and the hexa-D forms in the Sandoz Research Institute at East-Hanover, U.S.A. Water was purified by ion exchange (Milli-Q; Waters). All reagents and solvents were analytical grade. The perfusion fluid contained 0.8 mM Mg^{2+}, 1.1 mM Ca^{2+}, 2.9 mM K^+, 155 mM Na^+, 13.7 mM HCO_3^-, 0.249 $H_2PO_4^-$, 132 mM Cl^-, 0.249 mM SO_4^{2-} and also 5.92 mM glucose. Calibrators were prepared with the perfusion fluid, the concentration ranges for ENA and MET being 0.15-100 and 0.16-107 ng/ml; every solution contained i.s.-ENA and -MET, 99.8 and 98.4 ng/ml. Physostigmine hemisulphate (Sigma) was put in the rat plasma to obviate the enzymic hydrolysis of ENA.

Instruments.- FI and HPLC of dialysates was done with the Kontron (Zürich) #460 System, having three #460 pumps (0.01-2 ml/min) with pulse dampers and a #450 data system; an #ISS200 Autosampler (Perkin-Elmer; Ueberlingen, Germany) was linked to it. The TSP mass spectrometer (Finnigan Mat, San Jose, CA) was a model #TSQ-70 upgraded to #TSQ-700 status with a new dynode, and had the new ICIS software (version 7.0) for data acquisition and processing.

Microdialysis [cf. 11].- Experimental details have appeared elsewhere [16]. Briefly, the probes (0.5 mm o.d.; membrane tip 4 mm long) were implanted in the brain cortex of 4 male Wistar rats (250-300 g). The 2.0 µl/min perfusion began 1.5 h before the start of a 4-h infusion of 0.96 mg ENA (4 µg/min) into a femoral vein. Prior to the infusion, a bolus of 1 ml (0.4 mg of ENA) was injected in order to purge the infusion device and attain rapidly the steady-state conditions. Dialysate samples were collected 0.5 h before the start of the infusion, and at 30-min intervals over 8 h. In addition, blood samples were withdrawn (400 µl, from the femoral artery of the opposite leg) every 30 min from 2.25 up to 7.75 h after the infusion start. Blood was treated with heparin and physostigmine just after sampling, and plasma samples were prepared by centrifugation.

Extraction.- To the total volume of an unknown sample (30-60 µl exactly measured with a syringe), or to 100 µl of a calibrator, were added 250 µl of saturated sodium carbonate and 4 ml of tert.-butyl methyl ether in a glass extraction tube. The samples were extracted for 20 min on a horizontal shaker and centrifuged (5 min, 1400 **g**); then 3.5 ml of the organic layer was transferred into a new tube and evaporated to dryness under a gentle N_2 stream at 40°. The residue was taken up in 230 µl of a mixture (solution **A**) of 0.2 M ammonium acetate and acetonitrile (85:15 by vol.) with vortexing for 20 sec; the solution was transferred to an autosampler vial for injection.

MS-MS analysis.- FI transfer [see end of text] entailed injection *via* a 200 µl loop with a 1.2 ml/min flow of solution **A.** Water was added at 0.6 ml/min to obtain the optimal flow and ammonium acetate concentration for the TSP ionization. The source was held at 300° with the vaporizer at 105°. With the discharge and filament off there was a pure buffer ionization. The repeller voltage was optimized for each ion to be detected.

Positive-ion mass spectra were obtained by scanning the range 100-300 Da with the first quadrupole. In the MS-MS mode, ~1.7 mTorr (0.23 kPa) of argon was admitted into the collision chamber with a collision energy of -10 eV. The daughter ion spectra were m/z 206.1 for ENA and hexa-D-ENA and 121.1 for MET and hexa-D-MET. The scan time was 0.5 sec for each compound. The electron multiplier was operated at 1700 V.

GC-MS analysis of plasma was performed by an in-house procedure*, using 250 µl of sample per determination. The concentration of calibrators ranged from 0.5 to ~500 ng/ml.

Calculations.- Peak area ratios (compound/i.s.) were chosen as signals (= Y). Linear calibration parameters were calculated from the signals observed for the calibrators using

*Perhaj, Z. & Laplanche, R. (1989), Sandoz Pharma Document No. 303-263, Basle. Extractions (MTBE, with an acid back-extraction) and derivatization (propionic anhydride), then GC(EI) in SIM mode.

the program ELSFIT†. The following variance model was chosen for the weighting: variance = $P1^2 \cdot F^{P2}$ where P1 and P2 are constants and F is the value of Y calculated with the regression parameters (intercept a, slope b) for a concentration ratio X of the compound to the i.s. For either study-sample compound the concentration, where **X** = (Y - **a**)/b, is **X** • concn.$(i.s.)_{OUT}$, this concentration (for the corresponding i.s.) being calculated from the i.s. concentration in the calibrators, concn.$(i.s.)_{CAL}$ (99.8 and 98.4 ng/ml for ENA and MET) and from the median peak areas observed for the calibrators (S_{CAL}) and for the study samples (S_{SAMP}): concn.$(i.s.)_{OUT}$ = concn.$(i.s.)_{CAL} \cdot S_{SAMP}/S_{CAL}$

The relative recovery of the compound concerned (ENA or MET) was assumed equal to the relative loss (by retrodialysis) of the corresponding i.s. (hexa-D-ENA or -MET) in the dialysis probe [15]: rec. = [concn.$(i.s.)_{IN}$ - concn.$(i.s.)_{OUT}$]/concn.$(i.s.)_{IN}$. The i.s. concentration at the inlet of the probe, concn.$(i.s.)_{IN}$ (~100 ng/ml), was accurately calculated in the same way as concn.$(i.s.)_{OUT}$ but with samples taken before the start of the experiment. The concentration of each compound in the brain cortex ECF was given by the ratio concn./recovery.

RESULTS AND DISCUSSION

Mass spectrometry.- Fig. 1 shows the mass spectra for ENA and metabolite and the daughter-ion spectra of the protonated molecular mass peaks: m/z 251.1 for ENA, 166.1 for MET. Both parent spectra appear to be very simple, with little fragmentation and therefore good detectability of the compounds with the chosen daughter ions (m/z 206.1 and 121.0). The corresponding mass spectra of hexa-D-ENA and hexa-D-MET are very similar, giving rise to the same daughter ions by loss of the (labelled) dimethylammonium.

Assay performance.- The calibration curves were linear (r >0.999). Performance parameters are summarized in Table 1. The biases were between -5.8 and +1.7% for ENA and between -3.4 and +2.6% for MET. The overall variability (C.V.) was ~7% for concentrations of ENA down to 0.51 ng/ml and 9% for MET down to 0.55 ng/ml. The lower limit of quantification (LLOQ) with the usual definition (C.V. 20%) was 0.2 ng/ml for both compounds in artificial CSF. For GC-MS measurements on plasma samples the established LLOQ's were 1.6 and 2.0 ng/ml for ENA and MET.

Study results

The *in vivo* average recovery ±S.D. was 28 ±7% (range 21-37% for ENA and 19 ±9% (9-30%) for MET. It was fairly reproducible over the 4 experiments for ENA, but very variable with MET.

---

†Table 3 in L.B. Sheiner, ELSFIT User's Manual, Technical Rept. of Divn. of Clinical Pharmacology, Univ. of California, San Francisco (1985).

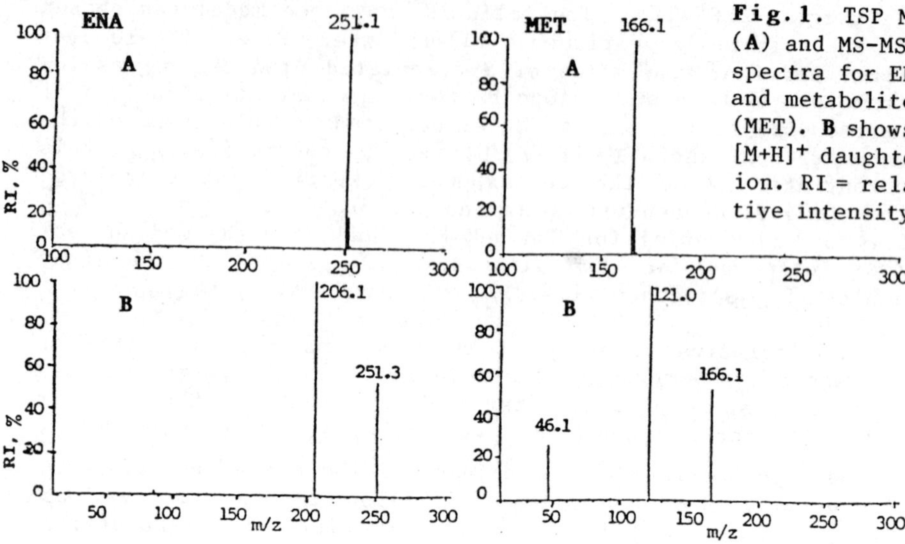

Fig. 1. TSP MS (**A**) and MS-MS (**B**) spectra for ENA and metabolite (MET). **B** shows $[M+H]^+$ daughter ion. RI = relative intensity.

Table 1. Method validation summary for determination of ENA and *(values in italics)* MET in artificial CSF by MS-MS(TSP) in the FI mode (n = 7 at highest concentration; otherwise n = 9).

Expected ng/ml	Observed ng/ml	Mean bias, %	C.V., %
0.15, *0.16*	0.15, *0.16*	0.9, *0.0*	17.2, *24.4*
0.50, *0.54*	0.51, *0.55*	1.2, *2.3*	7.5, *9.4*
1.01, *1.07*	1.00, *1.06*	-0.8, *-0.9*	3.7, *7.5*
4.03, *4.28*	3.79, *4.14*	-5.8, *-3.4*	4.2, *6.8*
20.13, *21.4*	20.43, *21.95*	1.5, *2.6*	1.8, *3.1*
100.65, *107.00*	102.38, *106.74*	1.7, *-0.2*	1.7, *1.1*

Variability within one experiment could not be assessed because of the variability of the absolute MS signals over the time. Attempted calculation of the ENA and MET sample concentrations with the absolute signals so as to assess a recovery at each sampling time resulted in a much poorer reproducibility of the estimates of the *in situ* concentrations within a single animal.

The **ENA** levels in plasma and in brain-cortex ECF (Fig. 2) showed steady-state attainment within 1 h of i.v. infusion and good reproduciblity across the 4 animals: plasma 112 ±11, brain 39 ±5 ng/ml (±S.D.). Taking into account that ENA is 45% bound in plasma, the ENA-in-ECF/plasma-free-ENA ratio was ~63%; moreover, the elimination half-life (calculated between 4 and 6 h) was similar in both compartments (Fig. 2). The high levels seen at the start of measuring ECF kinetics came from a pre-infusion i.v. loading dose. For MET the ng/ml values were: plasma 43 ±6, ECF 12 ±2. As for ENA, the steady state was rapidly attained and the elimination half-life was similar in both compartments (Fig. 3). A possible interpretation of the MET data is that

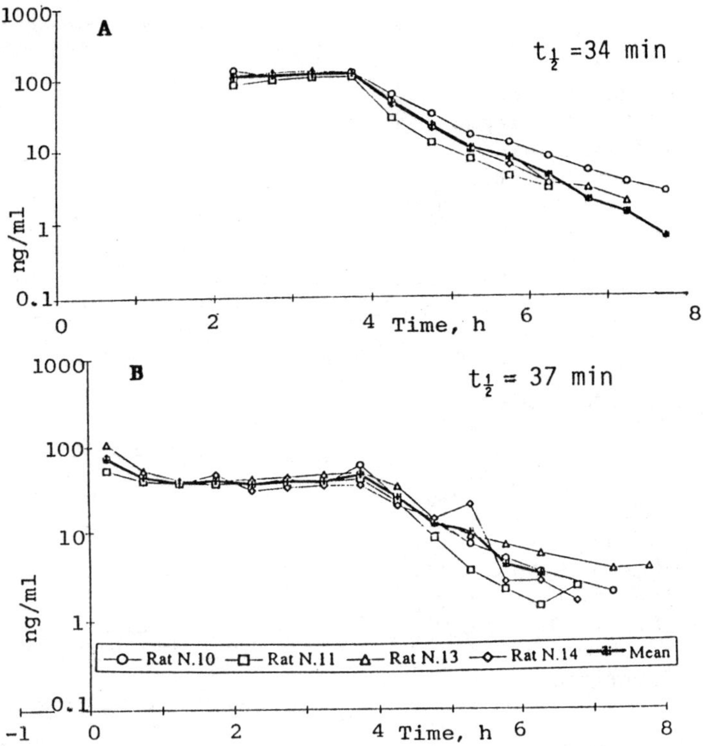

Fig. 2. Concentrations of ENA in plasma (**A**) and brain-cortex ECF (**B**).

its plasma concentration determined its kinetics in the brain. We must, however, keep in mind that MET is locally produced by the interaction with the target enzyme AChE.

CONCLUDING COMMENTS

This study showed the effectiveness of the methods used for quantifying the passage of ENA into the brain. No vascular blood contamination occurred. Using a suitable reference for recovery calculations the assay of brain cortex levels was quite accurate. The high sensitivity of the approach entailing brain microdialysis was amplified by the use of MS-MS(TSP), which is so specific that no chromatographic step is needed. It enables several compounds to be determined simultaneously (4 in our case), and is so sensitive that small sample volumes suffice. A further advantage of MS-MS with sample introduction by FI from an autosampler is that each sample took only 2.5 min to analyze, allowing ~100 to be injected in a single day.

Note on FI operation.- Connecting to the TSP interface is a 4-port valve (with a 'To waste' position; Valco), fed from a 20 cm capillary (0.25 mm i.d.) which receives water (see EXPERIMENTAL) passed through a 90 cm capillary and a T-valve that can receive, from a 3 m × 0.12 mm capillary ('loop'), the autosampler's input with 1.2 ml/min mobile phase.- Hence 1.8 ml/min enters the TSP.

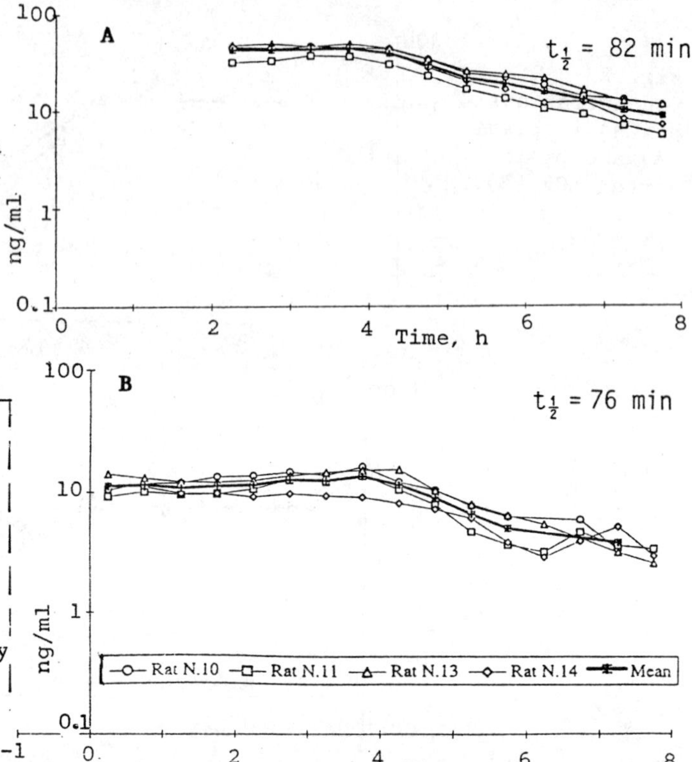

Fig. 3. Concentrations of MET in plasma (**A**) and brain-cortex ECF (**B**).

In *References,* Vol. 23 in 'this series': *Biofluid and Tissue Bioanalysis for Drugs Including Hypolipidaemics* (ed. E. Reid *et al.*), Royal Society of Chemistry, Cambridge.

References

1. Enz, A., Amstutz, R., Baddeke, H., Gmelin, G. & Malianowski, J. (1993) *Prog. Brain Res. 98,* 431-438.
2. Ungerstedt, U. (1987) in *Life Sci. 41,* 861-864 [Issue on *Measurement of Neurotransmitters in the Brain* (Pergamon)].
3. Di Chiara, G. (1990) *J. Neurosc. Meth. 34,* 29-34.
4. Hurd, Y.L., Kehr, J. & Ungerstedt, U. (1988) *J. Neurochem. 51,* 1314-1316.
5. Scott, D.O., Sorenson, L.R., Steele, K.L., Puckett, D.L. & Lunte, C.E. (1991) *Pharm. Res. 8,* 389-392.
6. Steele, K.L., Scott, D.O. & Lunte, C.E. (1991) *Anal. Chim. Acta 246,* 181-186.
7. Beneviste, H. & Huettemeier, P.C. (1990) *Prog. Neurobiol. 35,* 195-215.
8. Scott, D.O., Sorenson, L.R. & Lunte, C.E. (1990) *J. Chromatog. 506,* 461-469.
9. Sarre, S., Michotte, Y., Herregodts, P., Deleu, D., De Klippel, N. & Ebinger, G. (1992) *J. Chromatog. B 575,* 207-212.
10. Smith, S. & Sharp, T. (1994) *J. Chromatog. B 652,* 228-233.
11. Van Belle, K., Sarre, S., Ebinger, G. & Michotte, Y. (1994) in Vol. 23, this series, pp. 373-379.
12. Chilton, A. (1994) in Vol. 23, this series, pp. 381-388.
13. Caprioli, R.M. & Lin, S.N. (1990) *Proc. Nat. Acad. Sci. 87,* 240-243.
14. Menacherry, S.D. & Justice, J.B. Jr. (1990) *Anal. Chem. 62,* 597-601.
15. Wang, Y., Wong, S.L. & Sawchuk, R.J. (1993) *Pharm. Res. 10,* 1411-1419.
16. Alonso, M.J., Bruelisauer, A., Misslin, P. & Lemaire, M. (1995) *Pharm. Res. 12,* 291-294.

#C-2n

A Note on

A HIGHLY SENSITIVE GC METHOD FOR THE DETERMINATION OF DERAMCICLANE AND ITS *N*-DES METABOLITE IN RAT AND DOG PLASMA

Katalin Balogh Nemes†, Miklós Abermann, Erzsébet Bojti & Imre Klebovich

Egis Pharmaceuticals. Ltd., Dept. of Pharmacokinetics, Keresztúri Út 30-38, H-1475 Budapest, Hungary

Deramciclane is a new anxiolytic compound from our company. To determine the drug and its N-monodesmethyl metabolite (Fig. 1) in rat and dog plasma, a highly sensitive (LLOQ* 1 ng/ml) GC-NPD method has been developed, with initial SPE (Fig. 2) and

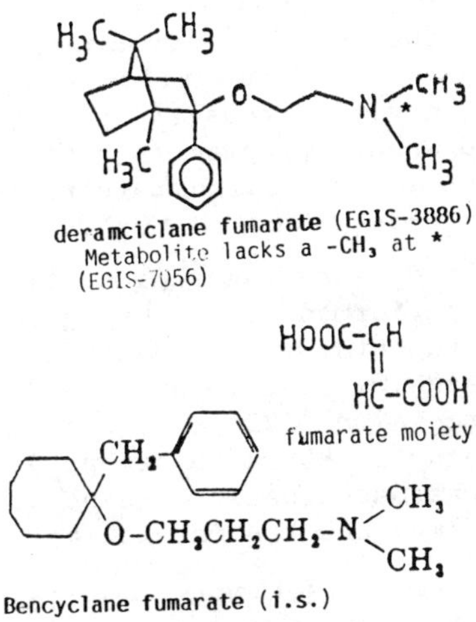

Fig. 1. Analyte formulae including i.s. Deramciclane: [1*R*, 2*S*, 4*R*]-(-)-*N*,*N*-dimethyl-2-[(1,7,7-trimethyl-2-phenyl-bicyclo[2,2,1]-hept-2-yl)-oxy]-ethanamine 2-(*E*)-butendioate (1:1).

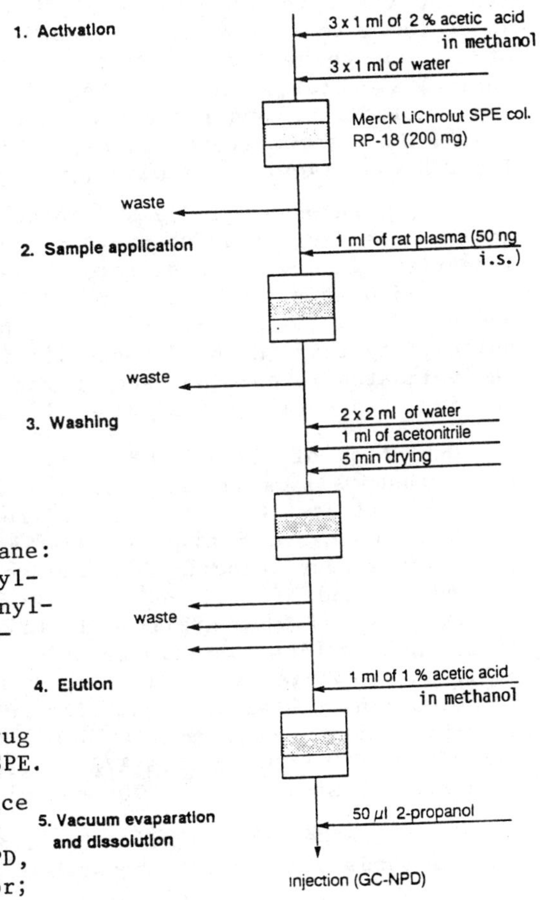

Right: **Fig. 2.** Drug and metabolite SPE.

---

†addressee for correspondence

**Abbreviations:* LLOQ, lower limit of quantification; NPD, nitrogen-phosphorus detector; i.s., internal standard.

Fig. 3. GC of extracts of rat plasma: **A**, spiked with drug and metabolite, 300 ng/ml (also i.s.); **B**, blank.

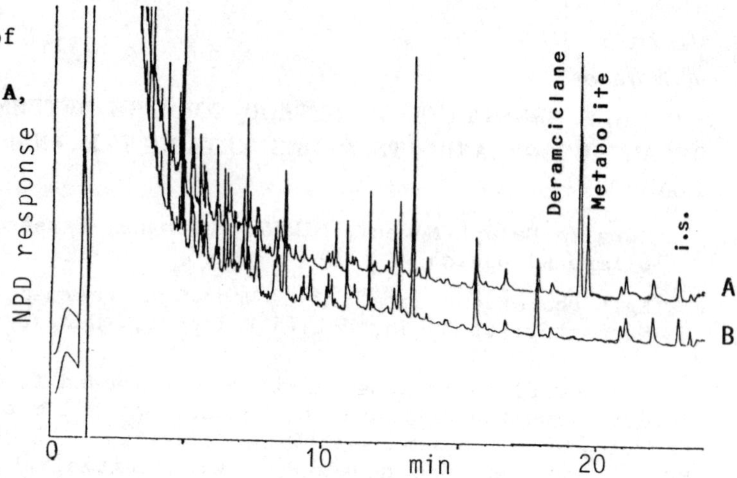

GC conditions as below. The built-in electronic pressure controller (EPC) is a new GC tool that allows 'pseudo-splitless' injection with a high head pressure in the column, fast elution of components not of interest, and elution of analytes at an optimal flow rate. Thereby there is benefit to sensitivity, R_T reproducibility, and run time. The pressure program is accompanied by a temperature program; they are optimized together. The SPE was likewise optimized, using ^{3}H-drug besides GC-NPD.

Fig. 3 shows GC profiles. Metabolite assay was semiquantitative. The drug assay was rugged and performed well: very good linearity (2 curves, on different instruments; r = 0.998) in the range 1-500 ng/ml; precision 8.16%/bias+4.07% inter-day (n = 7) and 4.17%/-1.41% intra-day (n = 6); precision 2.22% in a system suitability test (n = 6; 3 concns.); recovery 85.9% (5 concns.). The validated method was used in rat pharmacokinetic studies on the drug (one oral or i.p. dose, 10 mg/kg) and the metabolite.

Operation of the GC (HP #5890 series II with NPD; Supelco SPB-5 fused-silica column, 30 m × 0.25 mm × 0.25 µm).- Injector at 240°, NPD at 320°; oven at 100° for 0.4 min at start. Gases [& flow rates, ml/min].- Helium (99.995% pure) as carrier gas [0.7 with linear velocity 24.9 cm/sec during the 8-min isothermal phase] and NPD make-up gas [30.6]; H_2 [4.1]; air [106]; septum purge [1.0]. Split-mode injection [vent 8.0 at 100°]: liner packing (Serva) silicone SE30 Gas Chrom Q, 100-125 mesh; split ratio varies with the ramped PRESSURE PROGRAM.- 340 kPa for 0.05 min at start; then, for Levels 1/2/3: 500/400/30 kPa per min pressure change, to 170/110/180 kPa - held for 10/9/1 min. OVEN TEMP. PROGRAM.- Levels 1/2: 10/50° per min, to 210/280° - held for 8/3 min; Level 3, 70° per min drop, to 95° (end).

Why GC rather than HPLC? - Sensitivity is much better, and the compounds of interest are stable at high temperatures.

#C-3

A SENSITIVE, SPECIFIC AND RUGGED DETERMINATION OF FLESINOXAN IN PLASMA BY HPLC-EC†

D.J.K. van der Stel, P.H. van Amsterdam, K. Bijmholdt-Sierat, E. Jochem, G.Ph.D. Leeflang, C.M. Middlekoop & H.J.C. van Schie

BioAnalysis Dept., Pharmaceutical Division, Solvay Duphar bv, P.O. Box 900, NL-1391 CP Weesp, The Netherlands

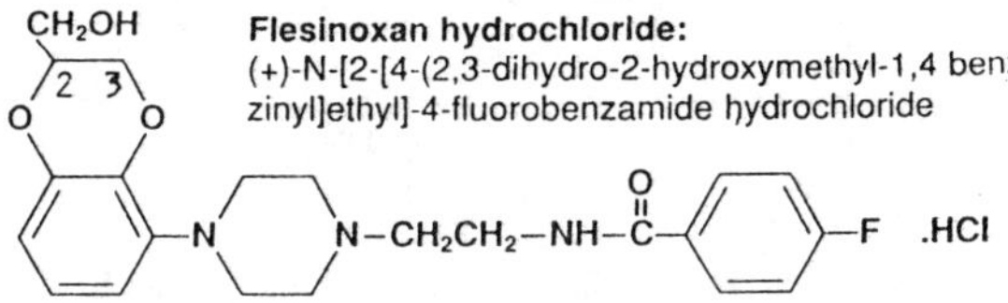

Flesinoxan hydrochloride:
(+)-N-[2-[4-(2,3-dihydro-2-hydroxymethyl-1,4 benzodioxin-5-yl)-1-piperazinyl]ethyl]-4-fluorobenzamide hydrochloride

3-Hydroxymethyl analogue as i.s.

Requirement — *A method more sensitive (aim: LLOQ ≦0.1 ng/ml with a 2 ml sample) than an early HPLC-fluor method, and superior in ease and robustness to a GC-ECD* method [1] that has risk of falsely high values through metabolite breakdown.*

End-step — *RP-HPLC-EC, with autosampler loading onto pre-column.*

Sample handling — *SPE (C-18), with sample concentration; solvent-extract and wash; dry down and redissolve in mobile phase.*

Comments — *The developed method performed well for 0.1-50 ng/ml and was specific; good matching with the GC-ECD method.*
The strategy and tactics are presented largely in 'panel' form together with the validation which finally included inter-laboratory comparison. Through the extraction and the washing steps, only apolar basic compounds reached the HPLC column.

Flesinoxan, a full 5-HT_{1a} agonist, is currently in development (Phase III clinical trials) for major depressive disorder (MDD) and generalized anxiety disorder (GAD). To support pre-clinical and clinical development in Phase I and early-Phase II a GC-ECD assay was developed; pre-treatment included derivatization and HPLC clean-up [1]. Ample sensitivity (LLOQ 0.1 ng/ml with a 5 ml sample) was shown during validation but seemed hard to maintain routinely with large batches and only ≈3 ml sample size - reduced since, in multi-session clinical pharmacological studies with volunteers, the total blood amount permitted to be taken was restricted.

†Two 'Notes' based on Forum Posters with 'panels' therefrom and already readied for press were amalgamated and reinforced when, too late, a full MS. arrived (from H. de Bree as 'correspondent').
* *[Further editorial point) Abbreviation* for electrochemical (detection) is EC (authors had ECD, now confined to GC: electron-capture detecion). Other abbreviations in Legends and Preface; i.s., internal standard.

Fig. 1.
The switching assembly. For components and operation see later in the text.

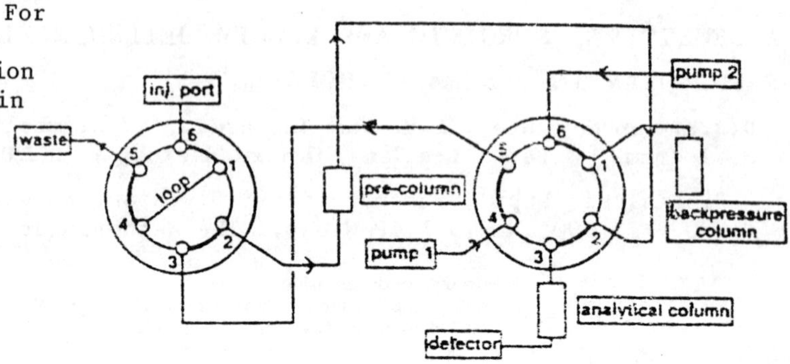

Additionally, the need for very skilled technicians and the sophisticated dedicated set-up for the GC-ECD method hampered efficient contracting out of assays. So the position we had reached, and the way forward, can be summarized as follows.

Need for a new method, and outline of its development

Since flesinoxan is taken by patients in sub-mg doses, its bioanalysis stretched the GC-ECD method that served at the outset. The trickiness of the method, including a derivatization step and ill suited to large batches, prompted an HPLC approach now that EC detection has grown in scope and reliability. Keeping a solvent extraction step as in the GC-ECD method but preceding it by RP-SPE, an effective procedure was developed (Scheme 1), using a switching assembly that enabled pre-column effluent to bypass the analytical column (Fig. 1). Amongst stratagems that were tried (Scheme 1), an NP-HPLC clean-up step as in the GC-ECD method did not warrant adoption, nor was NP-SPE helpful.

Amplification (referring to Scheme 1 *top left*)

The first step was assessing the sensitivity of HPLC-EC with the pure compound. A limit of detection (LOD) of at least 25 pg/ml was shown. The second step was performing a feasibility study, focused on sensitivity, with flesinoxan-spiked plasma. An LOD of at least 50 pg/ml, with reasonably clean chromatograms, was achieved using a **basic method** (Scheme 1) designed with the help of experiences using this compound type.

Sample preparation.- Spiked plasma blanks processed by the basic method revealed chromatograms with quite a number of peaks around the R_T of flesinoxan and the i.s. Moreover, the excessive presence of late-eluting compounds implied long run times, hampering a high sample throughput. Adaptations of the basic clean-up procedure, as follows, were tested one-by-one to enhance the selectivity of the procedure and to maintain a high analyte recovery and good performance of the HPLC system.-

Scheme 1. Method development and final protocol. Liquid-liquid extraction: L/L or (for analyte) LL_e or (to wash) LL_w. MIBK, methyl isobutyl ketone.

DEVELOPMENT

Basic set-up HPLC-EC assay

1. SPE (C_{18}) of plasma sample
2. L/L extraction of basified residue with MIBK
3. Wash MIBK with NaOH
4. Evaporate MIBK, dissolve residue in mobile phase, HPLC-EC

Tested adaptions of basic design (referring to the above numbers)

4. Acetone transfer of evaporated MIBK residue
1. Acidic wash of SPE residue (HCl and pentane/ether)
1. Neutral wash of SPE residue
1. Straight phase SPE of C_{18}-SPE residue
1. Straight phase HPLC clean-up of SPE residue
4. Column switching HPLC-EC

pre-1. Centrifugation of plasma sample

1. Acidic wash of SPE residue (HAc and MIBK/ether)

METHOD

Procedure

- Centrifuge plasma samples at ± 6000 x G for 20 minutes. Remove plasma surface layer with pasteur pipette/suction. Pipette 2 ml of sample into a 10 ml vial

SPE Add 10 ng of internal standard and 200 µl NH_4OH. Wash C_{18} cartridges 3 times 5 ml methanol and 3 times 5 ml water. Concentrate the sample on the cartridge. Wash 3 times with 5 ml water and with 3 ml of methanol/water (25/75). Elute the cartridge with 3 ml of a methanol/NH_4OH (98/2). Evaporate the solvent to dryness in a stream of nitrogen at 75 °C. Dissolve the residue in 500 µl of water.

LLw Add 5 ml of n-pentane and wash for 1 minute. Centrifuge for 1 minute at ± 3750 x G. Freeze the aqueous layer in a cryobath at -40 °C. Discard the n-pentane layer.

LLw. Add 1 ml of a 2% acetic acid solution in water. Add 5 ml MIBK/ether (50/50) and wash for 1 minute. Centrifuge for 5 minutes at ± 3750 x G. Freeze the aqueous layer and discard the organic layer.

LLe Add 1 ml of 1 mol/l NaOH solution. Add 4 ml of MIBK and extract for 1 minute. Centrifuge for 5 minutes at ± 3750 x G. Freeze aqueous layer and transfer organic layer into a new 10 ml vial.

LLw Add 1 ml of 1 mol/l NaOH solution and wash for 1 minute. Centrifuge for 5 minutes at ± 3750 x G. Freeze aqueous layer and transfer organic layer into a new 10 ml vial. Evaporate solvent in a stream of nitrogen at 75 °C.

HPLC Transfer the residue into a autosampler vial with two times 100 µl acetone. Evaporate the acetone in a stream of nitrogen at room temperature. Dissolve the residue in 35 µl of HPLC mobile phase and cap the vial. Inject 20 µl into the HPLC system.

HPLC-EC

Stationary phase	Pre-column	Brownlee RP-18, 7 µm, 15 x 3.2 mm
	Analytical	Brownlee RP-18, 5 µm, 220 x 2.1 mm
Mobile phase	water/acetonitrile/methanol/NH_4OH (600/300/100/0.5) + 5 g NH_4Ac/l	
Equipment	Waters and/or HP pumps, Gilson autosampler, ESA-Coulochem detector	
Program	Load pre-column at 1 ml/min for 1 min Connect pre-column to analytical column; flow 0.4 ml/min Wash pre-column with methanol/NH_4OH (1000/0.5) Condition pre-column for new injection	

- Acetone wash of the final residue to improve sensitivity.
- SPE (C-18) residue subjected to acidic and neutral washes and straight-phase (NP) SPE and (removing apolar components) HPLC.
- Centrifugation of plasma and acidic wash of SPE residue to remove polar compounds.

The steps that gave most benefit to selectivity and/or sensitivity were incorporated in the basic sample preparation set-up.

Chromatography and detection.- From previous experience with flesinoxan and other compounds within its class, it was known that RP-HPLC, especially with a C-18 phase, yielded reproducible chromatograms with good peak shape. In the feasibility studies a Hypersil ODS column (Shandon; 5 µm, 200 × 4.6 mm) was used. We changed to C-18 OD-222 (Brownlee; 5 µm, 200 × 4.6 mm) since this column proved to surpass the Hypersil ODS in lifetime and rate of conditioning. Although several options were tested to remove apolar plasma constituents from the processed sample, the overall result was that late-eluting compounds still precluded having a run time of ~10 min.

We added to the HPLC system a pre-column and Rheodyne valve (Fig. 1) that enabled us to bypass these compounds by column switching. After injection of the samples, as soon as analyte and i.s. have migrated from the pre-column onto the analytical column the pre-column is switched out of the primary flow path. The late-eluting compounds thus trapped on the pre-column are rinsed off by a fast gradient program, attaining 100% methanol. Shortly thereafter the pre-column is re-conditioned with mobile phase and switched into the primary flow path again. A back-pressure column is connected to the pre-column in the rinsing/conditioning phase to avoid drastic pressure changes in the chromatography system, so maintaining a stable base-line.

The **EC detectors** used were EG&C initially (amperometric) and ESA Coulochem II (coulometric). The former served well in the feasibility studies, but let us down when the need arose for long-lasting high performance. Seemingly the glassy carbon working surface had to be polished after each large injection series, and the sensitivity loss could not be kept below ~40% within a batch of ~50 samples. Moreover, it became clear that maintaining high constancy of mobile phase composition and surrounding temperature was crucial and it was hard to avoid post-column formation of vapour bubbles, resulting in detection cell malfunctioning that was not self-curing.

The ESA Coulochem II did not suffer from these drawbacks and had certain advantages: application of pre-oxidation which almost completely obviates contamination of the electrodes, and cell design that generates some back-pressure. Also relevant for a broad application of the method was that this EC detector seemed to be the standard one in many contract laboratories.

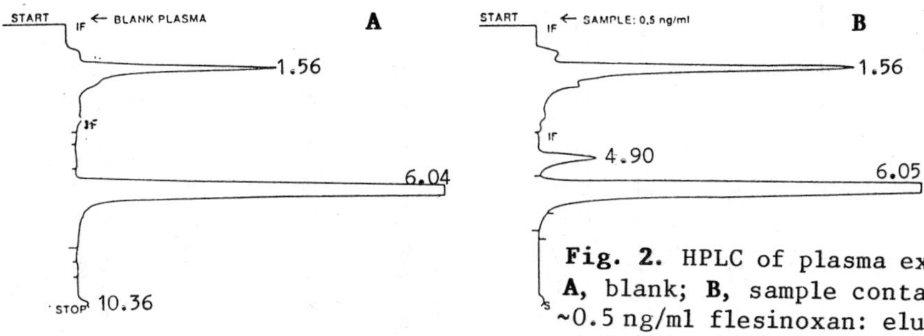

Fig. 2. HPLC of plasma extracts. **A**, blank; **B**, sample containing ~0.5 ng/ml flesinoxan: elutes at ~4.6 (i.s. 6.0) min.

FINAL ANALYTICAL METHOD

Materials.- Inorganic salts were of analytical grade from well established commercial suppliers. Solvents such as MIBK (extractant) and acetonitrile (HPLC eluant) were redistilled before use. This appeared to be of the utmost importance for sustained good performance of the column and detector cell.

Sample preparation is shown in Scheme 1 (and is also depicted later in relation to 'test factors'). Procedures for **HPLC** also appear in Scheme 1, and representative chromatograms are shown in Fig. 2.

VALIDATION*

Our validation strategy is first to perform a thorough basal validation to determine: system performance, calibration model, linearity, accuracy, precision and sensitivity. Validation parameters such as stability, specificity, ruggedness and inter-laboratory comparison are assessed whilst applying the method or by performing separate studies.

Basal validation was as shown in Scheme 2 which also summarizes our results. To assess **performance** a solution of pure drug and i.s. was injected 7 times evenly interspersed throughout an analysis batch. The maximum observed difference was <2% and the mean intra-run C.V. was 0.36%. **Calibration** standard (duplicates) responses gave the intra-run variance per level. C.V.'s being near-constant, a factor of $1/x^2$ was adopted for least-squares curve fitting [2] (typical curve: Fig. 3).

Fig. 3. A typical calibration curve (with non-linear scales; peak-height ratio basis) for flesinoxan in plasma. The quadratic model shown in Scheme 2 was used.

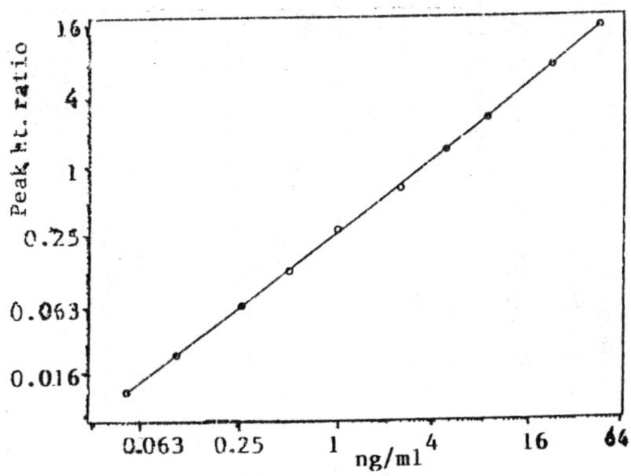

*Routine batches (30-50 test samples) have ≧9 calibrants, 3 × 2 QC's and 5-10 system performance samples (pure drug and i.s. in HPLC eluant).

Scheme 2. Features of the method. Lo, Me, Hi signify spike levels (5 in all).

VALIDATION

(Data statistically analyzed by usual descriptive univariate analysis procedures)

Design
7 runs with: 10 calibration standards in duplicate plasma blank in duplicate
QCS Lo", Lo', Lo, Me and Hi in duplicate 5 K-references i.e. 5 levels

System performance
HPLC repeatability: average CV = 0.36%
each run CV < 0.48%

Response function
Error structure: Constant CV (5.4%) from 0.1 - 50 ng/ml
Constant s at < 0.1 ng/ml
Conclusion: $1/x^2$ weighing

Models:	A	B	C	|%bias|	r
$y = A + Bx$	-0.001	0.117		5.2%	0.995
$y = A + Bx + Cx^2$	-0.001	0.113	0.0001	3.6%	0.997

Second model evidently preferable.

Accuracy

QC	Spike (ng/ml)	Mean found	%bias
Lo"	0.050	0.055	9.2
Lo'	0.10	0.097	-2.9
Lo	0.20	0.19	-3.3
Me	1.0	0.98	-2.1
Hi	10	10.5	4.6

Precision

QC	Spike (ng/ml)	Intra-day CV	Inter-day CV
Lo"	0.050	26	26
Lo'	0.10	13	15
Lo	0.20	7.1	9.0
Me	1.0	6.1	10
Hi	10	4.5	6.1

Sensitivity
LLOQ (based on inter-day s of low level QCs) = 0.08 ng/ml

Specificity ['LLOQ' = lower limit of quantification]
Matrices: No interferences observed regardless species and storage history of plasma sample
Metabolites: Less than LLOQ result for isolated metabolite fractions
Cross-corr: Perfect match in cross-correlating HPLC-ECD with former GC-ECD

Analyte stability
Storage: No decrease after 60 months at ≤ -10 °C
Analysis: Flesinoxan is stable in plasma at room temperature for at least 24 hrs.
Stock solutions in water stable for at least 3 months
Sterilization: Flesinoxan is stable at 50 kGy gamma irradiation of plasma

Ruggedness
Designed: *see later (text and Scheme 3).*

Solvents: Use of ultra pure (distilled) solvents is essential
In-use: More than 5000 samples analyzed from human (clinical and pharmacokinetic) and animal (toxico- and pharmacokinetic) studies within current quality demands. Assay has successfully been set up at 3 contract laboratories.

The quadratic model was slightly better than the other model tested (Scheme 2), and also allowed a good fit for incidental slightly concave or convex calibration curves.

Accuracy: at each of 5 levels (duplicate QC's) within each batch, agreement was within 10%. **Precision** was within 15% except at the lowest level. For **sensitivity**, defining the LLOQ as the level where C.V. becomes 20% implies that the actual LLOQ is 0.08 ng/ml (interpolating between Lo" and Lo' QC's); but for routine work the lowest calibration standard (0.05 ng/ml) is excluded and the practical LLOQ is ~0.1 ng/ml.

Further validation

Stability.- We routinely test whether an analyte survives γ-irradiation (50 kGy) aimed to kill life-endangering viruses, e.g. HIV, and so minimize risks to laboratory personnel [3]. Pilot experiments showed a 10^5-fold reduction of HIV-titre and no effect on drug plasma concentrations [4]. Flesinoxan-spiked blanks (0.2/2/10 ng/ml) and a pool of study samples (~4 ng/ml) showed no statistical differences during 5 years at -10°, with or without irradiation; nor was degradation found under other conditions (Scheme 2) including thawing.

Specificity verification is summarized in Scheme 2, and method comparison in Fig. 4. When 4 plasma samples of independent origins were assayed for flesinoxan both direct (a) and as fractionated HPLC peaks from metabolite profiling (b), matching was good: (a) 24.1, 18.1, 6.87, 7.70; (b) 23.5, 20.2, 8.33, 8.36 ng/ml. For processed metabolite fractions the values were <LLOQ. Testing of possible co-medications (Table 1) revealed interference only from moclobemide and prazosin at relatively high concentrations.

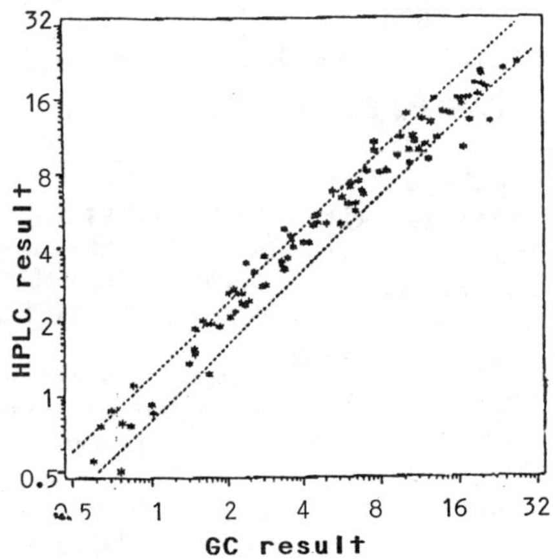

Fig. 4.
Results (log-transformed) for 101 samples from 3 subjects: GC-ECD compared with HPLC-EC, showing a perfect match (paired t-test; the two lines represent theory ±20%).

The perfect match between two assay methods differing in sample work-up and chromatography, and the metabolite-study evidence, firmly prove the HPLC-EC assay to be very specific.

Table 1. Flesinoxan HPLC-EC assay: specificity with respect to various drugs and metabolites (e.g. salicylic acid).

Drug	Level (ng/ml)	Drug	Level (ng/ml)
Acenocoumarol	200	Fluoxetine hydrochloride	1000
Acetylsalicylic acid	100000	Fluoxetine plasma	100
Alprazolam	200	Glybenclamide	2000
Alprazolam plasma	50	Haloperidol	50
Amitriptyline hydrochloride	1000	Hydralazine hydrochloride	1000
Amitriptyline plasma	100	Ibuprofen	20000
Antipyrine plasma	15000	Imipramine hydrochloride	1000
Bromazepam	250	Lorazepam	50
Buspirone	1000	Lorazepam plasma	50
Buspirone plasma	10	(±) Metoprolol tartrate	500
Caffeine	10000	Moclobemide *	3000
Carbamazepine	10000	Norfluoxetine	1000
Cimetidine	10000	Paracetamol	10000
Desogestrel	10	Phenprocoumon	5000
Diazepam	1500	Prazosin hydrochloride °	1000
Digitoxin	5	Propranolol	500
Digoxin	5	Ranitidine hydrochloride	2000
17β-Estradiol	0.1	Salicylic acid	100000
Estrone	0.5	Theophylline	50000
17β-Ethynylestradiol	0.25	Tolbutamide	200000
		(±) Warfarin	10000

* interferes ≥ 1000 ng/ml,
° interferes ≥ 100 ng/ml

RUGGEDNESS: Strategy (Scheme 3, A & B)

The foregoing outline (Scheme 2) is now amplified, firstly with consideration of the strategy. In summary, experience has indicated factors (61) that could possibly change performance of the assay if changed intentionally or inadvertently: 48 are part of the sample work-up (as re-capitulated in Scheme 3B) and the remaining 13 are in the chromatography part. Work-up has 6 logical parts: plasma preparation, SPE, L/L washings 1 & 2, L/L extraction and L/L washing 3. Parts containing ≤4 factors were investigated with (full) factorial designs; when >4 were identified, Plackett-Burman designs were used [5]. Application of this general strategy to flesinoxan is now considered.

RUGGEDNESS: Results for the flesinoxan assay (Scheme 3C)

Scheme 3 concludes with 'retrospective guidance'. Anticipating this, the **Panel** on the right summarizes the CONCLUSIONS reached.

Ruggedness testing of bioanalysis method using different factorial and Plackett-Burman designs was suitable for the determination of the effects of factors from the sample work-up and HPLC system.

Dividing the assay into sample work-up and HPLC was permitted because the variances of the measurement errors could not be pooled using the F-test.

Separation of the sample work-up in different parts was permitted because no interactions between factors from different parts were measured, at least for the parameters that are most significant for quantification: peak height ratio and peak height.

The assay of flesinoxan in plasma is (acceptably) rugged.

Scheme 3. Ruggedness testing, (A).

THEORY

Definition ruggedness test:
to examine the behaviour of an analytical process when small changes in the environmental and/or operating conditions are made, akin to those likely to arise in different laboratories.

The ruggedness test consists of three phases:
(a) pre-experimental phase
(b) experimental phase
(c) post-experimental phase

The most commonly selected experimental designs are:
(i) factorial design
- in general m^k experiments, k=number of factors. m=number of levels
- advantage : calculation of effects of all factors and all interactions between factors
- disadvantage : too many experiments for $k>4$

(ii) fractionated factorial design
- in general m^{k-p} experiments, $p=(1/2^p)^{th}$ degree of fractionation
- advantage : calculation of effects of all factors and some interactions between factors
: less experiments to perform
disadvantage : not all possible interactions between factors can be calculated

(iii) Plackett-Burman design
- in general n·4 experiments, n=1,2,3 etc.
- advantage : many factors can be calculated within few experiments (in x experiments x-1 effects can be calculated)
- disadvantage : each effect of (main) factor is confounded with interactions between factors

Example of a Plackett-Burman design for 8 experiments with confounding effects

Experiment number	Factor settings						
	A	B	C	D	E	F	G
1	+1	+1	+1	-1	+1	-1	-1
2	+1	+1	-1	+1	-1	-1	+1
3	+1	-1	+1	-1	-1	+1	+1
4	-1	+1	-1	-1	+1	+1	+1
5	+1	-1	-1	+1	+1	+1	-1
6	-1	-1	+1	+1	+1	-1	+1
7	-1	+1	+1	+1	-1	+1	-1
8	-1	-1	-1	-1	-1	-1	-1
Confounding effects	BF CD EG	AF CG DE	AD BG EF	AC BE FG	AG BD CF	AB CE DG	AE BC DF

METHOD

The tested method is flesinoxan in plasma. Quantification is based on the internal standard (i.s.) method.

Pre-experimental phase:
- selection of factors: total number of 71 factors. Divide method into parts.
- selection of designs: ≤ 4 factors per part use factorial
otherwise use Plackett-Burman

Scheme 3.
Ruggedness testing, (B).

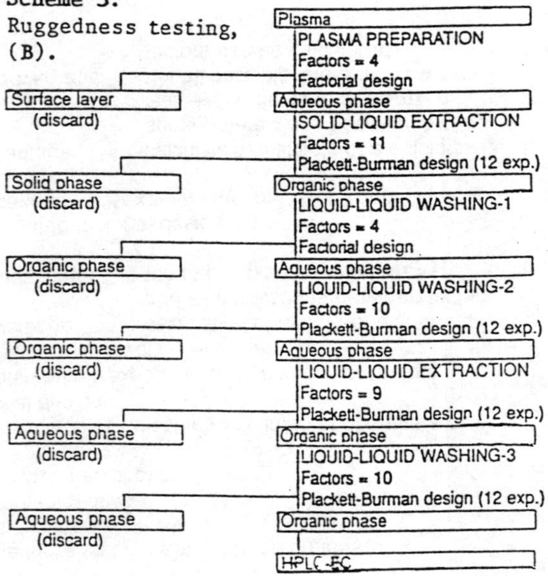

Block scheme of the bioanalytical method with selection of the number of factors and experimental designs

After completion of the experiments of sample work-up per parts, factors which showed relative large variances within parts were retested in a new design: sample work-up in total.
Purpose: search for possible interactions between factors originated from different parts of sample work-up. Selected number of factors was four and the design was factorial.

Experimental phase

- selection of response variables:
- • peak height ratio, peak area ratio, peak height and peak area
- • peak width, retention time, plate count, peak symmetry, resolution between analyte and i.s.
- • height of unretained peak (1.5 minutes) and height of matrix peak (4.0 minutes)

Post-experimental phase:

- reduction of response variables
- variances of measurement errors
- influences and effects of factors
- significance and relevancy of factors
- improving method or testing procedure
- system suitability criteria

Inter-laboratory comparison (see text)

Results from two laboratories compared

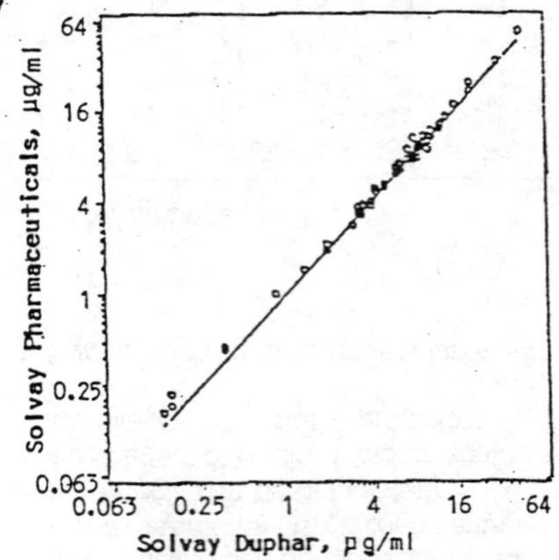

∘ observed level, —— theoretical correl^{n}.

Scheme 3. Ruggedness testing, (C).

RESULTS

(PH denotes peak height)

Measurement errors and the mean values of the selected response variables with 36 resp. 16 degrees of freedom for sample work-up resp. HPLC

Analysis on	Response variable	s of error	C.V. of error (%)	Mean
Sample work-up	PHRatio	0.0495	4.3	1.149
	PHeight	51200	14	367000
	plate count	266	17	1570
	PH(4.0 min)	158000	84	189000
HPLC	PHRatio	0.0123	1.0	1.200
	PHeight	33300	9.5	349000
	plate count	209	14	1530

Relevant significant factors in sample work-up per part and in total for the relevant response variables

Sample work-up		Response variable			
		PHRatio	PHeight	plate count	PH(4.0 min)
per part	1	volume of plasma	-	-	volume of plasma
	2	-	speed of elution	-	-
	3	-	-	-	-
	4	-	-	-	-
	5	-	waiting step	-	-
	6	-	-	-	-
in total		-	volume SPE-column speed of elution	2-factor interaction	

Relevant significant factors in HPLC for the relevant response variables

Response variable		
PHRatio	PHeight	plate count
volume of water	voltage of oxidation	batch of analytical colomn
amount of NH_4Ac		time before switching
batch of analytical column		pre-column from
batch of EC cell		analytical column
dummy factor		after injection

System suitability criteria based upon the three liquid-liquid washing steps

Response variable	analyte	i.s.
Retention time (min.	4.97 - 5.29	6.20 - 6.65
Peak width (min.)	0.30 - 0.36	0.36 - 0.46
plate count	1375 - 1936	1343 - 1907
Resolution	2.09 - 2.39	

CONSEQUENCES

What do these results mean for processing samples?

Sample work-up:
- volume of plasma; because the tested range (0.5 and 5.0 ml) was large, it might not be significant any more if a new design would be done with narrower ranges.
- speed of elution of SPE-column: treat the samples with equal speeds.

HPLC:
- if HPLC system functions properly within one series the variations of the factors are not expected to appear. If not, the user is warned about possible effects. Furthermore, within each series calibration standards are analysed, so it would not automatically mean that the quantification would be influenced.

The response variables chosen to assess ruggedness were those showing the largest variance, namely peak-height, plate count for the analyte, and peak height of a plasma constituent sporadically manifest at ~4 min. Obviously the statistical analysis also embraced peak-height ratio, the key response for quantitative analysis. Scheme 3(C) shows results obtained. It is noteworthy that the C.V. for peak-height (PH) ratio (4.3% in work-up and 1.0% in HPLC) is quite low and within the observed precision of the assay. Since this value represents the inter-experiment spread and is the result of all tested factors, it is clear that the impact of changes in steps is near-trivial.

Within each work-up step the factors that showed a significant (α = 0.05) and relevant influence were: the plasma sample volume affecting peak-height ratio and peak height for endogenous compound at ~4.0 min; speed of elution of the SPE cartridge affecting analyte peak height; and a pause during L/L extraction affecting analyte peak height. The established high/low levels of these factors were far from a normally occurring spread by chance; e.g. we tested processing 0.5 *vs.* 5.0 ml plasma.

Scheme 3(C, top-right) shows that relatively more influential factors apply in the HPLC step: for a given analytical run with its own set of calibration standards the tabulated factors are constant and cannot be of any influence. In our opinion, the ruggedness test showed without any doubt that the HPLC-EC assay of flesinoxan is rugged and will yield the same quantitative data under any circumstances not differing substantially from the pre-set sample work-up and chromatography procedure.

Inter-laboratory comparison.- In each of 4 participating laboratories, 40 plasma samples were assayed in duplicate. The observed levels and intra-sample, inter-day C.V.'s were:

- Solvay Duphar:	8.92 ng/ml;	C.V. 5.4%
- Solvay Pharmaceuticals:	9.15	6.8
- Laboratory **x**:	10.1	7.5
- Laboratory **y**:	9.61	2.8

Solvay Duphar, being the originator of the method, was regarded as being the reference laboratory. Comparing the result sample by sample and calculating the mean relative difference from Solvay Duphar showed that each laboratory was within the pre-set requirement of ±15%. Scheme 3(B) shows a typical plot of one laboratory's results *vs.* Solvay Duphar.

In-study validation.- During application of the assay, >10,000 plasma samples originating from clinical and toxicokinetic studies have been analyzed. In each of the hundreds of analysis batches QC's at low/medium/high levels were assayed in duplicate. Accuracy and precision data from these various studies, 'in-study validation', showed essentially the same values as during the basal validation; in fact, performance somewhat improved, as might be expected since the laboratory technicians had to pick up the routine at the outset.

CONCLUSIONS [See also Scheme 3(C), 'CONSEQUENCES']

The outcome of the investigation is summarized thus:-

System performance	Chromatography repeatability, C.V. < 0.5%
Response function	Constant C.V. from 0.1-50 ng/ml
	Best model $y = A + Bx + Cx^2$ with $1/x^2$ weighing
	%bias < 5% and mean r = 0.997
Accuracy (Lo, Me, Hi)	Each level (Lo, Me and Hi) < 5% bias
Precision (Lo, Me, Hi)	Intra-day mean C.V. = 6% and each level < 7%.
	Inter-day mean C.V. = 9% and each level < 10%.
Sensitivity	LLOQ about 0.08 ng/ml (in practice set at 0.1).
	ULOQ set at 50 ng/ml.
Specificity	No interferences from matrix nor from metabolites.
	Perfect match with GC method.
	Specific with regard to at least 40 drugs and/or metabolites.
Ruggedness	Rugged method. No influence of any factor for a specific analysis batch.
Inter-laboratory	Method operates equally well at 4 laboratories.
In-study validation	Performance is maintained over at least 10,000 samples.

We reckon overall that the HPLC-EC method described is a very reliable tool for determining flesinoxan concentrations in plasma and that it meets all current requirements set for bio-analytical methods [6].

References

1. van Berkel, M.P., de Bree, H. & Sierat, K. (1988) in *Bioanalysis of Drugs and Metabolites, Especially Anti-Inflammatory and Cardio-vascular* [Vol. 18, this series] (Reid, E., Robinson, J.D. & Wilson, I.D., eds.), Plenum, New York, pp. 209-212.
2. Massart, D.L., Vandeginste, B.G.M., Deming, S.N., Michotte, Y. & Kaufman, L. (1988) *Chemometrics: A Textbook* Elsevier, Amsterdam; see pp. 75-92.
3. Hiemstra, H., Tersmette, M., Vos, A.H.V., Over, J., van Berkel, M.P. & de Bree, H. (1991) *Transfusion 31,* 32-39.
4. de Bree, H. & van Berkel, M.P. (1990) in *Analysis of Drugs and Metabolites including Anti-infective Agents* [Vol. 20, this series] (Reid, E. & Wilson, I.D., eds.), Roy. Soc. Chem., Cambridge, pp. 221-225.
5. Box, G.E.P., Hunter, W.G. & Hunter, J.S. (1987) *Statistics for Experimenters*, Wiley, New York; see pp. 291-434.
6. Shah, V.P., Midha, K.K., Dighe, S., *et al.* (1992) *Pharm. Res. 9,* 588-592.

Editor's notes:

#Concerning 'EC' (electrochemical; see footnote on 1st p.): the authors had 'ECD' in their Posters; but the late MS. now amalgamated with them had 'ED' - editorially altered to 'EC'.

#Forum discussion points centered on ruggedness (robustness), and so have been appropriately put near the end of Section #F.

#C-4

DETERMINATION OF THE 5-HT RECEPTOR AGONIST 311C90 IN HUMAN PLASMA BY LC-MS-MS

S. Pleasance†, I.J. Fraser, A.E. Jones, J.A. Allanson & P. Sadra
Glaxo-Wellcome, Group Development Operations, Division of Bioanalysis and Drug Metabolism, Beckenham, Kent BR3 3BS, U.K.

An MS-based method is described for determining the 5-HT receptor agonist 311C90 and its desmethyl metabolite (183C91) in human plasma, with a deuterated analogue as i.s. The method employs SPE and LC-MS-MS with APcI and SRM. It offers increased sensitivity, selectivity and speed of analysis compared with an existing method using fluorescence detection (HPLC-fluor).*

Migraine, characterized by a throbbing headache which is usually unilateral and often associated with nausea, vomiting, photophobia and phonophobia, is a debilitating and recurrent disease affecting ~9-16% of the Western population [1].

The neurotransmitting and vasoactive amine 5-HT, widely distributed in the CNS and periphery, has a central role in migraine pathophysiology [2] although its precise pathogenic role remains undefined. The compound 311C90, *S*-4-[(3-(2-dimethylaminoethyl)-1H-indol-5-yl)methyl]2-oxazolidine [formula in Fig. 2 below], is a synthetic analogue of 5-HT that displays potent, selective agonist actions at the 5-HT_{1D} receptor subtypes [3]. Desirable properties of 311C90 include a physicochemical profile which, in animals, favours access of the molecule to central components of the trigemino-vascular system [4]; its ability to act at central as well as peripheral elements of this system, together with its high potency, may offer an anti-migraine profile surpassing that of existing oral therapies.

To support the clinical development of 311C90 an HPLC method has been validated for its determination in human plasma. Using fluorescence detection the method also allows determination of its desmethyl (183C91) and indole acetic acid metabolites. As shown in representative chromatograms (Fig. 1) for a spiked QC sample and plasma from a dosed volunteer (receiving propranolol too), the parent drug is well resolved both from its metabolites

---

**Abbreviations* [others in Preface; some introduced or varied by Editor].- APcI, atmospheric pressure chemical ionization; SRM, selected reaction monitoring; fluor, fluorimetric detection (in HPLC; authors' text had FLD); AmFm, ammonium formate; diAmH-phos, diammonium hydrogen phosphate; MeCN, acetonitrile; NaFm, sodium formate; QC, quality control; i.s., internal standard.
†addressee for any correspondence

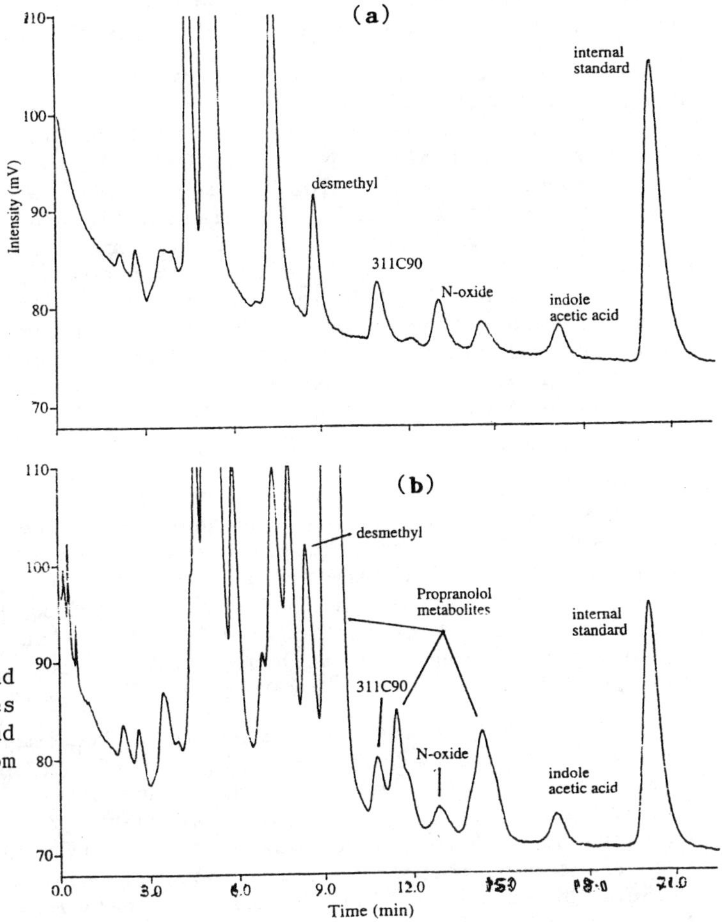

Fig. 1. HPLC-fluor profiles: extracts of (**a**) control plasma spiked with 311C90 and its metabolites (15 ng/ml), and (**b**) plasma from a volunteer dosed with 311C90 and propranolol.

and from endogenous plasma components. The method (description below) gives an LLOQ of 2 ng/ml for 311C90 and its metabolites. As the clinical development progressed it became apparent that pharmacokinetic data would be required at lowered oral doses (≤5 mg) and therefore a more sensitive method was required.

EXPERIMENTAL

Chemicals.- Super Purity acetonitrile was obtained from Romil Chemicals (Cambridge); HiPerSolv water and other chemicals came from BDH/Merck (Poole). The drug, its metabolites and the deuterated i.s. ($^2H_6$-311C90) were synthesized at Glaxo-Wellcome Research Laboratories, and primary stock solutions were prepared in 25% (v/v) ethanol.

HPLC-fluor method.- Plasma samples were analyzed by automated SPE and HPLC-fluor using a Gilson ASPEC along with a Jasco #PU-980 pump, #LG-980-02 ternary gradient controller, #CO-860 column oven and a #FP-920 fluorescence detector. Plasma samples were initially centrifuged for 10 min at 3000 rpm.

To a tube containing 500 µl water and 100 µl i.s. solution (2 µg/ml) was added 200 µl of sample. After vortex mixing, a 650 µl aliquot was loaded onto a 100 mg Bond-Elut C-18 cartridge using (typically 1 ml) MeCN, water (500 µl) and 10 mM NaFm pH 4 for pre-conditioning and, for washing, NaFm, water (750 µl) and 5% aqueous MeCN (v/v; 500 µl). The sample was eluted with 30% MeCN in 17 mM diAmH-phos (pH 3; 300 µl), and 200 µl was injected onto a Zorbax TMS (C-1) column (250 × 4.6 mm) operated isocratically (40°) using 15% MeCN in 17 mM diAmH-phos (pH 3), 1 ml/min. The column was protected by a 2 µm in-line filter and after each 17 min run was flushed by ramping the mobile phase to 80% MeCN. The output from the detector (nm settings: ex, 290; em, 350) was processed by a Fisons Multichrom data acquisition system.

LC-MS-MS method

A Hewlett-Packard Series-II liquid chromatograph was linked to a PE-Sciex API-III$^+$ triple quadrupole mass spectrometer. For manual SPE, 500 µl centrifuged plasma was added to a tube that contained 500 µl 10 mM AmFm and 100 µl i.s. (10 ng/ml). After vortexing, the sample was loaded onto a 100 mg Bond-Elut C-18 cartridge, using (1 ml portions) MeCN, water and 10 mM AmFm for pre-conditioning and, for washing, AmFm, water and MeCN. The eluent was 300 µl of 50% (v/v) MeCN in 150 mM AmFm (pH 4).

Eluate (20 µl) was injected onto a Phenomenex SCX column (30 × 4.6 mm) operated isocratically at 40° using 70% MeCN in 25 mM AmFm (pH 5) at 1 ml/min. The column was coupled direct to a heated nebulizer (N_2) interface at 500°, and positive ions generated by APcI for tandem mass spectrometry with SRM; the m/z transitions were respectively 288 → 58, 274 → 182 and 294 → 64 for 311C90, 183C91 and $^2H_6$-311C90 (i.s.). Instrument operation was at unit mass resolution, with 150 msec SRM dwell time per transition and with argon as the collision gas.

Calibration curve construction: 311C90 along with i.s. was spiked into control plasma, to get calibration standards (duplicate) or independently prepared QC's: respectively 0.2/0.5/0.75/1/2/5/10/15 (and, for initial validation, 0.1/16) ng/ml and 0.3/2/12 (and, initially, 0.2/0.5/5/15) ng/ml). LC-MS-MS peaks were integrated by the PE-Sciex MacQuan processing software and calibration curves were constructed by plotting peak area ratios of 311C90 and 183C91 to i.s. against concentration using a weighted (1/X) linear regression model. Concentrations in unknown and QC samples were subsequently interpolated from this line.

RESULTS AND DISCUSSON

APcI-MS-MS seemed, from previous bioanalytical experience [5], the most likely avenue to increased sensititivity without a significant increase in sample preparation or analysis times.

Fig. 2 shows the full-scan positive ion APcI spectra of 311C90, the base peak being the protonated molecule (MH^+) at m/z 288 with little fragmentation observed. The spectra were similar for the active metabolite 183C91 and the i.s., with MH^+ ions at m/z 274 and 294 respectively. Fig. 2 also shows the MS-MS ion product spectra of these MH^+ ions, all three yielding adundant product ions suitable for use as SRM transitions. The product ion at m/z 58 due to $[N(CH_3)_3 + H]^+$ and the corresponding ion 6 amu higher at 64 were selected as the 'Q3' ion in the case of 311C90 and the i.s. respectively. Despite an equivalent ion at m/z 44 in the spectrum of 183C91, the choice fell on the more abundant ion m/z 182 arising from the concerted loss of $N(CH_3)_2$ and NH_3CO_2H (with ensuing oxazoadinone ring rearrangement).

HPLC and extraction.- The selectivity of MS-MS reduces the need for complete chromatographic resolution of individual components. The emphasis in LC-MS-MS method development veers to shortening analytical run-time so as to improve throughput and to use of API-compatible mobile phases. So the HPLC-fluor method with a 25 min cycle time and involatile phosphate in the mobile phase seemed an unsuitable basis for LC-MS-MS. We have had some success in developing a 'generic' method for LC-MS-MS based on short, strong cation exchange (SCX) columns. Although not operated in a true ion-exchange mode, for several series of compounds and their metabolites they have furnished retention characteristics apt for LC-MS-MS.

A particular problem with APcI is that thermally labile metabolites, e.g. N-oxides, tend to fragment in the high temperatures of the interface/source to give a signal in the MH^+ SRM channel of the parent drug and hence must be chromatographically resolved. With RP operation of the SCX column a parent drug and metabolites of varying polarity are resolvable by mobile-phase optimization without significantly increasing the run time.

The SPE procedure used for HPLC-fluor was modified: AmFm was used rather than NaFm and conditions (volumes, molarities, pH) were re-optimized for extraction of just 311C90 and 183C91. Fig. 2 shows the LC-MS-MS chromatograms for an extract of a 100 pg/ml calibration standard. With a mobile phase of 70% MeCN in 25 mM AmFm at pH 5, both compounds and the 1 ng/ml i.s. elute in <4 min with good peak shapes and excellent signal-to-noise.

Method validation was formally done prior to the analysis of study samples, in respect of calibration model, precision and accuracy, specificity and recovery. Although the stability of the analytes* in analytical stock solutions, frozen human plasma (including freeze-thaw cycles†) and processed extracts had been assessed already in the HPLC-fluor method validation, confirmation was sought by further experiments at lower concentrations.

*Relevant arts. in Vols. 22/23, and 20: p. 403 relates to †.- *Ed.*

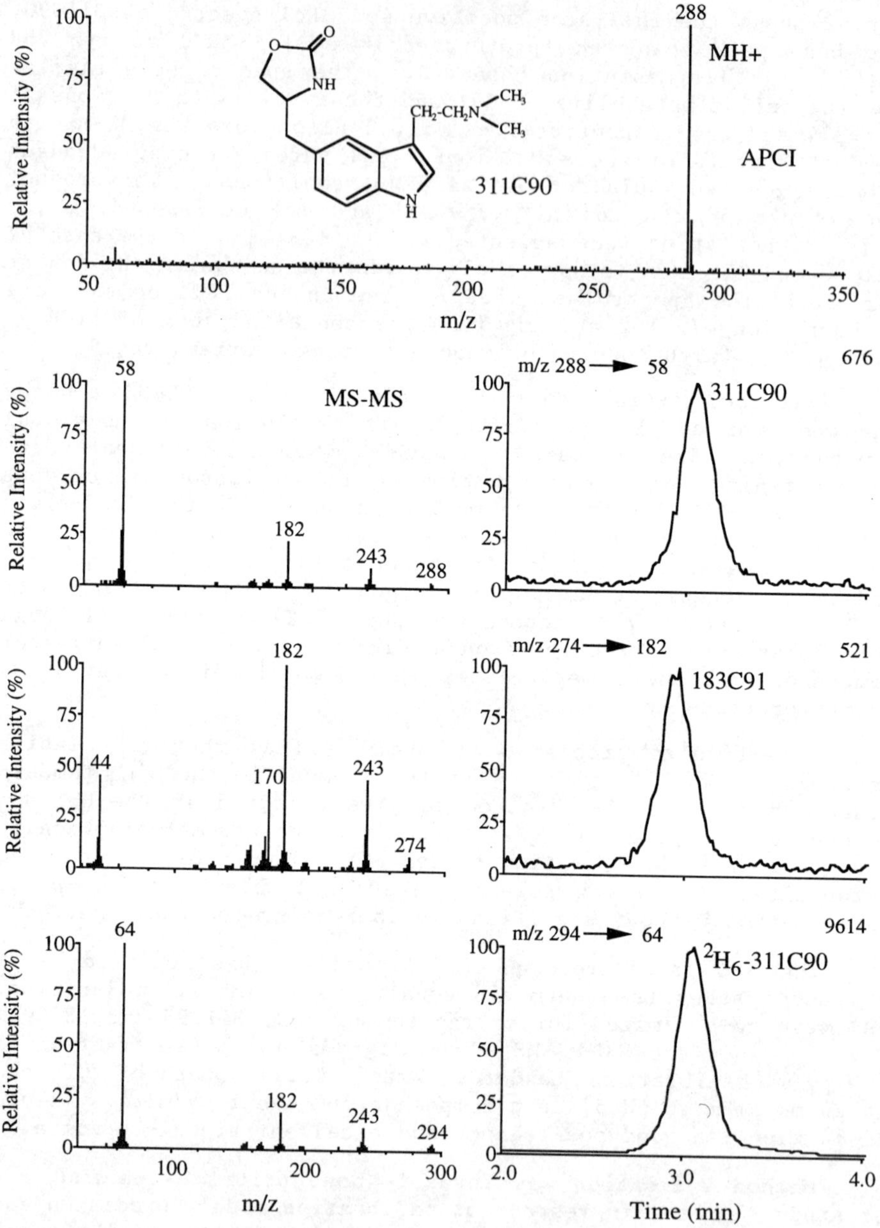

Fig. 2. APcI (positive ion) spectra of 311C90 [top panel], and MS-MS product ion spectra [left-hand panels] and corresponding LC-MS-MS SRM chromatograms [right-hand panels] of 311C90, 183C91 (each 100 pg/ml) and i.s. (hexadeuterated 311C90, 1 ng/ml).

The aptness of a weighted (1/X) linear regression model for both analytes was confirmed by examining standardized residual plots of calibration data from 4 inter-assay experiments (see below). There was no evidence of curvature and the 1/X weighted linear regression provided the most homoscedastic plot (constant variance with concentration) and minimized residuals magnitude.

The method's precision (% C.V.) was determined by assessing the agreement between replicate analyses of mixed spiked control samples independent of the calibration standards: 4-fold replicates at 4 concentrations were analyzed on 4 separate occasions. The data for both 311C90 and 183C91 were analyzed by an ANOVA macro within Excel (Microsoft Corp.) to give estimates of inter- and intra-assay precisions. Both values were <15% at all spike concentrations within the limits of quantitation (0.2-15 ng/ml; Table 1). Method accuracy (% bias) was determined by assessing, for the spiked control samples, the agreement in concentrations - measured (means from the precision assessment) *vs.* nominal. Accuracy was good (Table 1), the greatest divergence being 14%.

Table 1. LC-MS-MS precision and accuracy from the initial validation.

	311C90				183C91			
Spike Control (ng/ml)	0.20	0.81	5.07	15.21	0.20	0.81	5.08	15.23
Mean	0.21	0.92	5.44	16.33	0.20	0.91	5.47	16.34
Precision (%CV) Intra-assay	6.4	5.4	2.6	3.4	7.8	5.4	2.6	3.5
Inter-assay	6.3	5.7	neg.	3.1	10.5	neg.	4.4	1.8
Accuracy (% bias)	1.0	13.7	7.2	7.4	-1.5	12.1	7.9	7.3

The specificity of the method was assessed by visual examination of SRM chromatograms generated in the analysis of human plasma from several sources - pre- and post-dose samples from 311C90 clinical trials - and of control plasma spiked with metabolites and analogues of 311C90. The retention times and twin mass-selection stages confer superb specificity, as illustrated by the lack of interferences in any of the SRM traces in Fig. 2. The 6 amu difference between the MH^+ ions and respective product ions of 311C90 and i.s. ensures no between-channel interference.

Method application.- The method was first used to support double-blind, placebo-controlled dose escalation studies aimed at determinining tolerability, pharmacokinetics and pharmacodynamic effects of 311C90 in healthy male volunteers after single and multiple dosing. Samples were also analyzed by HPLC-fluor, providing an opportunity to cross-validate the methods. Fig. 3 shows, with two oral dose levels, representative 311C90 plasma profiles which indicate a good between-method correlation and

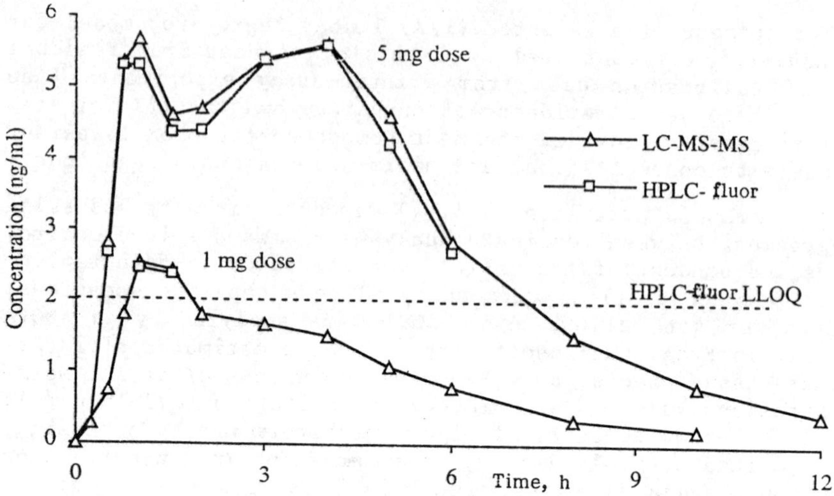

Fig. 3. Representative 311C90 plasma profiles obtained by LC-MS-MS and **HPLC-fluor after an** oral dose of 1 or 5 mg.

demonstrates the ability of LC-MS-MS to provide pharmacokinetic data below the LLOQ of the HPLC-fluor method. The correlation between HPLC-fluor ('test method', Ct = concentration found) and LC-MS-MS ('reference method', giving Cr) is further illustrated for plasma 311C90 levels in Fig. 4, wherein Ct/Cr is plotted as a function of Cr. We concluded that the methods give equivalent results: for the Ct/Cr ratio the calculated confidence limits (upper, 0.9425; lower, 0.9235) lie between 0.8 and 1.2, and the value of zero is contained within the calculated 95% confidence intervals of the regression line's slope. The performance of the LC-MS-MS method is illustrated by a control chart (Fig. 5) and summary QC data (Table 2), generated in support of a single-dose study representing 23 individual assays.

Table 2. LC-MS-MS QC data during support to a volunteer study on 311C90.

	311C90			183C91		
QC concentraton (ng/ml)	0.32	2.10	12.61	0.28	1.85	11.11
Mean	0.30	1.97	12.36	0.25	1.71	10.96
Std. Dev.	0.03	0.14	0.64	0.02	0.15	1.08
%CV	9.49	6.93	5.15	8.75	8.72	9.88
% bias	-5.81	-6.02	-1.98	-10.36	-7.55	-1.34

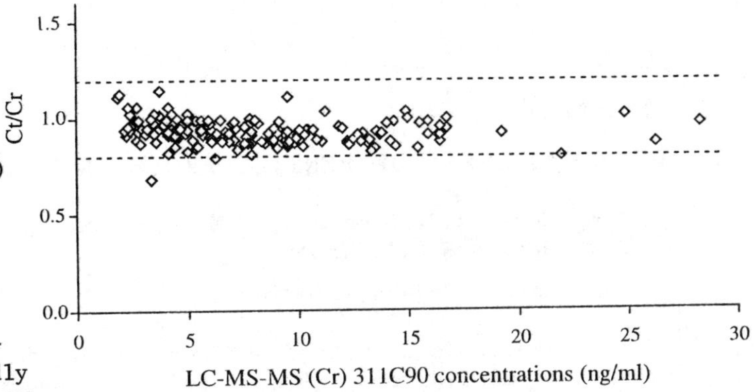

Fig. 4. 311C90 ng/ml: HPLC-fluor ('Ct') compared with LC-MS-MS ('Cr).

Below:
Fig. 5. Control chart for medium-level QC repeatedly assayed by LC-MS-MS.

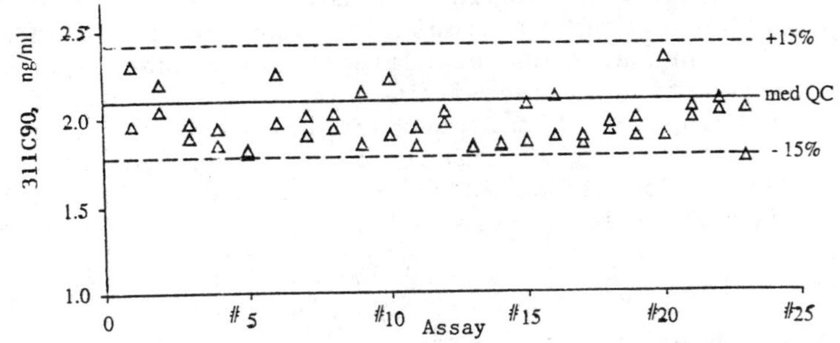

The selectivity of LC-MS-MS has since proved invaluable in studies on the effect of potential concomitant medication on 311C90 pharmacokinetics - a problem illustrated by an HPLC-fluor chromatogram (Fig. 1b) arising from co-administration of 311C90 and propranolol. The metabolites of propranolol can be seen to mask both 311C90 and its desmethyl metabolite, not resolvable by modification of the HPLC conditions. No such interferences were observed in the SRM traces from LC-MS-MS, which served without modification for the propranolol interaction study.

CONCLUSIONS.- The LC-MS-MS method developed for plasma 311C90 and its active desmethyl metabolite provides equivalent results to an HPLC-fluor method, at least for 2-15 ng/ml. It is more sensitive (LLOQ 200 pg/ml), HPLC takes only 2 (*vs.* 25) min albeit the manual extraction is labour-intensive, and the high specificity reduces many problems encountered with matrix variations.

Acknowledgement is made to Ms C. Dean for technical assistance, and thanks given to colleagues and project-team members at WRL.

References

1. Goldstein, M.E. & Chen, T.C. (1982) *Adv. Neurol. 33,* 377-390.
2. Lance, J.W. (1991) *Eur. Neurol. 31,* 279-281.
3. Martin, G.R. (1994) in *New Advances in Headache Research 4,* 3-4 [Clifford-Rose, F., ed.; Smith-Gordon, London].
4. Goadsby, P.J. & Edvinsson, L. (1994) *Headache 24,* 394-399.
5. Fraser, I.J., Clare, R.A. & Pleasance, S. (1995) *J. Chromatog. 663 B,* 275-287.

#C-5n

A Note on

LC-MS-MS ASSAY OF FENTICONAZOLE IN HUMAN PLASMA

T.A. Enos, R.J. Simmonds, J.M. Long, W. Speed, J.J. Dervan & C.A. James†

Pharmacia & Upjohn, Fleming Way, Crawley RH10 2LZ, U.K.

Fenticonazole nitrate is used as an antifungal in treatment of vaginal candidiasis. A new product is under development for treating mixed vaginal bacterial and candidiasis infections - a cream having fenticonazole combined with clindamycin. Due to the topical route of administration and relatively low systemic absorption, a sensitive assay was required to support further development of this product line. Existing HPLC-UV methods had LLOQ's* ~5 ng/ml. The present objective was to quickly develop and validate an HPLC-MS-MS method with a sub-ng/ml LLOQ in female plasma. Here we complement a published description [1].

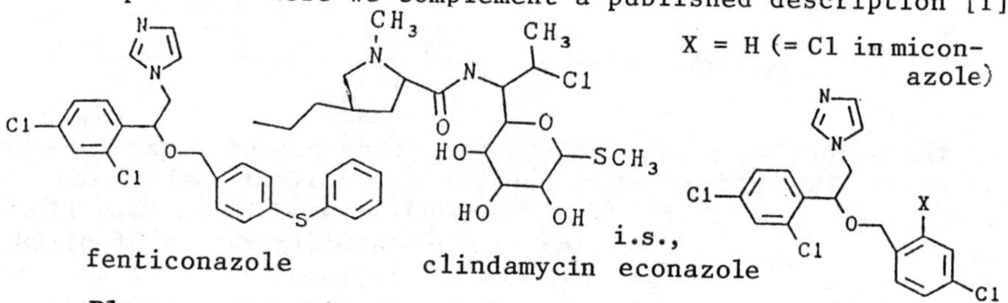

X = H (= Cl in miconazole)

fenticonazole clindamycin i.s., econazole

Plasma extraction.- Human plasma aliquots (50 µl) were precipitated with 100 µl MeCN containing econazole nitrate as i.s. (2.016 ng/ml), and centrifuged (2 min; 11,500 **g**). Using a polypropylene fine-tip 'pastette', the supernatant was transferred to autosampler vials with low-volume inserts.

Instrumentation and chromatographic conditions

The LC-MS-MS system consisted of a Kratos Spectroflow 400 HPLC pump, Waters Ultra WISP 715 Autosampler, Shimadzu CTO-6A column oven and PE Sciex API III+ MS-MS system. A Shandon BDS 3 µm C-8 column (50 × 4.6 mm), at 45°, was used with a Brownlee C-18 guard column (15 × 3.2 mm; Anachem); 20 µl of extract was loaded. The mobile phase comprised MeCN/10 mM pH 3.5 aqueous ammonium formate (70:30 by vol.) with 1.0 ml/min flow rate; the run time was 2.5 min.

**Abbreviations:* LLOQ, lower limit of quantification; i.s., internal standard; MeCN, acetonitrile. MS-MS, tandem MS (LC = HPLC); APcI, atmospheric pressure chemical ionization.

†Editor the channel (will forward) for any contact with authors

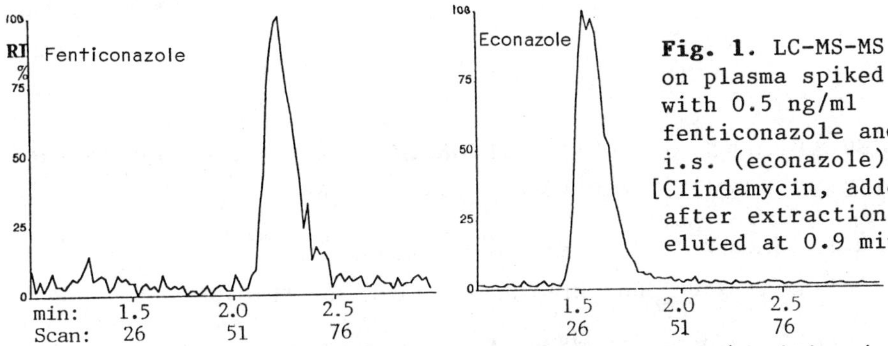

Fig. 1. LC-MS-MS on plasma spiked with 0.5 ng/ml fenticonazole and i.s. (econazole). [Clindamycin, added after extraction, eluted at 0.9 min.]

The MS-MS detector was operated in the APcI (positive ion) mode utilizing the heated nebulizer interface, its probe being set at 500° and the gas pressure 70 psi. Curtain, nebulizer and auxiliary gas (N_2) flows were 0.8, 0.6 and 2.0 L/min respectively. Multiple reaction monitoring (MRM) was performed; the collision gas was argon (~330 × 10^{12} molecules/cm; collision energy 35 eV). The parent-to-daughter transition, with 400 msec dwell time, was m/z 455>199 for fenticonazole, 381>125 for i.s. and 425>126 for clindamycin. Peak area ratios for the selected ions were determined by the PE software package MacQuan.

Results and Discussion (Fig. 1 shows MRM peaks)

The assay was linear over the 0.5-20 ng/ml range and was accurate and precise; three appraisals (each n = 9) gave:-
ng/ml set: 1.53/4.025/13.99; found: 1.54/4.035/14.72
C.V., %: 5.1/ 6.1/ 4.9; bias, %: +0.7/ +0.2/ +5.2
There was good recovery (~108%) and stability at -20° or after three freeze-thaws. In comparison [1] with HPLC-UV the method gave a lower LLOQ, with a smaller sample, and was faster since the extraction was simpler and the column shorter. Miconazole was disfavoured as i.s. because of molecular ion complexity.

Use of LC-MS-MS (APcI preferable to electrospray) has wide applicability and usually needs little change in sample preparation and HPLC conditions that have served us for many compounds including those assayed in the past by HPLC-UV. LC-MS-MS, even for compounds with good UV chromophores, is quick, relatively simple, and conducive to rapid method development as needed in drug discovery; high throughput may be the primary advantage in large-scale studies. Very short run times result from use of short (3 or 5 cm) columns with 3 µm packings and a choice of phases, relying on MS-MS specificity and selectivity to resolve components. Increasing bioanalytical use of LC-MS-MS will come soon with the advent of relatively low cost bench-top systems coupled to powerful PC-based data systems (e.g. the Finnigan LC-Q 'ion trap' system offering LC-MS^n).

Reference: 1. Speed, W., Long, J.M., Simmonds, R.J. & Enos, T.A. (1995) *Rapid Comm. Mass Spectrom. 9,* 1452-1456.

#C-6

DEVELOPMENT AND APPLICATION OF LC-MS-MS BIOANALYTICAL ASSAYS - MYTHS AND REALITIES!

R.J. Simmonds†, T.A. Enos & W. Speed

Pharmacia & Upjohn, Fleming Way, Crawley RH10 2LZ, U.K.

API interfaces for MS or MS-MS, coupled with HPLC, are so advantageous, e.g. in shortening run times to <3 min, that it is tempting to use them for all applications and to rely on MS-MS selectivity for obviating interferences which in the past would have been overcome by chromatographic and sample-preparation techniques whose importance for overall assay performance may now be less appreciated.- The idea that they may be no longer vital for selectivity and sensitivity may somewhat shock an established bioanalyst and entails a change in perspective, a 'paradigm shift'. Really, in our experience, they may remain crucial for developing robust and sensitive assays even with MS detection.*

Exemplified by suleptanate, a method-development strategy is now presented. It is akin to that for 'conventional' procedures where UV or fluorescence is used for HPLC detection. Its success gives us some confidence that bioanalytical skills developed over a number of years are not now completely redundant!

HPLC with on-line triple-quadrupole MS-MS* is a burgeoning technique. Approaches to its bioanalytical use will inevitably vary, reflecting the background expertise and philosophies of users. LC-MS-MS is a maturing discipline and there is as yet no one accepted approach to solving the problems and meeting the challenges. Accordingly, a description of our experience when coming into this field and working with the Perkin Elmer Sciex API III+, after many years of conventional bioanalysis, may be instructive, together with our current approach to LC-MS-MS development, taking a particular drug analyte as an example. We give a personal perspective, without literature citations, of the 'learning curve' we had to go through to reach our present level of competence. It may not reflect the reactions and experiences of others.

†addressee for any correspondence; but send to Editor (will forward)

**Abbreviations include:-* DEA, diethylamine; EC, electrochemical (detection); MeCN, acetonitrile; QC, quality control; TFA, trifluoroacetic acid. MS-MS, tandem MS [prefix LC = HPLC]; AP(c)I, atmospheric pressure (chemical) ionization; ISP, ionspray; RI [in Figs.], relative intensity.

Technical advances.- Though on-line LC-MS interfaces have been available for 10-15 years, the difficulties and drawbacks associated with the techniques first available, e.g. the moving belt or chain interfaces, severely limited their application. LC-MS could detect and quantify analytes largely independently of any physico-chemical characteristics, e.g. UV absorptivity which, if poor, renders the analyte 'difficult' by conventional HPLC assays. However, LC-MS had always seemed a very promising approach whose time had not quite arrived.

Thermospray techniques were a step forward, but had little application to bioanalysis of drugs, due in the main to ion instability and limited sensitivity. Hence not till the advent of API interfaces with ISP or, as in the Sciex instrument, cI (chemical ionization), did the problems seem to have been solved and a 'universal detector' was seemingly available at last to bioanalysts. Other articles in this and the preceding volume deal with operation of such instruments, wherein key challenges such as mobile-phase removal and effective analyte ionization have been elegantly overcome. Especially notable features are the potential sensitivity and selectivity, robustness, and the ease of coping with previously intractable analytical problems.

Myths

Whilst news of these features was welcomed by bioanalysts, there did seem to be a downside. Thus, exquisite and interesting techniques such as pre- and post-column derivatization, or column switching and 'heart-cutting', would not be required as such, if at all, to achieve adequate sensitivity with difficult analytes. This could be considered a 'plus'; but it came as rather a shock to learn that, based upon the experience of earlier users within the industry, established bioanalytical techniques such as extraction, column selection and solvent optimization would no longer be critical to the performance of an assay, since all selectivity would be ensured by parent and daughter ion MS (MS-MS). Sample preparation, we were assured, did not need to be more than simple plasma protein precipitation with an organic solvent ('dilute and shoot'), and very short columns would suffice for chromatography, with analytes just clear of the solvent front. In short, we were led to understand that method development would be a trivial exercise and validation and application hardly more onerous. Some manufacturers' literature and presentations emphasized these points and, in some instances, they became 'received truths'.

These are the 'myths' referred to in the title, the term in this context not being meant pejoratively or to refer to definite untruths. It is used more in a Greek sense, 'stories about the activities of heroes and Gods'- to which we mere mortals should aspire. So great a change in the way biofluid assays are developed and applied has been called (mindful of

Thomas Kuhn) a 'paradigm shift'. As established and experienced analysts we felt the ground shift under our feet. What now remained for us to do? Paradigm shift it may be; but for us it was (pardon the pun) a case of 'Paradigm lost'.

Reality

Whilst accepting that some of the myths were perhaps 'sales hype', we nevertheless tried, in the first months of working with the PE Sciex API III+, to fulfil the expectations associated with the machine and to run what could be called 'quick and dirty' assays of development compounds. Maybe we were unlucky, but we were largely unsuccessful. Assays seemed to fall to pieces during validation, and whilst good sensitivity could sometimes be obtained, reliability and reproducibility were very poor. Some of our problems were undoubtedly due to the HPLC pump and injector hardware - so we replaced these by equipment we were more familiar with; but we also felt that more emphasis needed to be placed on sample handling and chromatography so that in a sense we were asking the machine to do less and so were putting less stress on it. With some reluctance we went back to the bioanalytical techniques that we knew, viz. developing SPE for the samples if feasible and adapting, for MS-MS detection, our established HPLC techniques - which generally involved 15-25 cm RP analytical columns with 20-40% organic mobile phases containing 0.05-0.1% TFA.

Some success was achieved with assays thus developed. Our approach was to treat the MS-MS instrument (working in the APcI - heated nebulizer - mode) as a somewhat temperamental detector that had to be fed low volumes of clean samples free of interferants or co-eluted substances that could 'quench' ionization at the interface. Extracts appeared quite 'clean' by HPLC-UV and, with our usual mobile phases and analytical columns, run times ('cycle times' in LC-MS-MS parlance) were ~5 min. We strived to obtain good chromatographic efficiency as an aid to reliability of peak integration and to enhance sensitivity. We were able to analyze up to 60-80 samples in a batch reliably, though we were still seeing some calibrants with C.V.'s >15%.

Many related methods were developed by us for a variety of research compounds from three different chemical classes. The assays fulfilled their brief, but the promise of faster method development and sample throughput was, in the main, not met. We were now more confident with our own and the instrument's capabilities, but we felt that perhaps we were working still within the old paradigm.

A CASE HISTORY: suleptanic acid (from an administered ester)

It was at this point, when we were still in the early learning phase of our LC-MS-MS experience, that we were faced with a very real analytical challenge. In order to enable registration of Promedrol®† in a European country, regulators

†Assay in authors' lab.: p. 71 in Vol. 22, this series (1992).- *Ed.*

Fig. 1. Promedrol: metabolism in man

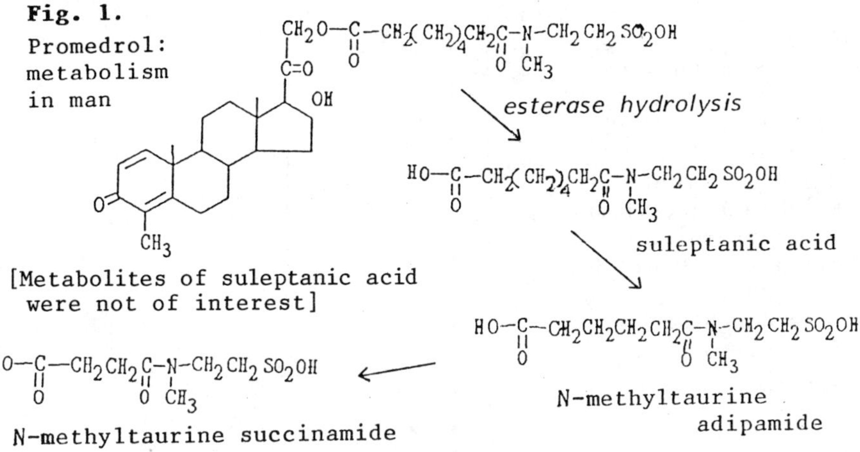

required pharmacokinetic data, within a couple of months, on the suleptanic acid (or 'suleptanate') moiety. The suleptanate ester of methylprednisolone facilitates solubility of the formulation, and the active component is released by plasma esterases (Fig. 1). Whilst we had copious concentration data on methylprednisolone, we had no knowledge of the disappearance of suleptanate *in vivo*, and it had been assumed that it was inactive and did not accumulate in the body. We had plenty of clinical samples, but no bioanalytical assay.

There had been several attempts in the past to develop a sensitive bioanalytical assay using a variety of techniques. These had not been successful, largely because of the unhelpful characteristics of the compound in respect of conventional HPLC or GC assay. Suleptanic acid has no usable UV chromophore and is involatile for GC assay, and there is no selective reagent for derivatization; it is very polar (with a negatively charged group at each end) and can act as a detergent, both features having implications for extraction and chromatography. Given the above, only LC-MS-MS techniques could readily provide an assay with the requisite sensitivity, ~5 ng/ml or better.

MS characterization

Suleptanate generated a pseudo molecular negative ion, $[M-H]^-$, at m/z 294.6 which fragmented to a characteristic daughter ion at m/z 79 (Fig. 2). This boded well for selectivity. However, efficient ionization at the interface, and hence good sensitivity, required strongly alkaline conditions (at least pH 10) and at least 30% organic solvent (MeCN) in the mobile phase. Below pH 10, where the two acidic groups are still fully ionized, a much poorer signal was obtained. A further difficulty, which could not have been predicted and which was encountered when investigating anion-exchange HPLC (SAX), was that cations appeared to suppress ionization at the

Fig. 2. MS(APcI) of suleptanate (100 ng/ml, introduced at 10 μl/min in 15% MeCN containing 0.01% TFA and [not routinely used] dimethyloctylamine. **A**: MS pattern (Q1 scan), showing the $[M-H]^-$ ion at m/z 295. **B**: its daughter ion (after MS-MS), of m/z 79.

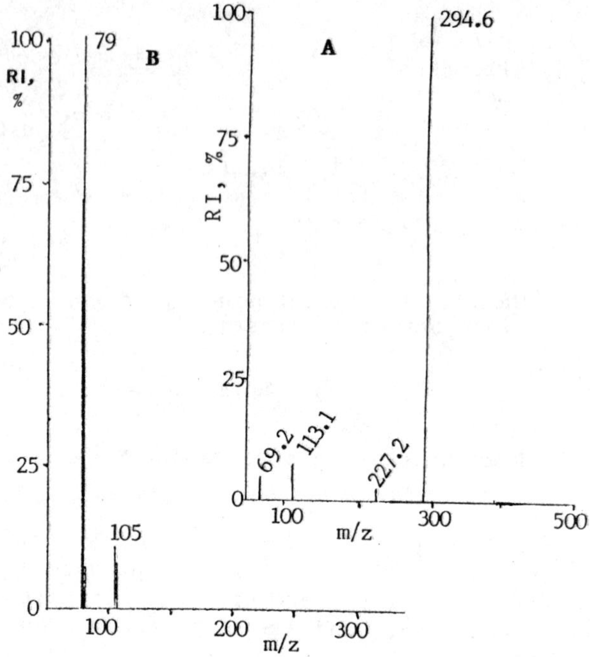

LC-MS interface. Method development had to address potential problems.- Sample work-up had to largely remove proteins and cations but maintain reasonable extraction efficiency for the analyte, and HPLC entailed presenting the analyte in a very alkaline medium at the LC-MS interface free of any co-elutants that could quench ionization. The development of the HPLC and the feed-in for MS was therefore subject to constraints as difficult as those seen with any ultra-sensitive HPLC assay with UV, fluorescent or (more pertinently) EC detection.

Sample clean-up

Clean-up was multi-step, with initial protein precipitation and two SPE steps, the first to remove cations and the second to remove other contaminants, suleptanate being retained.

(1) To 0.25 ml plasma add 0.5 ml MeCN; mix and centrifuge.
(2) Apply supernatant, water-diluted to 2 ml, to a 100 mg Bond-Elut SCX cartridge, and collect the break-through.
(3) Apply to a 100 mg DEA cartridge, rinse with 1 ml water, and elute with 0.25 ml 20% '880' ammonia solution.
(4) Dry down at 50° under air and reconstitute in 50 μl 25.5 mM ammonium acetate buffer pH 10.

Up to 20 μl could be injected on column without degradation of peak shape. All the above steps appeared necessary to achieve the sensitivity target. Thus, the SPE eluting solution could not be injected on column without some loss of MS signal.

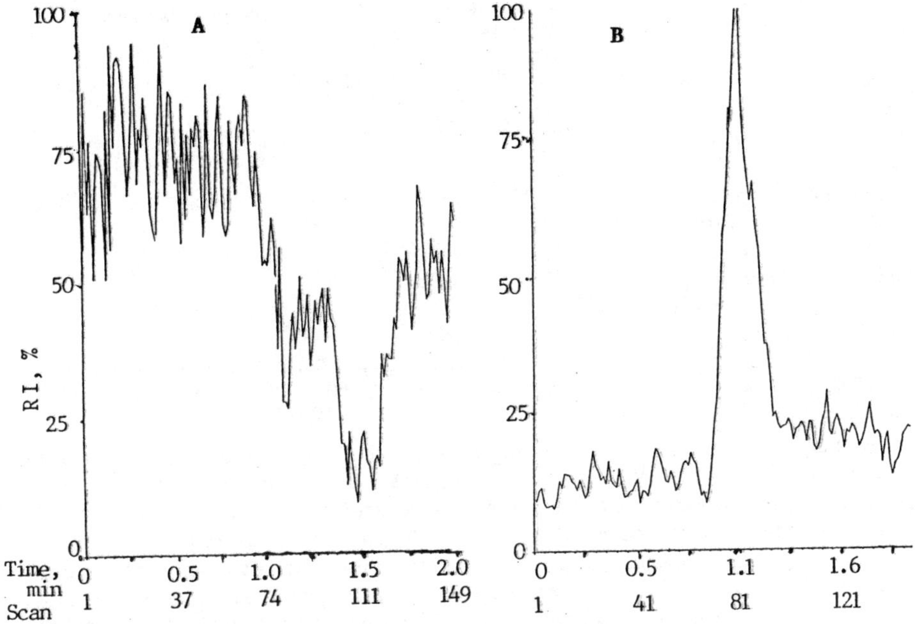

Fig. 3. HPLC-MS-MS patterns for m/z 79 (from m/z 275). **A**: Extract of 0.25 ml blank plasma (peak 'exaggeration' due to the RI basis). **B**: Plasma containing 11 ng/ml suleptanate.

Chromatography

The high pH of the mobile phase limited analytical column choice, since silica-based stationary phases were likely to dissolve in use and give unreliable results. Chromatography was, however, obligatory, not only to resolve suleptanate from co-extractants that quenched ionization but also to enhance sensitivity. We had looked at various options for ion-exchange packings, but obtained consistent and repeatable results using RP-HPLC with Shandon 'Hypercarb' (graphite; 150 × 4.6 mm) - a phase resistant to alkaline pH, and retentive enough to move the very polar suleptanate well away from the solvent front (R_T 1.1 min) even with the relatively high proportion of MeCN in the mobile phase (30%; 70% pH 10.0 ammon. acet.; 1 ml/min). Some remaining problems of poor signal-to-noise at the lower end of the 5-150 ng/ml calibration curve were ameliorated by use of an SCX guard column (Brownlee 15 × 3.2 mm) to scavenge for any cations still present in the extract.

When the various parts of the assay - extraction, HPLC and detection - were put together and briefly validated, we were somewhat relieved to find adequate selectivity (Fig. 3) and sensitivity (Fig. 4), down to 5 ng/ml - which, since 20 µl was loaded and

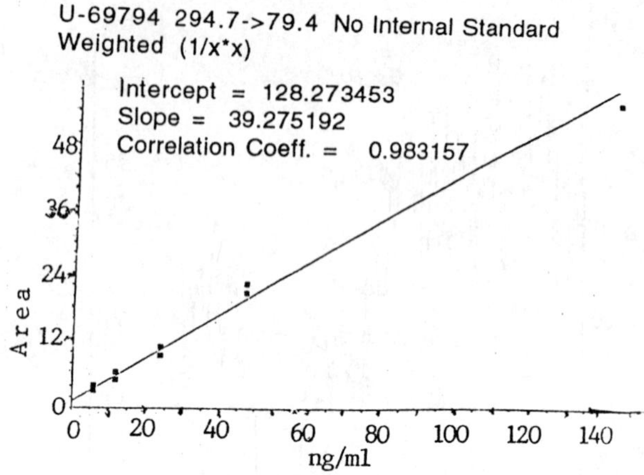

Fig. 4.
A representative calibration curve for suleptanate in plasma extracts, covering the range 5-148 ng/ml.

and recovery was ~50%, represents ~250 pg on-column. Quite good precision (C.V. ~25% at 5 ng/ml) and linearity were found despite the absence of an internal standard (none was available in the short development time). With some precautions, e.g. replacing the guard column and flushing the column and interface, up to 50-60 samples could be analyzed in a single run. The assay enabled analysis of clinical samples and, for the first time, estimation of suleptanate pharmacokinetics.

Benefit from developing the assay

Though this is an assay that would have been impossible by any other means, it is by no means a 'typical' LC-MS-MS assay, and contains some 'traditional' aspects. For example, the plasma extraction, although extensive for an LC-MS-MS assay, would have been unremarkable for a highly sensitive HPLC assay with conventional detection. However, success in this complex assay, which needed much 'bioanalytical guile', greatly increased our confidence in our approach to LC-MS-MS assay development. We were also more confident in the machine and our own capabilities with it. We became less concerned that our approach was somewhat idiosyncratic, and were content to develop assays in our own way. We subsequently generalized and refined the strategy that was successful with the suleptanate assay.

LC-MS-MS METHOD DEVELOPMENT STRATEGY

#1. Define context of the assay. What is it to be used for? What sensitivity is required and what is the projected sample throughput? Since all assays are compromises, the assay context remains all-important. While some LC-MS-MS assays are potentially of exquisite sensitivity (pg/ml), demanding the greatest sensitivity possible will have repercussions on the development time. Thus, the sample work-up and chromatography necessary to remove 'interference' (whether peak contamination or quenching at the LC-MS interface) may need to be especially complex. Though the well documented problems with the traditional ultra-sensitive (HPLC) assays may shrink because of the selectivity inherent in MS-MS, they will not have gone away completely. Achieving a 'blank' control sample will still be necessary and call for sample work-up. Compromises have to be made, and assay goals set, realistic in relation to the time available. Ideally, the capabilities of the LC-MS-MS instruments should be understood by the 'customers' for bioanalytical data. If they have heard only the 'myths', then their expectations may be unrealistic!

#2. Establish MS and MS-MS parameters. Initially sensitivity, linearity and precision should be checked with standard solutions of analytes and simple HPLC (whilst we are happy to use mobile phases with TFA in the first instance, buffers are preferable as TFA causes corrosion problems in the long term). The parent and daughter spectra of the analyte(s) and internal standard are chosen with both selectivity and sensitivity in mind, and choice can also be limited by the analogs available as internal standards. Achieving efficient ionization may put constraints on the pH or organic content of the mobile phase and hence the choice of stationary phase; to get efficient and reliable ionization at the interface the samples may have to be injected only in organic media, or at a high or low pH (maybe detrimental to analyte stability). Whatever on-column sensitivity is achieved with standard solutions will affect how small a biofluid sample can be. LC-MS-MS assays generally require less sample than typical HPLC assays. Unexpectedly poor sensitivity with biological extracts will require investigatation of both HPLC and sample extraction.

It is sometimes difficult to discriminate between poor extraction efficiency and possible suppression of ionization, and extracts of blank plasma can reveal where the problems lie. Here it should be noted that sample extraction schemes that have been successfully used for a sensitive assay with UV or fluorescent detection may not be viable for assay of the same analyte with MS-MS detection, and contamination of the interface or ion suppression may be found. Different HPLC may solve this, but more often than not a different sample work-up will be required, e.g. a change from liquid/liquid to SPE. Again, greater problems will be found at high sensitivity. MS parameters set up and optimized with standard solutions may need re-optimization for extracts.

#3. To establish aptness for batch analysis, analyze extracts and controls in a ~50-60 sample batch. Such a run may reveal problems not seen with small batches. Interferences and MS signal diminution may be seen as the run progresses. Again this bears on the extraction and chromatography required.

#4. Characterize methods before use. Full 'validation' might not be appropriate; but analyte stability, sensitivity/ selectivity and some statistics (precision, accuracy, etc.)

should be established before real use. Whilst actual analysis of extracts can be accomplished faster than for a typical HPLC-UV assay, checking analyte stabilities upon storage or freeze-thawings, or preparing and characterizing calibrants at the LLOQ or QC's, will need as much time as in any other assay.

#5. Include all appropriate QC's and checks in actual assays. We have found outliers to be rather more common with LC-MS-MS assays than with HPLC-UV. This may reflect the better sensitivities we are working at, but it does argue for having more points in a calibration curve, or including extra QC's. Routine QC's cannot be skimped, and more controls may well be needed than for conventional assays. As cycle times are short, even with extra 'control' samples, throughput should not suffer.

CONCLUDING COMMENTS

We do not claim any originality for the scheme, which we trust will be uncontentious - a point that needs stressing. The scheme, for what it is worth, is similar to established schemes used in bioanalytical development. These revolve around the need to ensure that detection selectivity and sensitivity are adequate, whereas the MS-MS scheme revolves around achieving and maintaining adequate ionization at the LC-MS-MS interface. This is a key, as we see it, to a successful LC-MS-MS assay.

This may be a somewhat naive point of view, but it has facilitated the development of effective assays where extraction, chromatography, etc., are developed in step with each other. We are now quite happy with 'dilute-and-shoot' sample work-up, or using short analytical columns, especially for assays of limited application, if sensitivity and especially reliability are not compromised.

With hindsight, we can see that in trying to develop fast assays on a par with established LC-MS-MS practitioners we were trying to run before we could walk. To develop a style of assays that one is comfortable with, and produce acceptable data with a new technique, does take time. We were perhaps too eager to match the standards apparently achieved by other groups, without having the necessary expertise. However, the whole learning experience has been good, especially as we found that our bioanalytical skills, so necessary in the development of the suleptanate assay, were not so redundant after all. For us it has been a case of (and this is irresistible) 'paradigm regained'.

#C-7

A STRATEGY FOR THE DEVELOPMENT OF BIOANALYTICAL METHODS USING LC-MS

C. Mason†, J. Moore, M. Churchill, P. Solanki & W.A. Galloway

Department of Cellular Biochemistry and Bioanalytical, British Biotech Pharmaceuticals, Watlington Road, Cowley, Oxford OX4 5LY, U.K.

LC-MS enables sensitive, reliable and robust bioanalytical methods to be quickly and routinely developed. The approach now described can help early compound selection including bioavailability testing, and any follow-on registration studies. After investigating analyte solubility and LC-MS conditions, devising a simple extraction procedure and finding a remedy for any matrix instability of the analyte, variability is appraised and plasma or blood samples from a few drug-dosed rats are analyzed. Using this approach, any delays in the formal validation process are tolerable, suitability for real study samples can be assured, and pilot information to aid drug selection can be gained.

Using MS and LC-MS-MS, methods that are sensitive, reliable and robust can be developed quickly and routinely. Bioanalysis and pharmacokinetics are becoming involved at an ever earlier stage in compound development, and analytical methods are often required, as part of compound selection, before the start of formal toxicology and clinical programmes. These methods have traditionally been of indifferent quality due to the limited development time available. The approach now described allows methods to be developed rapidly to support compound selection activity, whilst being robust enough for subsequent validation in the support of registration studies.

DEVELOPMENT STRATEGY

Solubility.- When a compound is first encountered, information on its physico-chemical properties may not be available; but knowledge of the compound's solubility in organic solvents provides the basis for the HPLC mobile phase.

Internal standard (i.s.).- The use of an i.s. is strongly recommended since the final method will be all the more robust if it is included. A high purity (>95%) analogue of the test compound selected should suit, provided there is no possibility of its metabolic formation from the analyte.

†addressee for any correspondence

Chromatographic systems can be quickly devised when a standard approach is used for development. Methods should in general satisfy the following criteria.-

Run time: should be >1 min and <6 min to ensure that the peaks are suitably resolved from any interferants and that the run time does not limit the productivity of the method.

Isocratic mode: gradient elution systems should be avoided since they require lengthy equilibration times.

Resolution of analyte from a suitable i.s. is desirable.

Columns: methods are more robust and reproducible if conventional stationary phases packed into commercially available columns are used.

Mobile phase: this should be a simple mixture of methanol or acetonitrile and water; ion-pair or other inorganic modifiers should be avoided, to simplify development and to give a more robust system.

Mass spectrometer.- Optimizing the analytical conditions for analyte detection by MS involves applying techniques that can be easily mastered by a non-MS specialist. The MS is tuned by infusion of a methanolic solution of the compound directly into the LC-MS interface using a syringe pump. In order to enable the compound to be detected with linearity from the low ng or pg region up to the low μg region, it is important to consider the following parameters.-

Choice of interface, APcI or ESP: APcI is strongly recommended for rapid and robust methods.

HPLC flow rate: the APcI interface will handle conventional LC flow rates whereas ESP necessitates techniques such as a column splitter or a micro-bore column to reduce the flow into the MS to <0.1 ml/min.

MS optimization of sensitivity and analyte stability entails modifying capillary and vaporizer temperatures and gas flows.

Extraction schemes

The commonest sample pre-treatments in bioanalysis are protein precipitation with centrifugation and direct injection, SPE and solvent extraction.

Protein precipitation and direct injection.- Precipitation of plasma proteins with agents such as acetonitrile or trifluoroacetic acid (TFA) inevitably results in a dilution of the sample which can limit the sensitivity of the method, and moreover there may be low analyte recovery that further limits sensitivity. In some applications the sensitivity of the MS compensates for this limitation; but in general the adoption of protein precipitation as the sole means of sample preparation prior to analysis by LC-MS is best avoided.

SPE has two main advantages over solvent extraction: the technique is easily automated and on-line methods are feasible.

However, the very short analysis times achievable for LC-MS methods mean that off-line SPE methods are preferred since even a simple procedure can take up to 10 min per sample to execute. SPE schemes are developed in a sequential manner, thus:

- conditioning of the cartridge with methanol;
- solvation of the cartridge with buffer of set pH;
- aliquot sample (say 0.5 ml) and add i.s. and buffer (1 ml);
- loading of the cartridge;
- washing with buffer (further washes with methanol/water may be required);
- elution with mobile phase.

Experiments are required to establish which pH gives the best recovery and to optimize the wash steps. [D.L. Massart deals with such aspects, and an 'expert system', in Vol. 23.- *Ed.*].

Solvent extraction is generallly less specific. By screening a range of pH's (e.g. 4, 7 and 10) and of solvent polarities, a simple extraction can be quickly developed. In practice ethyl acetate suits for extracting most analytes.

ASSESSMENT OF METHOD ROBUSTNESS AND INFORMATION ABOUT ANALYTE

Matrix stability of the compound is determined in human, rat and marmoset plasma at room temperature and 37°. A pool of each matrix is spiked, and aliquots (n = 4) are assayed after storage for 0, 0.5, 1, 2, 3, 4 and 6 h. Concentrations are calculated using a calibration curve derived only from human plasma.

In the example shown (Fig. 1), the analyte was unstable in rat plasma at 37°. A method of circumventing this instability was required. The successful approach was, using acetonitrile,

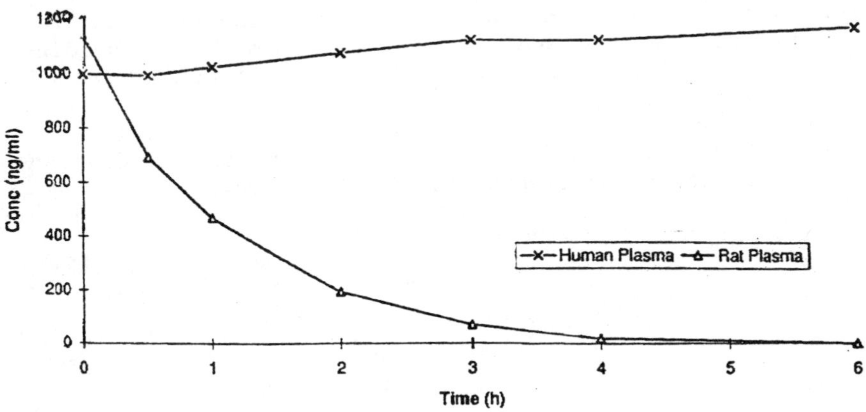

Fig. 1. Stability results for the compound stored at 37° in different matrices.

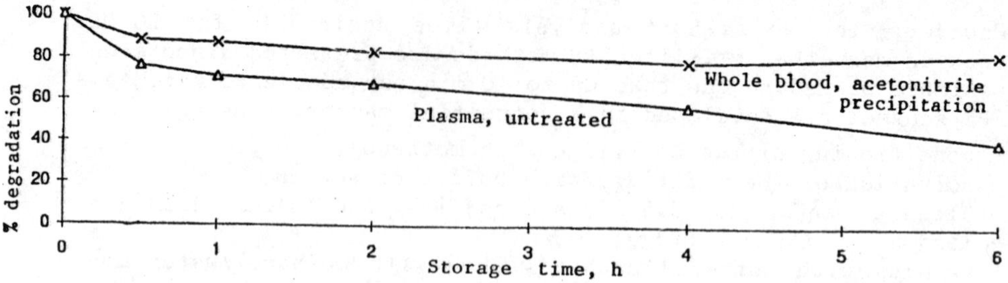

Fig. 2. Comparison between acetonitrile-precipitated whole blood samples and plasma samples, all kept at room temperature.

precipitation of whole blood samples, put straight into tubes containing it; the acetonitrile layer was removed following centrifugation, dried down and re-constituted in buffer of appropriate pH prior to SPE. As Fig. 2 shows, this approach overcame the analyte instability problem.

Intra-batch variation for assay in each matrix, which must be <20% (C.V.), is determined after completion of the stability experiments. This involves 6 calibration curves and 6 quality controls (QC's).

Method applications are now feasible, the method being suitable for support of research studies and early toxicokinetic work or, following formal validation, registration studies.

CONCLUDING COMMENTS

The foregoing approach minimizes the problems caused by delays during formal method validation, while ensuring that the method is suitable for analysis of study samples, providing useful range-finding and bioavailability information to aid compound selection.

In a companion article, #B-4, we describe method development for two matrix metalloproteinase inhibitors.

#C-8n

A Note on

ULTRA-TRACE BIOANALYSIS: THREE APPROACHES FOR ASSAY OF A NOVEL PHARMACEUTICAL AT THE pg/ml LEVEL

M. Bayliss†, V.J.C. Willson, J.R. Preston, K.E. O'Neill, J.L. Furness & D. Wilkinson

Astra Charnwood, Dept. of Physical Chemistry,
Bakewell Road, Loughborough LE11 ORH, U.K.

The trend towards increasing potency in newer drugs calls for bioanalytical methods of ever-increasing sensitivity. For both animal and human plasma samples, assays at the pg/ml level had to be developed in relation with initial therapeutic indicators for a novel drug, a moderately hydrophobic amino compound.

Three different analytical procedures are summarized here: (1) HPLC-MS; (2) column-switching HPLC combined with fluorescence derivatization; and (3) RIA. Each entailed a simple SPE step with C-18 cartridges prior to analysis.

(1) Since the initial preparation for RIA of a radiolabel and specific antiserum typically needs at least 6 months and there is no guarantee of success, an HPLC-MS method was used to generate analytical results from early studies. It employed an ionspray interface and multiple reaction monitoring. The LLOQ*, as pg/ml (and equivalent amount injected on column, pg), was 1000 (33) for rat samples, 100 (33) for dog and 75 (50) for human; the sample sizes were 0.1, 1 and 2 ml respectively.

(2) HPLC with fluorescence detection was evaluated as an alternative approach to be used if the other techniques failed. Derivatization was performed using a novel two-reagent system comprising phenyl isothiocyanate and, as fluorescent label, Lissamine (Rhodamine B sulphonyl chloride). Resulting extracts were applied to a 3-column system: a C-1 column removed reagent excess and was back-flushed onto a base-deactivated C-8 column; the eluate was heart-cut (loop transfer mode) to a 'hydrophobic' column (Primesphere HC C-18). The LLOQ was 25 pg/ml (5 pg load).

(3) A high-titre antiserum was successfully raised in sheep against a cationized BSA conjugate of the analyte. When combined with a high specific activity ^{125}I-labelled derivative of the analyte, this gave an extremely sensitive RIA: for human plasma the LLOQ was 10 pg/ml (equivalent to 2 pg per tube).

---

†addressee for any correspondence
*LLOQ = lower limit of quantification

#C-9n

A Note on

DRUG BIOANALYSIS MADE EASY?

W.John Lough[a], **Sara-Jane Groves**, ***Brian Law**,
†John Maltas, **Malcolm J. Mills** & **Mansoor Saeed**

School of Health Sciences. University of Sunderland, Sunderland SR1 3SD, U.K.
*Zeneca Pharmaceuticals, Alderley Park, Chesh. SK10 4TG
†Glaxo-Wellcome Group Research, Greenford, Middx. UB6 OHE

By the very nature of the problems posed by the determination of drugs in biological fluids, method development times in bioanalysis are frequently lengthy and the methods developed are often complex. Accordingly, three approaches were investigated which offered the prospect of reduced method development times and/or simpler methods.

1) Assays with chiral distinction.- Despite the increasing trend towards the development of homochiral drugs, there is still much demand for the determination of drug enantiomers in biofluids. This is particularly the case where some benefit might be derived from re-investigating a drug which has been marketed as a racemate. Studying its enantiomeric distribution following administration of the racemate is often informative. In such instances the sample pre-treatment developed for study of the racemate may be utilized for a chiral assay by simply treating the 'cleaned up' sample with an achiral fluorescent π-basic derivatizing agent and carrying out HPLC of the product on a microbore column containing a π-acidic chiral stationary phase (CSP). This should allow minimal method development, guaranteed enantiomeric separation, selective detection and low limits of detection (LOD).

2) Microbore columns may, furthermore, be exploited in achiral methods by selecting a suitable combination of SPE stationary phase and HPLC stationary phase such that large volumes of eluate from an SPE cartridge may be injected into a microbore column and a peak compression effect brought about. Method development for this approach might be more protracted; but the resulting method ought to offer the prospect of simplicity, no evaporation stages, ease of automation and low LOD's.

3) Immunoaffinity clean-up, using immobilized antibodies (Ab's).- Whilst the generation of suitable Ab's might be rather time-consuming, subsequent method development employing this approach ought to be straightforward, and the clean-up stage ought to be simple and highly effective.

[a]correspondent

Studies that we have undertaken with particular analytes have shown these three approaches to be effective, but with practical issues some of which are now considered in relation to foregoing paragraphs: 1), 2) [focus on 'compression'] & 3).

Instrumentation.- All HPLC was done with modular systems using Shimadzu instrumentation (Dyson Insts., Hetton-le-Hole) including CR5A computing integrators. For the chiral work, 1): pump, LC-6A; detector: UV, SPD-6AV variable wavelength; fluor, RF 551 - likewise for microbore LC (pump, LC-10AD) and for 3), IAf studies (LC-4A chromatograph). All injectors were Rheodyne.

1): deriv (fluor aCDApBs), then LC with a CSPpAc*

With Pirkle's popular R-DNBPG [1] as the CSP, a wide range of agents, fluor aCDApBs (Fig. 1), has been used to evaluate the approach; the outcome was an optimized implementation strategy.

i) Wherever possible, sample preparation to furnish a residue from an extract should use, unrefined, a published method for the racemate. If none is available, the procedure should be simple: protein precipitation, or liquid-liquid extraction and evaporating down the organic layer.

ii) Naphthyl isocyanate should then be used, being the most convenient choice for the aCDApBs although most will in fact give rise to LC separation on a R-DNBPG column. However -

iii,a) If better selectivity of detection is required, a phenanthrene-based aCDA should be used since it will give a longer wavelength for emission than the equivalent naphthalene-based agent. iii,b) If LOD has to be lower, then a fluorene-based aCDApBs should be used. Alternatively, a microbore [2] R-DNBPG column should be used to take advantage of the greater mass sensitivity that such columns offer. iii,c) On the rare occasions that better resolution is required, it is worth considering π-acid deriv (aCDApAc) in conjunction with a π-basic CSP (CSPpBs); on reciprocity grounds this would not be expected to give better results, but the CSPpBs tried did in fact prove useful, being urea- rather than amide-bonded to the support.

In general it was found that the aims of ease of method development, selective detection, low LOD's and good resolution were obtainable. However, as well as the obvious drawback that not all drugs possess easily derivatizable functional groups, it was found that in using microbore columns there was a reduction in resolution either when performing deriv on biological extracts rather than standards or when using fluor detection - most conventional fluor flow cells being slightly larger than

---

* *Abbreviations* [see also previous p.].- Ab, antibody; aCDApBs, an achiral derivatizing agent of π-basic type; CSPpAc, a chiral stationary phase of π-acid type; IAf, immunoaffinity [with an immobilized Ab; authors' text had AME = Ab-mediated extraction]; deriv, derivati|ve/zation; fluor, fluorescence: R-DNPBG, *R*-dinitrobenzoylphenylglycine. LC signifies HPLC.

those used in UV detectors. Fig. 2 illustrates some results, with experimental details. For the iii) approach the main thrust of future work will be to establish whether methods developed are sufficiently robust to cope with the rigours of multiple runs with numerous samples.

2) Microbore LC with peak compression [2]

Microbore columns may be exploited in achiral methods by selecting a suitable SPE packing/LC stationary phase combination such that large volumes of eluate from an SPE cartridge may be injected onto a microbore column, achieving a peak compression effect. Method development for this approach may be more protracted; but the outcome ought to offer the prospect of simplicity, no evaporation stages and low LOD's.

The use of this approach was studied for a range of drugs and biological fluids*- namely NSAID's and anti-leprosy drugs in urine, plasma and aqueous humour. However, before this could take place, much preliminary work using indomethacin pure drug standard solutions in aqueous/organic solvents was needed to identify the optimal practical conditions to apply to the subsequent attempts at 'off-the-shelf' bioanalysis. Thus, it was found that even when using 'inert' materials, adsorption of analyte is likely to take place in the injection system [3, 4]. However, the adverse effect of this on quantitation may be circumvented by using partial loop filling.

Further practical issues such as injector loop i.d., buffer purity and optimal detector flow cell were also addressed [5]. Ultimately it was found that in RP-HPLC on C-18 columns, peak compression could be very well achieved with as little as 25% difference in methanol content between injection solvent and mobile phase. Further it was found that the maximum tolerable injection volume was considerably higher than would have been expected from the literature [e.g. 6]. Accordingly an equation was developed which allowed calculation of the maximum tolerable injection volume:

$$V_{pc(max)} = V_{Io(max)} \cdot \frac{(k'_0 + 1)^2}{(k' + 1)}$$

where $V_{pc(max)}$ = allowable increased injection vol.† due to peak compression (limit), $V_{io(max)}$ maximum tolerable injection vol. in the absence of peak compression; k'_0 = capacity factor of solute, when eluted by the sample solvent; k' = same, when eluted by the mobile phase.

The actual derivation of this equation (relevant to art. #D-1 by D.J. Anderson - *Ed.*) will be reported elsewhere by us. However, it may be explained qualitatively by considering that there are two stages in the peak compression. (1) The band is compressed at the head of the column due to the fact that upon injection the mobile phase is effectively the very weakly eluting sample solvent' (2) further compression occurs when the mobile phase eventually displaces the injection solvent at the columnn head, just as occurs with a gradient - here one-step.

*mainly human

†at $k_o = 0$

Following the systematic groundwork study the SPE/microbore LC/peak compression approach for drugs in biofluids allowed much improved methods (as in Fig. 3), including a method for a quite polar analyte (more difficult) which, compared with an existing method, was more compatible with LC-MS, had better C.V.'s at all concentrations and a lower (×5) LLOQ, and suited robotic operation [M.J.M.]. But the approach still has a few impediments to universal applicability.- (1) 'Unclean' samples do not allow the theoretical V_p, doubtless because of mass overload, matrix interaction and column-head displacement effects. (2) Sometimes the SPE eluate may need its water content raised so that peak compression occurs; but with automated systems this addition hardly affects error or analysis time. Work in hand may well overcome these two drawbacks and simplify method development.

3) IAf (Ab) extraction/clean-up [Overleaf: an example, Fig. 4]

A working model allowing in-depth systematic study (not on the affinity-strength aspect) has been aimed at. For ICI 215001 in plasma removal of endogenous material was very effective and no possible co-administered drugs co-eluted; recovery/precision /between-cartridge reproducibility/linearity were good provided cartridge elution volumes were sufficiently large (warranting peak-compression LC conditions to get a good LOD). Re-use was feasible with the wash/elution conditions used; but due to the vacuum strength the Sepharose beds often needed repacking. The Ab amount/specificity (Ab and assay) relationship needs study.

Conclusions.- The three approaches already offer promise for drugs in biofluids, and could be refined to enhance their appeal to bioanalysts - especially with combination use: e.g. 3) followed by 1), giving cleaner extracts and so minimization of LC resolution problems. Rather than become redundant with the increasing use of LC-MS, the approaches may bring benefit through their minimization of sample preparation.

Acknowledgements.- We thank Zeneca Pharm'ls, Glaxo Group Research and Sunderland Univ. School of Health Sciences for funding.

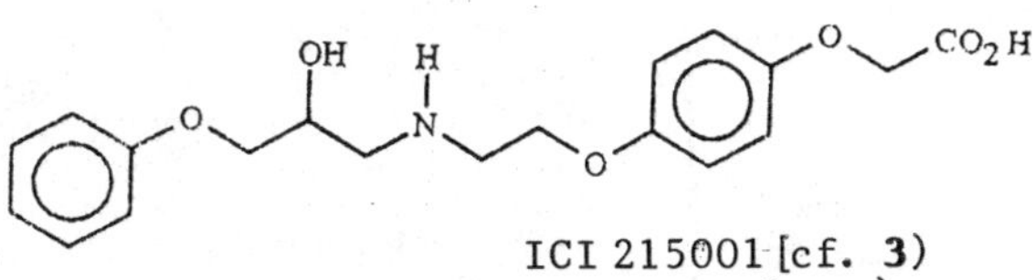

ICI 215001 [cf. **3**) above)

References [*Chr* denotes *Chromatography*]

1. Pirkle, W.H., House, D.W. & Finn, J.M. (1980) *J Chr 192*, 143-158.
2. Cited by Chervet, J.P. *et al.* (1992): *LC-GC Int. 5*, 33-38.
3. van der Water, C., Tebbal, D. & Haagsma, N. (1989) *J. Chr 478*, 205
4. Lough, W.J., Mills, M.J. & Maltas, J. (1996) *J Chr A 726*, 67-75.
5. Hambleton, P., Lough, W.J., Maltas, J. & Mills, M.J. (1995) *J. Liq. Chr 18*, 3205-3217.
6. Guiochon, G. & Colin, H. (1984) in *Microcolumn....* [*J Chr Library* Vol. 28; Kucera, P., ed.], Chap. 1, Elsevier, Amsterdam.

Added by Ed.: in Vol. 16, this series (1986; Plenum; ed. Reid, E. *et al.*) I.W. Wainer surveys CSP's useful for drug studies.

Fig. 1 (RIGHT).
A range of achiral derivatizing agents (see text).

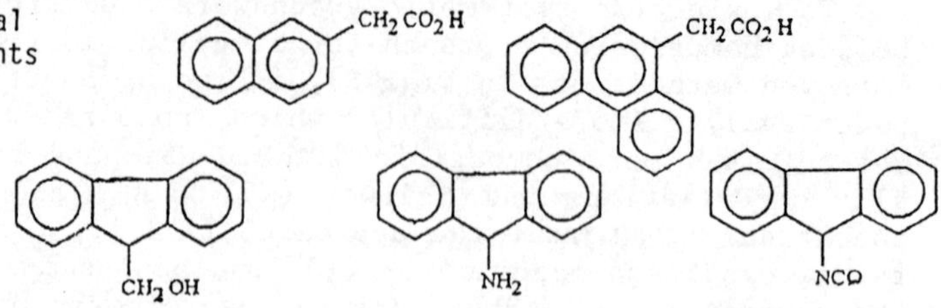

Fig. 2. Chiral HPLC of amide from propylamine and (d,l)-naproxen. **A**, pure drug/conventional column/UV: baseline resolution (just!). **B**, extract from urine/microbore col./fluor detection: poor resolution despite use of 3 µm particles.

BOTH: 20 µl loaded; propan-2-ol/hexane, 30:70 by vol.
A: Hichrom (R)-3,5-DNBPG col., 250×4.6 mm; 1.0 ml/min; 225 nm detection.
B: Chiralcap (R) 3 µm col., 125×2.1 mm; 0.2 ml/min; ex 265, em 335 nm.

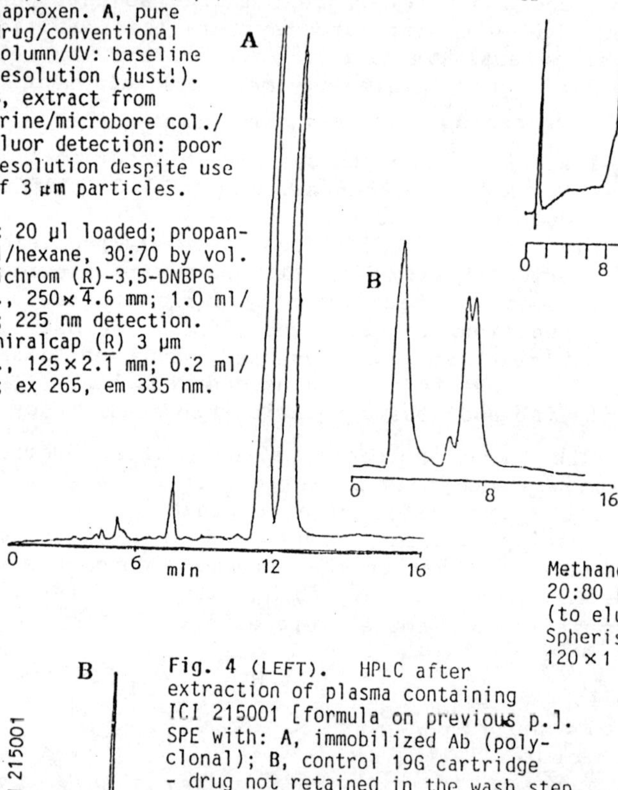

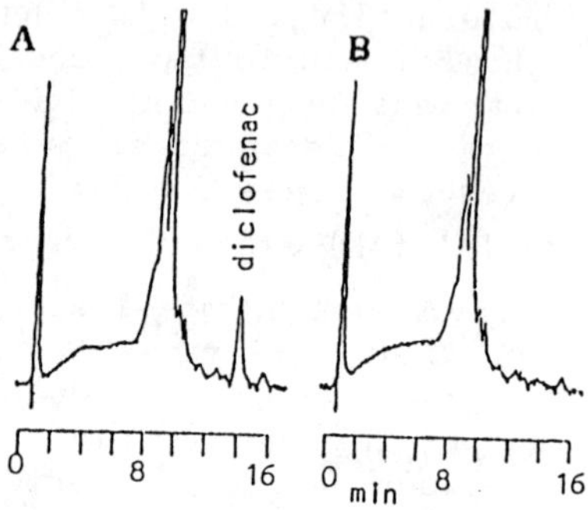

Fig. 3 (ABOVE). Microbore HPLC-UV, aqueous humour: SPE on 50 µl; 500 µl eluate injected. A: diclofenac, 100 ng/ml; B, blank. At such low levels conventional columns do not allow assay.

Methanol/20 mM pH 8 phos., 20:80 (to load) or 50:50 (to elute; 71 µl/min); Spherisorb ODS2 5 µm col., 120×1 mm; detector, 283 nm.

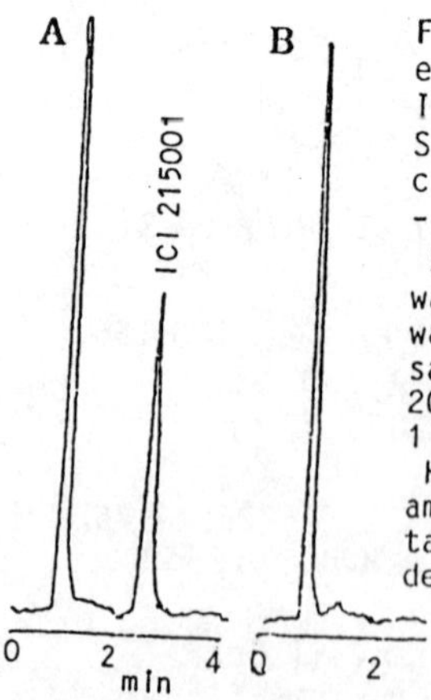

Fig. 4 (LEFT). HPLC after extraction of plasma containing ICI 215001 [formula on previous p.]. SPE with: A, immobilized Ab (polyclonal); B, control 19G cartridge - drug not retained in the wash step.

SPE cartridges (4 ml reservoir) prewashed with 1 ml methanol, then 1 ml water or buffer (pH 8.0 phos.); 200 µl sample loaded; then 1 ml additions: 20% aq. methanol (x 3) to wash; 1:99 1 M HCl/methanol (x 3-5) to elute.

HPLC: Spherisorb-5-SCX column; 20 mM ammon. formate/methanol, 20:80, containing 0.1% (v/v) TFA; fluorescence detection (ex 276, em 320 nm).

EDITOR'S NOTE: the Figs. and experimental details had to be fitted into a single page (beyond which the book had already been paged), and 'tight' layout turned out to be requisite.

#C-10n

A Note on

FLUORIMETRIC DETECTION OF NON-FLUORESCENT QUINOLONES AFTER POST-COLUMN PHOTOREACTION

K. Borner*, E. Borner, H. Hartwig & †H. Lode

Institute for Clinical Chemistry and Clinical Biochemistry, Universitätsklinikum Benjamin Franklin
Free University, D-12200 Berlin, Germany

and †Pneumonological Dept. 1, City Hospital Zehlendorf-Heckeshorn, Free University, D-14109 Berlin, Germany

Require-ment *An HPLC assay, highly sensitive and specific, for non-fluorescent quinolone antibiotics and quinolone metabolites in biofluids from human volunteers and patients.*

End-step *RP-HPLC; post-column photochemical reaction; fluorimetry.*

Sample handling *Dilution of sample with mobile phase.*

Comments *The purchasable photoreactor is cheap and trouble-free to operate. Its application is limited by low energy output and only 2 available wavelengths (254 & 360 nm). The approach offers desired sensitivity and specificity.*

Most quinolones can be determined in biofluids and tissues by HPLC with fluorimetric detection, with good sensitivity [1]. However, some quinolone metabolites yield little or no fluorescence, as do some newer quinolones (or naphthopyridines). Non-fluorescent quinolones can be degraded to fluorescent products by post-column UV irradiation, so allowing fluorimetric determination [2]. We now exemplify this, using a commercial photoreactor, for sparfloxacin glucuronide [3].

METHODS

Sources of materials.- Sparfloxacin (purity 98.8%): Rhône Poulenc Rorer, Cologne. Its glucuronide: from urine of a volunteer (HPLC-UV assay) after one 400 mg oral dose. Other quinolones: gifted (by manufacturers), or bought. Other chemicals and solvents: E. Merck, Darmstadt.

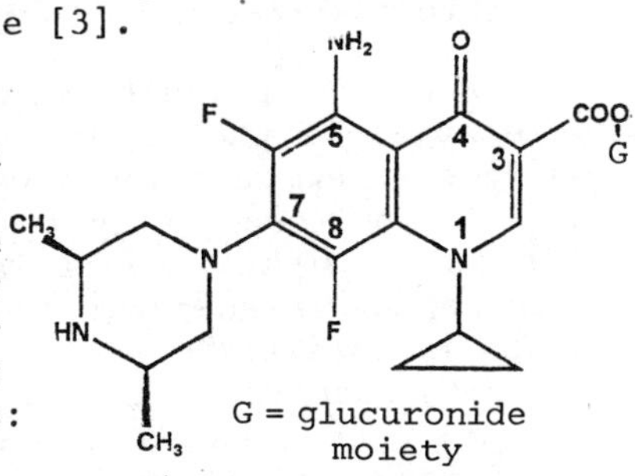

G = glucuronide moiety

HPLC [4].- The isocratic assembly comprised conventional modules (Fig. 1) with a photoreactor inserted between column and fluorescence detector (#RF-535; Shimadzu Co.). The mobile phase was acetonitrile/0.1 M phosphate buffer pH 2.92 (NaOH to adjust; 23 mM Na^+),¤ 3:1 by vol. The packing was Nucleosil SA,

*addressee for any correspondence

¤apparent pH of mixture 3.82

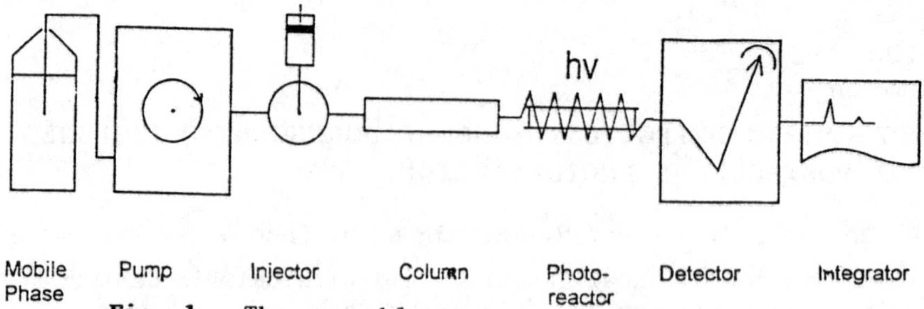

Fig. 1. The assembly.

5 μm, 250 or 125 × 4 mm, protected by a pre-column (30 × 4 mm) filled with Perisorb RP18. The flow rate was 0.7 ml/min, and 5-20 μl was injected. Excitation and emission were generally at 295 and 470 μm. The photoreactor (Beam Boost; ICT Co., Bad Homburg, Germany) had the reaction coil wrapped round a mercury lamp (254 nm; vol. 8.1 ml; irradiation generally for 5.66 min). To get spectra off-line (#DF-540 spectrofluorimeter; Hitachi) the sample was put manually into the coil and irradiated.

Samples of urine, from volunteers who ingested a 400 mg sparfloxacin dose, were protected from light and stored at -80° pending HPLC, when they were diluted 1:4 with phosphate buffer. The concentration of glucuronide (identified through alkaline or enzymic hydrolysis) was measured in the reference sample by HPLC-UV [4]; other quinolones were dissolved in mobile phase.

RESULTS

Effect of post-column irradiation.- When using the fluorimetric detector alone, urine showed only the parent substance (i.e. sparfloxacin; Fig. 2A), whereas UV detection indicated a second peak identifiable from previous work [5] as sparfloxacin glucuronide. Irradiation of the eluate converts the glucuronide into fluorescent material as shown in Fig. 2B.

Reaction product peaks (not illustrated nor identified) were multiple when mobile-phase solutions of sparfloxacin and various other quinolones were subjected to HPLC after off-line irradiation (especially if for >5.6 min), or even when kept for 1 h in sunlight. Irradiated sparfloxacin showed only minor changes in UV spectrum and (for the glucuronide likewise) no shift in excitation spectrum; but there was a marked shift in emission spectrum, from 525 to 470 nm. The glucuronide showed a continuous rise in detector response when, by varying flow rate, irradiation time was increased from 0.5 tO 11.1 min.

Glucuronide assay.- The response-concentration curve was parabolic. The detection limit was ~0.44 μg/ml (1.6 ng column load); for intra-series precision in the 0.29-99.2 μg/μl range the C.V.'s ranged from 5.0 to 9.3%.

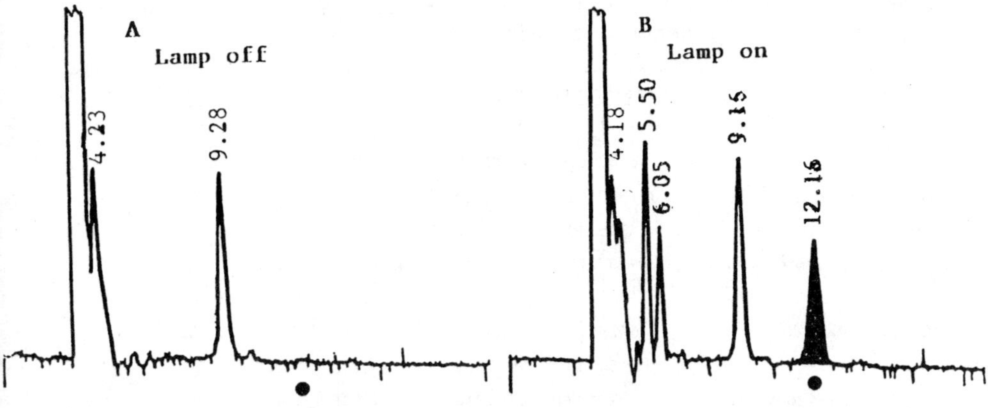

Fig. 2. HPLC (giving peak R_T's as shown, min) of human urine containing sparfloxacin (20.9 μg/ml) and its glucuronide (cf. peak •; 87.3 μg/ml), without (A) and with (B) irradiation.

Peak broadening (3-fold) resulted from inclusion of the photoreactor, with 0.7 ml/min flow rate and 5.66 min residence (radiation) time, using an open tubular (not crocheted) coil.

Other quinolones and metabolites (cf. [1]) also showed signal enhancement with irradiation (Table 1) and gave linear response curves, unlike sparfloxacin glucuronide (above).

Hydrogen peroxide addition to the mobile phase led, in preliminary experiments with sparfloxacin, to enhanced fluorescence; but the glucuronide's fluorescence yield diminished.

DISCUSSION

Fluorescence is an inherent property of the quinolone ring, but is quenched by some types of chemical change: a) esterification of the 3-carboxyl; b) oxo- or N-sulpho-substitution on the piperazine ring; c) substitutions at positions 5 and/or 8 in the quinolone ring; d) replacement of the piperazine ring by a different heterocyclic ring. Whereas a) and b) are due to biotransformation, c) and d) result from attempts to increase effectiveness against gram-positive organisms. All four types of varied structure allow degradation to fluorescent products, conducive to analytical sensitivity and specificity [2, 3, 6, 7].

The photochemical reactor now used, commercially available, is a low-cost instrument and is operable without major difficulties. It suffers mainly from its offering only two wavelengths (254 & 360 nm) and its low radiation energy such that irradiation for 5.6 min or longer was required with the model analyte sparfloxacin glucuronide. (The manufacturer recommends crocheted reaction coils to minimize peak broadening, but we prefer simple wrapped coils that generate less back-pressure.)

Table 1. Effect of irradiation (5.6 min; open tubular coil; 254 nm) on fluorescence intensity.

substance	slope (area units/mg)			detection limit (µg/ml)		
	lamp off	lamp on	ratio on/off	lamp off	lamp on	ratio on/off
ciprofloxacin	6138087	6388145	1,04	0,03	0,01	0,48
ciprofloxacin desethylen metabolite	327339182	73144645	0,22	0,01	0,01	1,00
ciprofloxacin oxo metabolite	305410	773243	**2,53**	0,21	0,12	0,55
ciprofloxacin sulfo metabolite	37279	618931	**16,60**	0,31	1,32	4,22
enoxacin	995703	1335507	**1,34**	0,31	0,25	0,80
enoxacin oxo metabolite	6424522	7136593	**1,11**	2,98	2,13	0,72
fleroxacin	5950918	6689380	**1,12**	0,01	0,01	1,77
fleroxacin desmethyl metabolite	1320192	8163050	**6,18**	2,46	1,35	0,55
fleroxacin N-oxide metabolite	49923197	52159737	1,04	0,34	1,50	4,41
lomefloxacin	43371151	47530740	1,10	0,01	0,02	1,32
nalidixic acid	16022400	3308340	0,21	(insufficient data)		
norfloxacin	1696514	1710290	1,01	0,45	0,40	0,89
ofloxacin	6760635	7038291	1,04	0,05	0,03	0,57
pefloxacin	1416754	1403430	0,99	0,57	0,57	0,99
sparfloxacin	1558648	1475200	0,95	0,24	0,63	2,60
sparfloxacin	329076	1440347	4,38	0,23	0,93	4,05
sparfloxacin glucuronide	0	20604	>100		1,71	
sparfloxacin glucuronide 11.1 min irradiation	0	91079	>100		1,62	

Further disadvantages of the reactor were peak broadening and, peculiar to the model analyte, a non-linear response curve - its surmised cause being the formation of multiple reaction products. Most of these disadvantages could be overcome by heating the column eluate, as in the photochemical reactor designed by Scholl *et al.* [2] although this is not commercially available. Sensitivity should also be improvable by use of a more sensitive fluorimeter.

CONCLUSION.- Post-column photoreaction offers HPLC sensitivity and specificity for non-fluorescent quinolones and metabolites.

References

1. Present authors (1990), pp. 131-144 in Vol. 20, this series (Reid, E. & Wilson, I.D., eds.) [publisher as now] - *Analysis for Drugs and Metabolites Including Anti-infective Agents.*
2. Scholl, H., Schmidt, K. & Weber, B. (1987) *J. Chromatog. 416,* [321-330.
3. Present authors (1994) *Lab. Med. 18,* 384.
4. Borner, K., Borner, E. & Lode, H. (1992) *J. Chromatog. 579,* 285- [289.
5. as for 4. (1993) *Drugs 45, Suppl. 3,* 303-304.
6. Schaefer, H.G. (1993) *J. Chromatog. 616,* 87-93.
7. Aoki, H., Ohshima, Y., Tanaka, M., Okazaki, O. & Hakusiu, H. (1994) *J. Chromatog. B 660,* 365-374.

#C-11n

A Note on

COMPOUNDS POSING AN ENDOGENOUS BLANK PROBLEM: RETINOIDS

Terence A.G. Noctor[▫], **Martyn J. Hemsley & Elizabeth Sharp**

Corning Hazleton, Dept. of Biopharmaceutical Analysis, Otley Road, Harrogate, N. Yorks. HG3 1PY, U.K.

Tretinoin (all-*trans*-retinoic acid) and isotretinoin (13-*cis*-retinoic acid) - oxidative metabolites of vitamin A (retinol) with cell differentiation/proliferation roles - have extreme lability, especially photochemical. Measurement of endogenous levels is difficult because of this and other factors including autoxidizability, instability to even moderate heat, the very low *in vivo* levels, the presence of other physico-chemically similar vitamin A metabolites, and avid protein binding. The inability to obtain blank control matrices is a major problem.

The most acceptable approach for accurate quantification of the retinoids in biological samples is direct introduction of the matrix concerned into the analytical system. Retinoid methods have been reviewed by Wyss [1]†. We sought to use new technology to develop a rapid, simple and robust method, as now achieved using HPLC-MS-MS(API)* and an ISRP 'pre-column'.

EXPERIMENTAL

Chemicals were procured thus: tretinoin and isotretinoin, USP grade: Promochem (Welwyn Garden City); MeCN, methanol and absolute ethanol, all Super Purity: Romil (Cambridge); ammonium acetate, AnalaR: Merck/BDH (Poole).

HPLC-MS-MS(API).- On-line SPE and separation/measurement were done using different systems coupled by a 6-port switching valve. System A for pre-concentration and clean-up, comprising an autosampler (Gilson 231 XL), pump and gradient controller (Jasco, PU-980 & LG-980-02), and ISRP cartridge (Regis SPS C-18; Fisons, Loughborough), was linked to the Universal Switching Unit (Anachem, Luton) such that in position 1 the effluent from the cartridge was lost to waste. System B, for separation and quantitation, comprised a pump (Perkin-Elmer, 410) supplying an Interpak ODS-80 column (150 × 4.6 mm; Capital HPLC, Broxburn).

▫addressee for any correspondence
†Elsewhere in this vol. D.J. Anderson considers findings by Wyss - who contributed to Vol. 22 of this series (1992).- *Ed.*
**Abbreviations*.- In MS (MS-MS = tandem): API, atmospheric pressure ionization; MRM = multiple reaction monitoring. MeCN, acetonitrile; QC, quality control; i.s., internal standard; ISRP, internal surface reverse phase [cf. art. #D-1]

The valve was configured so that analytical column effluent fed to the detector in both positions 1 and 0:- respectively, the mobile phase from pump B went directly through the analytical column to the detector/was back-flushed through the cartridge before reaching the head of the column, with restoration to position 1 after a 1-min transfer period. Whilst the separation was proceeding in system B, the pre-concentration cartridge was washed with MeCN and, for loading, returned to HPLC-grade water.

For MS-MS detection (API III+; Sciex, Thornhill, Ont.) the HPLC system was interfaced (55°) using a Sciex heated nebulizer probe kept at 400° with N_2 at 800 psi as nebulizer gas - also used as curtain gas (0.6 L/min) and auxiliary gas (2 L/min). The orifice was at -70 V and the multiplier at 3600 V. API was in the negative-ion MRM mode with argon as collision gas (300 × 10^{13} atoms/cm^3). The transitions measured were m/z 299 ⟶255.4 (tretinoin and isotretinoin) and 325.0⟶266.3 (i.s.). The dwell time was 200 msec and the pause time 30 msec. Data acquisition was done by Sciex RAD software with MacQuan for manipulation. The total time for the entire analytical cycle was 15 min.

Sample preparation was done entirely in the dark or under subdued yellow lighting. To each 1.0 ml aliquot of plasma was added 200 µl i.s. solution (50 ng/ml). After vortexing, 1 ml MeCN was added. The tubes were vortexed and centifuged at 4° (20 min, 3500 rpm), and 1.0 ml of supernatant transferred to a clean round-bottomed tube containing 1.0 ml 10 mM ammonium acetate pH 3.5. After mixing, 1.2 ml was transferred to amber-glass crimp-topped injection vials. These were loaded into an autosampler tray kept at 4°; 800 µl was injected onto the SPS cartridge, in an HPLC-grade water mobile phase, with a flowrate gradient ramping from 0.4 to 0.6 ml/min over 4.8 min. After this washing stage the analytes of interest were back-flushed from the pre-concentration to the analytical column with MeCN/acetic acid (100:0.05 by vol.; 0.6/ml min) as mobile phase.

Quantification of tretinoin and isotretinoin.- Calibration plots show large positive intercepts because of endogenous retinoids whose levels were determined by the standard-addition method. Second calibration plots were constructed using the actual levels of the retinoids in calibration standards, and these served to get the retinoid levels in test samples.

BEHAVIOUR OF THE METHOD

The following ng/ml values for isotretinoin|tretinoin were obtained with plasma from 10 healthy volunteers: 2.73|2.26, 2.39|1.75, 1.73|1.64, 7.16|1.57, 1.88|1.42, 2.03|1.97, 1.99|1.21, 1.93|1.86, 1.58|1.94, 2.10|1.51; mean 2.55|1.71. Fig. 1 shows typical HPLC patterns. The method is rapid, simple and specific for the compounds of interest. With spikes the linear range was 0.25-12.5 ng/ml. Inter-day precision was at least 11.5% for isotretinoin and 8.2% for tretinoin.

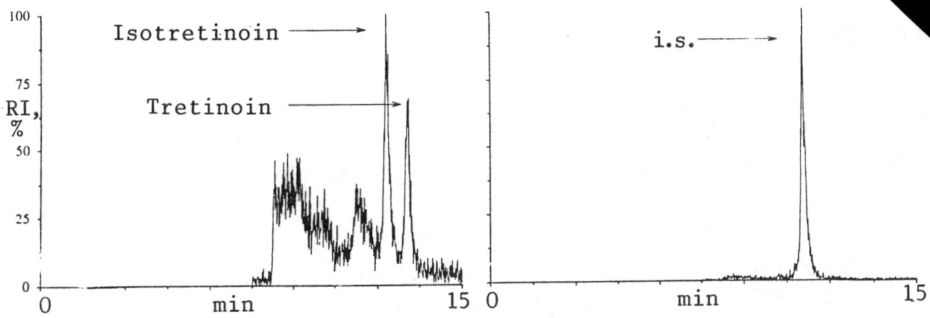

Fig. 1. HPLC-MS-MS(API) with MRM (R.I. = relative intensity) of plasma from an undosed volunteer, containing 2.36 ng/ml isotretinoin and 1.75 ng/ml tretinoin with spiked i.s.

Attempts to obtain analyte-free plasma.- Two methods were examined and found unsuitable. Charcoal treatment of control plasma apparently decreased tretinoin and isotretinoin levels. However, the levels in a particular sample varied greatly on subsequent analysis; this was thought to be due to variable release of analytes from a microparticulate suspension of charcoal that remained in the matrix even after centrifugation. Secondly, a batch of control plasma was UV-irradiated (254 nm) for 24 h with stirring. Initially the levels of both retinoids were found to be greatly reduced. However, on storage (-20°) the levels of isotretinoin and especially of tretinoin rose, even beyond their original levels. The most likely cause was initiation of an autocatalytic oxidation reaction, giving as end-products the analytes of interest.

SUMMARY

An assay procedure employing **HPLC-MS-MS**(API) is described for tretinoin and isotretinoin. Endogenous levels were determined using a standards addition approach [cf. hirudin studies by D. Gygax earlier in this vol.- *Ed.*]. The true levels in calibration and QC samples prepared using that plasma could then be calculated. This approach should be applicable to all cases where it is difficult to obtain a blank matrix.

Reference

1. Wyss, R. (1990) *J. Chromatog. 531,* 481-508.

Added by Editor.- Besides having published briefly on retinoid-type drugs (Vol. 18), R. Wyss describes an on-line/switching SPE-HPLC-UV method for retinoids [formulae shown] in Vol. 22 (1992). As now, spiked-in standards took care of the blanks problem; likewise in the GC-MS method of A.H. Parton *et al.* for assaying a guanidine-type agent that is native to plasma (p. 249, Vol. 23).

RESOLUTION OF CARBOXYIBUPROFEN, ARY METABOLITE OF IBUPROFEN

S.C. Tan, †N. Stevens, †V. de Biasi & A.J. Hutt [¤]

Department of Pharmacy, King's College London, Manresa Road, London SW3 6LX, U.K.

and †SmithKline Beecham Pharmaceuticals, New Frontiers Science Park, Third Avenue, Harlow CM19 5AW, U.K.

(R,S)-Ibu* is an important NSAID used in various musculo-skeletal disorders, showing stereoselectivity in both action and disposition as much investigated in recent years due to the metabolic chiral inversion of the weakly active R-enantiomer to its active S-antipode [1-4]. Various chromatographic methodologies have been reported for determining the drug's enantiomeric composition in biofluids after racemate administration; but the analysis of the stereochemical composition of the major metabolites has been much less studied. Ibu becomes oxidized *in vivo* in both animals and man to yield two main products, hIbu and cIbu (Fig. 1) [5-7]. Their urinary excretion together with Ibu and their glucuronides accounts for ~80% of an oral dose following administration of the drug to man [6, 7].

The formation of cIbu involves introducing an additional chiral centre into the molecule; hence four stereoisomers are possible. To date an examination of the 'total' stereochemical

Fig. 1. Oxidative pathways of ibuprofen in man (with Editor's *abbreviations*; R or S may precede)

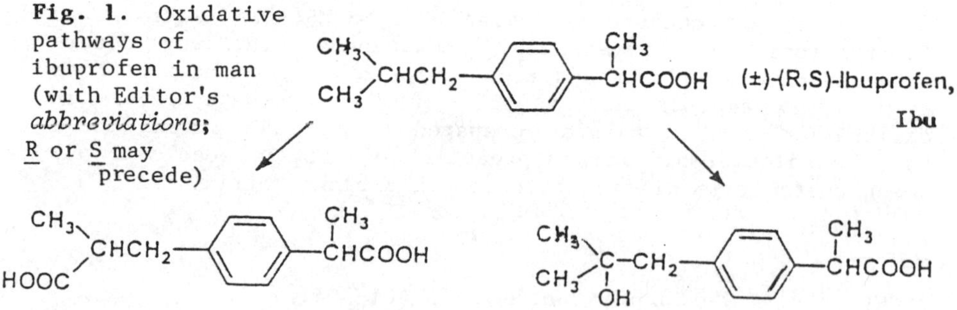

Carboxyibuprofen, **cIbu** Hydroxyibuprofen, **hIbu**

**Abbreviations* (by Editor) in Fig. 1; also TFA, trifluoroacetic acid. All mobile phase compositions are by vol. (CSP: see next p.)

¤addressee for any correspondence

outcome of Ibu metabolism has been hampered by lack of appropriate methodology for chromatographically resolving all four stereoisomers, not achievable with the indirect approach in the original packed-column GC assay [8] or capillary GC-MS assays [9, 10]. The application of a Pirkle-type chiral stationary phase (CSP), (R)-N-(3,5-dinitrobenzoyl)phenylglycine, to their resolution achieved separation of the two pairs of diastereoisomers but not in enantiomeric resolution [11]. The only successful method reported to date is based on the indirect approach: derivatization of the di-acid with (S)-2-phenylethylamine followed by HPLC with a C-8 stationary phase [12]. So as to examine the stereochemistry of Ibu in man, we have investigated the use of a polysaccharide CSP for cIbu isomers.

Experimental.- (R)-, (S)- and (RS)-Ibu and hIbu were the generous gifts of the Boots Co., Nottingham; cIbu amd its di-esters (methyl, ethyl, isopropyl) were prepared at KCL as will be described elsewhere. HPLC was done with an LDC Constametric #3000 pump/#3100 Spectromonitor/#CI 4000 computing integrator, or a Perkin Elmer Integral #4000 system. Fig. legends specify mobile phases (to suit the analyte) and columns - eventually 250 × 4.6 mm Chiralpak AD, 10 µm [amylose tris(3,5-dimethylphenylcarbamate)], obtained from J.T. Baker (Milton Keynes, U.K.)

RESULTS

Initial attempts to resolve the cIbu stereoisomers were indirect: derivatization using (R)-1-(naphthen-1-yl)ethylamine, then RP-HPLC (C-18). These failed, two of the stereoisomers co-eluting (Fig. 2); no benefit came from stationary-phase or mobile-phase composition changes. There are reports [13, 14] that enantiomers with a carboxyl group at the chiral centre can after conversion to neutral derivatives (e.g. amides or esters) be resolved using CSP's with a polysaccharide moiety. Hence a range of cIbu di-esters was tried with Chiralpak AD. Whilst the dimethyl ester's four stereoisomers showed near-baseline resolution (Fig. 3), retention and resolution were poorer with the di-ethyl and -isopropyl esters (data not shown), the latter giving only three peaks. As NP conditions were used, poorer retention with increasing hydrophobicity of the ester moiety would be expected, and evidently resolution suffered with increasing steric bulk of the ester functionality.

Raising the mobile phase's ethanol content to 10% (v/v) along with TFA addition (to 0.05% v/v) allowed the free acid to be used; all stereoisomers were baseline-resolved (Fig. 4). To date we have been unable to obtain stereochemically characterized samples of the di-acid's individual isomers; so elution order cannot be assigned. However, HPLC metabolite examination on a volunteer's urine after (S)-Ibu ingestion indicated the presence of the two later-eluting stereoisomers. We therefore

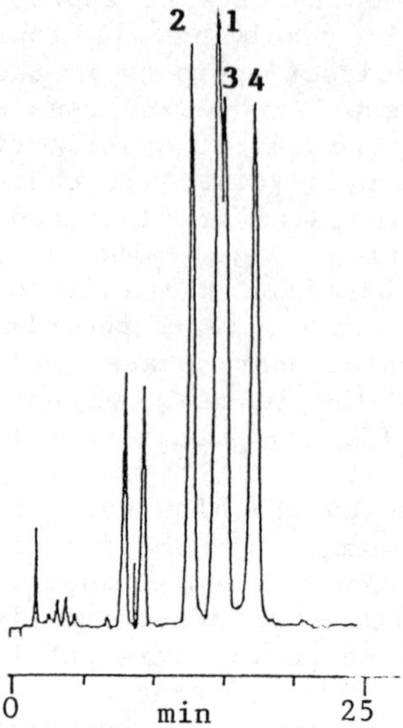

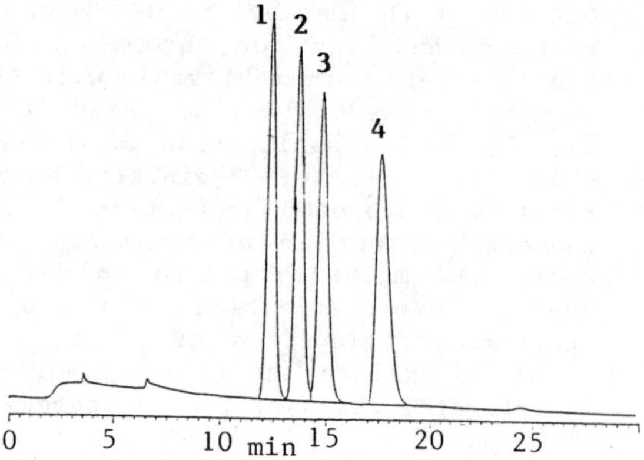

Fig. 3. Resolution of cIbu isomers as Me esters on a Chiralpak AD CSP (250 × 4.6 mm, 10 μm) column, with hexane/ethanol, 99.5:0.5; 1 ml/min. Detection at 213 nm. Peak R_Ts, min: 1, 12.2; 2, 13.5; 3, 14.6; 4, 17.6.

Fig. 2. RP-HPLC separation of cIbu isomers, pre-derivatized (see text), on a Resolve C_{18} column (150 × 3.9 mm, 5 μm). Mobile phase: acetonitrile/pH 3.5 10 mM phosphate, 50:50; 1.5 ml/min. Detection at 254 nm. Peak R_Ts, min: 2, 13.2; 1, 15.1; 3, 15.6; 4, 17.9.

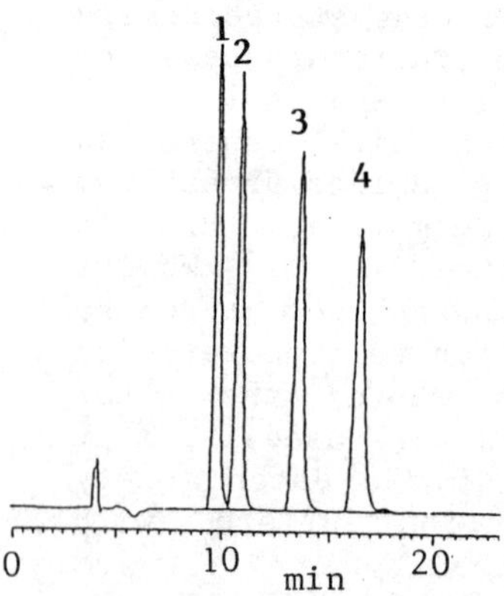

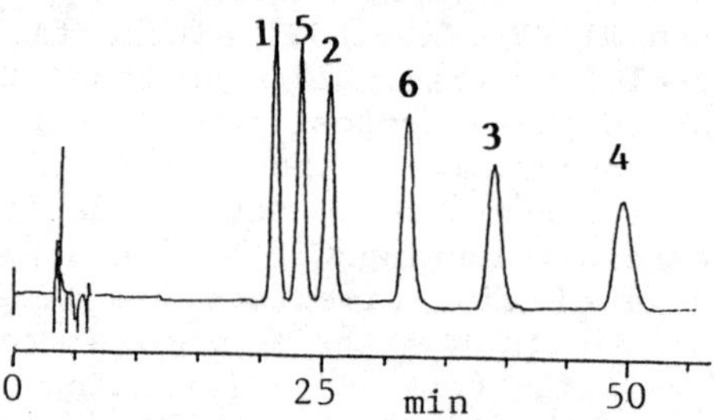

Fig. 5. Isomer resolution, on Chiralpak AD, for admixed cIbu (peaks 1-4) and hIbu (5, R-; 6, S-), using hexane/methanol/ethanol (95:3.5:1.5) having 0.05% (v/v) TFA; 1 ml/min; 220 nm. R_T's, min: 20.9/23.0/25.4/31.6/38.9/49.4.

Fig. 4. As for Fig. 3 but underivatized cIbu and 90:10 eluent mixture containing 0.05% v/v TFA; 220 nm. Peaks 1 (R_T 9.0 min) and 2 (10.6) are reckoned to be, in respect of the propionic acid moiety, R-cIbu, and 3 (13.9) and 4 (16.3) S-cIbu; peak identification in respect of the second chiral centre has not been made.

conclude that peaks 1,2 and 3,4 in Fig. 4 correspond respectively to the stereoisomers with the R and S configurations in the propionic acid moiety.

As also shown in Fig. 5, a single run allows stereochemical analysis of both metabolites, hIbu besides cIbu. Previously, with TFA but no methanol (hexane/ethanol 92:8), (R)-hIbu had been found to elute before (S)-hIbu (α = 1.15; R_s = 1.54).

In conclusion, the above data show that the cIbu stereoisomers are resolvable either as such or as their methyl esters using a Chiralpak AD CSP, whose use will allow isolation and so facilitate stereochemical characterization of the individual isomers. The stereochemical composition of both major metabolites of Ibu may be determined in a single HPLC analysis.

References

1. Hutt, A.J. & Caldwell, J. (1983) *J. Pharm. Pharmacol.* *35,* 693-704.
2. Caldwell, J., Hutt, A.J. & Fournel-Gigleux, S. (1988) *Biochem. Pharmacol.* *37,* 105-114.
3. Williams, K.M. (1990) *Pharmac. Ther.* *46,* 273-295.
4. Evans, A.M. (1992) *Eur. J. Clin. Pharmacol.* *42,* 237-256.
5. Mills, R.F.N., Adams, S.S., Cliffe, E.E., Dickinson, W. & Nicholson, J.S. (1973) *Xenobiotica* *3,* 589-598.
6. Lockwood, G.F., Albert, K.S., Gillespie, W.R., Bole, G.G., Harkcom, T.M., Szpunar, G.J. & Wagner, J.G. (1983) *Clin. Pharmacol. Ther.* *34,* 97-103.
7. Evans, A.M., Nation, R.L., Sansom, L.N., Bochner, F. & Somogyi, A.A. (1990) *Biopharm. Drug Disp.* *11,* 507-518.
8. Kaiser, D.G., Van Giessen, G.J., Reischer, R.J. & Wechter, W.J. (1976) *J. Pharm. Sci.* *65,* 269-273.
9. Young, M.A., Aarons, L., Davidson, E.M. & Toon, S. (1986) *J. Pharm. Pharmacol.* *38,* 60P.
10. Baillie, T.A., Adams, W.J., Kaiser, D.G., Olanoff, L.S., Halstead, G.W., Harpootlian, H. & Van Giessen, G.J. (1989) *J. Pharmacol. Exp. Ther.* *249,* 517-523.
11. Nicoll-Griffith, D.A., Inaba, T., Tang, B.K. & Kalow, W. (1988) *J. Chromatog.* *428,* 103-112.
12. Rudy, A.C., Anliker, K.S. & Hall, S.D. (1990) *J. Chromatog.* *528,* 395-405.
13. Wainer, I.W. (1988) *A Practical Guide to the Selection and Use of HPLC Chiral Stationary Phases.* J.T. Baker, Phillipsburg, pp. 11-14.
14. Shibata, T., Mori, K. & Okamoto, Y. (1989) in *Chiral Separations by HPLC. Applications to Pharmaceutical Compounds.* (Krstulovic, A.M., ed.), Ellis Horwood, Chichester, pp. 336-398.

#C-13n

A Note on

AN HPLC ASSAY DIRECTLY ON PLASMA USING AN ISRP*

S. Braggio & R.J. Barnaby

Research Labs., Glaxo SpA, via Fleming 2, 37135 Verona, Italy

Injecting untreated plasma onto an RP-HPLC column is prone to cause blockage due to irreversible binding or accumulation of denatured protein on the packing, with rapid deterioration. This is avoidable by use of an ISRP* packing of pore size such that water-soluble proteins are unretained, not penetrating or interacting with the particles, while the smaller hydrophobic analyte molecules penetrate and are resolvable, the internal surface being of conventional RP character. A simple assay for a Glaxo development drug is now described (pore size 10 μm).

Assay was done with 10 μm Ultrabiosep C8 (Shandon) in the 100 × 4.6 mm column and 18 × 4.6 mm guard column, with a Waters 600E solvent delivery system and 712 (Wisp) autosampler and a Jasco 975 UV detector at 300 nm. After loading 10 μl of water-diluted (1:1) plasma, elution (at ~8 min) was effected by 67 mM pH 7.4 phosphate buffer/acetonitrile (65:35 by vol.; 1 ml/min).

Results.- The peak areas over the spiking range studied (0.5-100 μg/ml) showed a linear curve, unaffected if water was used instead of plasma; thus the drug's strong binding (>99.9%) to human serum albumin did not impair transfer into the pores and hence recovery. Precision and accuracy were evaluated at 7 concentrations, each with 6-fold replication: for 0.5, 10 and 100 μg/ml the C.V. (and bias) values were 5.88 % (-3.8%), 2.4% (+6.0%) and 4.9% (+2.7%) respectively. No change in column efficiency or back-pressure was noted after ~300 injections. Assays were then done on volunteer plasma samples, containing therapeutic concentrations of the drug, that had already been assayed in a pharmacokinetic study with a different validated method, more elaborate. The two methods showed good accord (Fig. 1). In general, the simple ISRP method, needing only a small sample, is apt for clinical monitoring if sensitivity need not be high and the drug is reasonably lipophilic and/or can be selectively detected. Throughput could be as high as 100 samples/day/analyst.

*Internal Surface Reversed Phase (external surface hydrophilic through modification with a 'biological' polymer).

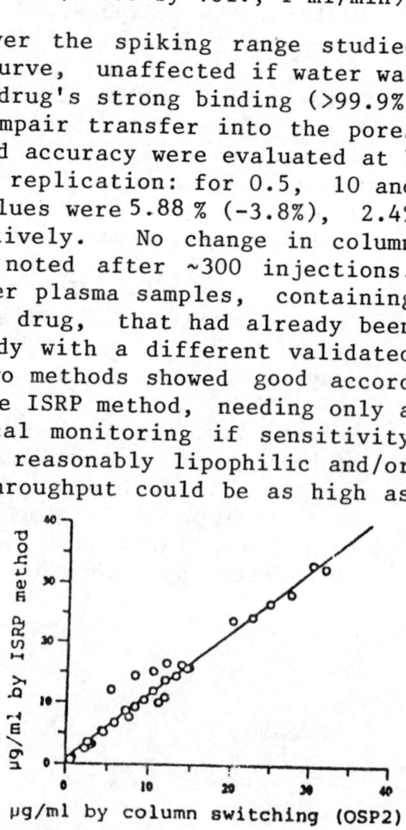

Fig. 1. Assays on volunteer samples by the two methods (r = 0.981). 'OSP2' signifies an On-line Sample Preparator (Merck).

Comments on #C-1, C. Nuez: MICRODIALYSIS AND MS FOR A CNS DRUG
#C-4, S. Pleasance: A 5-HT RECEPTOR ANTAGONIST*
#C-6, R.J. Simmonds: DEVELOPING LC-MS ASSAYS

C.D. Bevan asked Nuez whether % recovery could be a problem with compounds of mol. wt. >800. **Reply**: whilst our compound has mol. wt. ~250, it is ~800 for others studied by colleagues, who get adequate recovery. **H. de Bree asked** if the dialysis-with-i.s. approach suits even for highly protein-bound analytes. **Reply**: not suitable; only the free fraction can be measured.

Pleasance, replying to J. Moore.- (1) Addition of ammonium formate to the mobile phase does improve sensitivity over that with solvent/water mixtures; but a more robust separation is the main benefit. (2) Concerning applicability to peptides of the 'generic approach': rather than APcI we would use ESP (ESI; **Moore remarked** that this is more art than science).

Comment to Simmonds by D. Dell.- For compounds with very poor water-solubility it seems vital to avoid evaporating to dryness during sample preparation. Automatic switching methods allow injection of supernatants after protein precipitation, as done with the extremely lipophilic retinoids (Wyss & Bucheli, cited in art. #D-1). **J. Gilbert asked** whether derivatization might have helped in the suleptanate assay; **in reply, Simmonds** saw merit in the idea; but they hadn't tried it. **Answer to K. Jolley**: sample throughput is 300-400 per week.

I.D. Wilson, to Simmonds: our experiences with LC-MS-MS bear out your cautionary tales. **C.D. Bevan enquired** why the chosen packing for pre-MS clean-up was Hypercarb, to which one might think the molecule wouldn't stick. **Reply by Simmonds.**- It has good stability at pH 10 and gives an R_T of only 1 min.

Citations contributed by Senior Editor For journals, Chr denotes Chromatography

***In vivo* microdialysis** (cf. art. #C-4)

The narrow-bore HPLC approach mentioned by Y. Michotte et al. in their Vol. 23 art. [this series] as a way to achieve sensitive off-line measurement is documented in J. Chr: **B: 657**, 149-154 (1994) and, with description of a 70 µl flow cell, **672**, 97-102 (1995; Van Belle, K. et al.). For HPLC assay of paracetamol in plasma of freely moving rats, an on-line system with switching valves was effective: Steele, K.M. & Lunte, C.E. (1995) J. Pharm. Biomed. Anal. **13**, 149-154.

GABA in rat-brain microdialysates was measured by off-line HPLC-EC (C-18; 4.6 mm bore) after OPdiA-sulphite derivatization: Smith, S. & Sharp, T. (1994) J. Chr **B 652**, 228-233. For etoposide, microbore HPLC-fluor (LLOD 15 ng/ml; solvent extraction for serum) was done on 1-h collections: Burgio, D.E. et al. (1996) Biochem. Pharmacol. **51**, 987-992.

*being transferred from Glaxo Wellcome to Zeneca

Off-line conventional HPLC-EC (C-18) with prior SPE has also been performed on plasma taken by cannula (not microdialysis) from freely-moving rats given a dopamine D2 agonist, a thienyl-aminotetralin: Swart, P.J. & de Zeeuw, R. (1993) J. Pharm. Sci. **82**, 200-203.

Citations relevant to HPLC-fluor determination of **quinolone antibiotics** (cf. art. #C-10n) appear within #ABC, Group 6a. Especially pertinent is a method for a new quinolone (DU 6859a) that entails SPE, HPLC (C-18), photolysis and then fluorescent detection.- Aoki, H. et al. (1994) J Chr **B 660**, 365-374.

Material bearing on assay-development 'strategy' contributions in Sect. #C appears in Sections #D (SPE etc.), #E (notably LC-MS-MS) and #F (assay validation).

Retinoids, mainly endogenous (cf. arts. #C-11n & #D-1)

Retinol has been determined in capillary blood by HPLC-UV (C-18; 325 nm); with a Phenomenex instead of a Rainin column retinoic acid could be determined simultaneously.- Barua, A.B. et al. (1995) J. Liq. Chr **18**, 1459-1471. For retinoic acid and the isomer isotretinoin in plasma, SPE (C-18) was followed by NP-HPLC (360 nm detection).- Lefebvre, P. et al. (1995) J. Chr. **B 666**, 55-61.

For appraisal of retinol 'status', using deuterated retinol as i.s., ethanol-deproteinized plasma was subjected to subtle solvent treatment; acetonitrile-extracted material was put through HPLC (C-18) and, after derivatization, GC-MS (SIM).- Handelman, D.J. et al. (1993) Anal. Chem. **65**, 2024-2028. By a comparable extraction approach and then C-18 isocratic HPLC, the levels of retinol, α-tocopherol and individual carotenoids were ascertained in plasma with programming of the detection wavelength.- Bui, M.H. (1994) J. Chr **B 654**, 129-133. From a similar HPLC study a database helpful for identification was compiled.- Ben-Amotz, A. (1995) J. Liq. Chr **18**, 2813-2825.

Subtle solvent extraction has also been done on plasma/bile /liver samples from perfusions where acitretin, etretinate and metabolites were measured by gradient HPLC (C-18; 350 nm).- Decker, M.A. & Zimmerman, C.L. (1995) J. Chr **B 667**, 105-113. A comparable but simpler approach was used to measure a novel plasma component, 9-*cis* -retinoic acid.- Marchetti, M.N. et al. (1994) Anal. Lett. **27**, 1847-1862.

Hypolipidaemics, as featured in Vol. 23

Their efficacy in the prophylaxis of heart disease had not been proved. Now (early-1996) a 5-year trial in Glasgow by the West of Scotland Coronary Prevention Study has ended: pravastatin given to hypercholesterolaemic subjects brings firm benefit.

In the Assay Compendium (#ABC) certain hypolipidaemics are included in Group 4.

#ABC

COMPENDIUM *by* **E. Reid** (Guildford)

RECENT PAPERS ON DRUG ASSAY OR METABOLITE DELINEATION

mostly HPLC/GC; therapeutic-group listing, as in Vols. 22 & 23

1. **MUSCULO-SKELETAL AGENTS** incl. analgesic [anti-pain/-pyrexia in #5a], anti-inflammatory [cf. #2], skeletal-muscle relaxant
2. **RESPIRATION-EASING AGENTS/ANTI-ALLERGICS**: bronchodilator, mucolytic [expectorant: #7]; anti-histamine/-inflammatory [cf. #1]
3. **(ANTI)ADRENERGICS/CHANNEL MODULATORS/ACE INHIBITORS** ['straddles' #2, #4 & #5; (anti-)dopaminics are in #5b]
4. **CARDIOVASCULAR AGENTS** incl. anti-thrombotic/-cholesterol, diuretic, platelet/AII-directed, inotropic; some cytokines
5. **NEUROACTIVE AGENTS** [& see #1 for some analgesics, #3 for (anti)adrenergics, #7 for smooth-muscle relaxants]:-
   5a: 'damping-down' action, esp. anaesthetic (incl. local), narcotic/analgesic, hypnotic/sedative; anxiolytic/nootropic; **anti**-convulsant/-emetic/-insomnia/-migraine/-spasmodic
   5b: various, incl. stimulatory/antidepressant, psychotic, (anti-)cholinergic/dopaminic; anti-MAO/-5-HT; memory-aiding
6. **ANTI-INFECTIVE/ANTI-CANCER AGENTS**:-
   6a: some structural groups [A-to-Z within each]: #cef/ceph/cillin/penem/β-lactamase-inhibitory; #oxacin; #nucleoside-type
   6b: various [immuno-modulatory in #7]
7. **UNCLASSIFIED**: e.g. anti-diabetic/-ulcer/-CCK [anti-emetic in #5a]; hormone-related, immuno-modulatory, smooth-muscle relaxant

----

Classes featured in **earlier vols.** [info. source on p. vi]: V.14: #6; V.16: #5 etc. [includes peptides]; V.18, #1, #3 & #4; V.20: #6; V.22: #2; V.23: #4 (hypolipidaemics).

At end of each A-Z listing are agents having a **company-related number; generic names** may later appear in Merck [new edn. with a CD-ROM option due mid-1996] or Martindale. A reference book, publ. BPC/HMSO (with supplements): *British Approved Names 1994*.

#ABC entries (mainly 1994/95), not 'quality-controlled', bear on automation, on drug efficacy/metabolism/safety and peripherally on therapeutic monitoring, but not on screening (e.g. forensic). Generally the sample is plasma/serum (**blood** = whole blood) and - unclear in some cited arts.! - from humans.

UP-DATING SEARCH recommendation.- For a listed paper, ascertain in *Science Citation Index* (by first author) whether any later paper cites it; also seek the analyte in *Anal. Abs.* and in post-**B 622** vols. (a Cumulative Index may appear) of *J. Chromatog.*

LAYOUT AND ABBREVIATIONS KEY (& see p. vi) *Mostly human plasma samples*

Prefixes: *: noteworthy, e.g. methods discussion or technique exploration (maybe a 'model' compound); †: pro-drug; °: peptide. For metabolites, **mtb(s)**, there may be an investigative slant (**invstg.**; **radio** if with isotope use); Phase I is denoted by [1], OR *a bold-type* **[1]**/[1] oxid. if including N-des/oxidation products; [2] denotes Phase II if conjugates studied intact, not routinely **hydrol.** = hydrolyzed. Enantiomer studies/agents: **chiral.**

LE, liquid-liquid extraction; **pptn.** = protein precipitation. (Commonly - in SPE too - dry down and redissolve; not listed.)

Automation denoted **auto;** **swtch** = switching, maybe implicit. Types of column (col.)/cartridge packing include: C-18, CN, AGP (α_1-acid glycoprotein); cIEx = cation-exchange, aIEx, anion-.

Solutions (aq., aqueous; **grad.** = gradient): e.g. AcOH, acetic acid; acet., acetate; iPrOH, isopropanol; MeOH, methanol; MeCN, acetonitrile; phos., phosphoric acid/phosphate; TEA, Et_3N.

Derivatization/conversion: **deriv** (post-column if so listed). Where **nm** value given, HPLC detection was by UV (or visible).

Approx. sensitivity (LLOQ or LOD): **sens.**; PK, pharmacokinetic use.

References: (95) = 1995, etc.; J = *Journal;* Chr = *Chromatog.* (thus J LqChr = *J. Liq. Chromatog.*); Ph = *Pharm(aceutical)* as in J PhSci or J PhBmAn (Bm, *Biomed.;* **An**, *Anal.*). AtMcCh, AtMcAgCh:[ᵅ] *Antimicrob. (Agents) Chemother.* Bch Ph'col, *Biochem. Pharmacol.;* Chr'ia, *Chromatographia;* HRC, *J. High Resolu. Chromatog.*

[ᵅ] was An... not At... in Vols. 22 & 23

1. MUSCULO-SKELETAL AGENTS incl. NSAID, analgesic [some in #5a], anti-inflammatory, skeletal-muscle relaxant

Tabulations of HPLC and MS behaviour for **NSAID's** accompany method descriptions for **flunixin** in horse urine (sens. 10 ng/ml). Stanley, S.M.R. et al. (95) J Chr **B 667**, 95-103.

Auranofin (rabbit; urine): C-18: aq. MeOH; -SH liberation, then deriv; 412 nm; sens. <0.1 µg/ml. Kizu, R. et al. (93) Chem Ph Bull **41**, 1261-1265.

Cyclobenzaprine (incl. urine): LE; C-18: MeCN/formate/acet.; MS-MS [& UV for comparison; inferior]; sens. 0.1(urine 10) ng/ml. Constanzer, M. et al. (95) J Chr **B 666**, 117-126.

° A **cysteine proteinase inhibitor** [peptidomimetic; carbamoyl groups], ester & free acid (mouse incl. muscle): pptn. (acetone); SPE; C-18: MeOH/NH_4-TCA; pH 3.1; MS(APcI). Kai, M. et al. (93) J Chr **B 621**, 181-187.

Diclofenac [& see **Nitrofenac**, BELOW].- (1): rapid method, incl. oxid. [1]**mtbs**: LE; C-18, grad.: MeCN/ THF/pH 7 phos.; 280 nm; sens. 5 ng/ml. Schmitz, G. et al. (93) J Chr **620**, 158-163. (2): pptn. (MeOH/Zn^{2+}); C-18: MeCN/pH 2.8 phos./SDS; 280 nm; sens. 30 ng/ml. Mason, J.L. & Hobbs, G.J. (95) J LqChr **18**, 2045-2058.

Ibuprofen.- (1; rabbit; erythrocytes similar to plasma): LE; C-18: MeOH/pH 3 PCA; 222 nm; sens. suffices for PK. Sochor, J. et al. (95) J PhBmAn **13**, 899-903; & same authors & method (but SPE; blood): J Chr (94) **B 654**, 282-286. (2): LE; **chiral** col. (β-cyclodextrin): MeCN/TEA pH 4; UV; sens. 1 μg/ml. Naidong, W. & Lee, J.W. (94) J PhBmAn **12**, 551-556. *(3; rat): SPE (diatomaceous earth); **chiral** deriv; C-18: MeCN/acetic acid; fluor; sens. ~0.3 μg/ml. Kondon, J. et al. (94) BmChr **8**, 170-174.

Ketorolac.- (1): LE; AGP (**chiral**): pH 7 phos./iPrOH; 320 nm; sens. 5 ng enantiomer/ml. Jones, D.J. & Bjorksten, A.R. (94) J Chr **B 661**, 165-167. *Cf. (2): Mills, M.H. et al., **B 658**, 177-182; similar (but pH 5.5 phos.); freezing steps during LE helped extract protein-bound drug.

Methocarbamol: LE; CN & **Chiracel** cols.: hexane/EtOH; 272 nm; sens. 0.5 μg/ml (each enantiomer). Naidong, W. et al. (94) J LqChr **17**, 3737-3758; also J Chr **B 654**, 288-292: LE; C-18: MeOH/pH 3 phos. (not chiral).

Mivacurium isomers & (with assay variations) [1]**mtbs**: SPE; C-18: aq. MeCN/ octanesulphonate; fluor; sens. ~10 ng/ml. Lacroix, M. et al. (95) J Chr **B 663**, 297-307.

Nitrofenac (rat; PK; similar for **Diclofenac**): pptn. (MeCN); C-8: MeCN/ pH 3 AcOH; 280 nm; sens. 0.1 μg/ml. Benoni, G. et al. (95) J PhSci **84**, 93-95.

Penicillamine [& **Captopril**] (incl. liver & kidney): pptn. (TCA); SPE if a tissue; C-18, grad.: phos. pH 2.8 at start/octyl sulphate/EDTA; EC with a graphite electrode; sens. unclear. Wakabayashi, H. et al. (94) J PhBmAn **12**, 1147-1152.

Pranprofen & mtb: pptn. (MeCN); C-18: MeCN/phos. acid; then **Chiracel** OJ col.: hexane/iPrOH/acetic acid; fluor; sens. 10 ng/ml. Nōmura, T. et al. (93) Biol Ph Bull **16**, 298-303.

Quercetin: listed in #6b class.

* **Rocuronium & [1]mtbs** (invstg; incl. tissue homogenates): LE; C-18: aq. dioxan/deriv agent; LE (auto); fluor; sens. 10 ng/ml. Kleef, U. et al. (93) J Chr **B 621**, 655-676.

Tenoxicam (light-sensitive): pptn. (Zn^{2+}); C-18: pH 2.8 phos./MeCN/ lauryl sulphate; 355 nm; sens. 40 ng/ml. Mason, J.L. & Hobbs, G.J. (95) J Chr **B 665**, 410-415.

* **Verkulast** [LT blocker]: pptn. (MeCN; **robotics**); C-18: MeOH/pH 4.5 phos.; fluor; sens. 0.1 μg/ml. Hsieh, J.Y-K. et al. (94) J Chr **B 661**, 307-312.

DuP 697 [thiophen ring] & [1]**mtb** (incl. urine): LE; C-18: aq. MeCN; fluor; sens. 3 ng/ml. Joshi, A.S. et al. (94) J Chr **B 660**, 143-150.

---

2. RESPIRATION-EASING AGENTS/ANTI-ALLERGICS: bronchodilator, mucolytic [expectorant in #7]; anti-histamine/-inflammatory [cf. #1]

* **'AZE'** [an azepin-phthalazinone; anti-histamine/-LT] used as model compound: pptn. (TCA); **chiral** (OV) col.: MeOH/pH 5 acet.; frit-FAB MS; sens. <1 ng/ml. Mano, N. et al. (94) J PhBmAn **12**, 557-567.

Chlorpheniramine: LE; C-18 (2 cols., swtch): pH 3 phos./MeCN & (1st col.) octanesulphonate; 210 nm; sens. 0.5 ng/ml. Yamaguchi, M. et al. (94) J Chr **B 661**, 168-172.

(+)-**'MPPB'** [a piperidinylbutyric acid] & [1]**mtbs**, invstg (rat, slow & fast metabolisers; urine & bile): TSK gel col.: acet./MeCN; MS(frit-FAB) to identify. Takahara, E. et al. (94) J Chr **B 658**, 154-160.

For **KEY** (including Journal abbreviations) see p. 162

Nedocromil sodium (urine; swtch): C-18 (to concentrate): acidified MeOH; then aIEx: pH 3 phos./MeOH; 253 nm; sens. 20 ng/ml. Baker, P.R. et al. (95) J Chr **B 668**, 59-65.

Quercetin: listed in #6b class

Salbutamol.- (1): SPE (NP); C-8: MeOH/MeCN/phos./octanesulphonate; fluor; sens. 0.2 ng/ml. Gupta, R.N. et al. (94) J Chr **B 654**, 205-211. (2; mouse, incl. urine): SPE; **chiral** col. (Chirex 3022): solvent mixture/TFA; fluor; sens. 0.25 ng enantiomer/ml. Boulton, D.W. & Fawcett, J.P. (95) J Chr **B 672**, 103-109.

* **BAY x 7195** [anti-LT_4D_4]: load direct, or (more sens.) pptn. (MeCN) & SPE; C-18, grad.: aq. MeCN (acidified); irradiate (deriv); fluor; sens. 10 μg/ml if direct loading. Heinig, R. (95) J Chr **B 667**, 137-147.

BN 50727 [anti-PAF; a triazolodiazepine] (auto, swtch): SPE; C-18, grad.: MeCN/pH 7 acet.; 250 nm; sens. ~3 ng/ml. Prunonosa, J. et al. (95) J Chr **B 668**, 281-290. Likewise anti-PAF: **BN 50730** [heterocyclic] & ^{1}**mtbs** (incl. urine): SPE; C-18: aq.MeCN; MS(spray; nCI); sens. 1 ng/ml. Girault, J. et al. (94) Chr'ia **39**, 228-238.

CDRI 73/602 [pyrrole & quinazoline rings; anti-allergic]: LE; C-18: MeCN/ pH 7 phos.; fluor; sens. 2 ng/ml. Singh, S.K. et al. (94) J Chr **B 658**, 198-201.

MK 0476 [light-sensitive]: pptn. (MeCN); C-18 (40°): MeCN/phos. pH 3.5; fluor; sens. 5 ng/ml (PK). Amin, R.D. et al. (95) J PhBmAn **13**, 155-158.

MKS 213-492 [a xanthine]: SPE (pre-col.; swtch); C-18: MeOH/phos./ citrate pH 4.5; EC; sens. ~0.2 μg/ml. Berliz, G. et al. (95) J Chr **B 666**, 368-373.

MX 2/120 [alkylated xanthine] (guinea-pig): SPE; C-18: pH 3.7 phos./ MeCN; 280 nm; sens. 50 ng/ml. Agostoni, O. & Toja, E. (94) J Chr **B 661**, 313-318.

----

3. (ANTI)ADRENERGICS/CHANNEL MODULATORS/ACE INHIBITORS

['straddles' #2, #4 & #5; (anti-)dopaminics are in #5b]

Alfuzosin: direct onto C-18; swtch; CN col.: MeCN/MeOH/pH 2.5 phos.; fluor; sens. 1 ng/ml. Carlucci, G. et al. (94) J LqChr **17**, 3989-3997.

Amlodipine: SPE; phenyl col.: pH 4 acet./MeOH/SDS/EDTA; EC; sens. <0.5 ng/ml. Jofesson, M. et al. (95) J Chr **B 672**, 310-313.

Atenolol (incl. urine): SPE; C-18: pH 3 phos./diMe-octylamine/MeCN; fluor (N/A when interference problems overcome) & simultaneous 224 nm; sens. <50 ng/ml. Chatterjee, D.L. et al. (95) J LqChr **18**, 791-806.

Captopril: SEE #2, Pencillamine entry

* **Diltiazem & ^{1}mtbs** (auto; lability/gel-formation/interference problems): SPE (ASPEC); C-8: pH 2.9 phos./MeCN/octylamine; 238 nm; sens. 2 ng/ml. Method robust. Ascalone, V. et al. (94) J Chr **B 657**, 133-140.

Dobutamine [a catechol]: pptn. (TCA) or LE; C-18: MeCN/pH 3 phos./ octanesulphonate; EC; sens. 1 ng/ml. Husseini, H. et al. (93) J Chr **B 620**, 164-168.

Felodipine & ^{1}mtb: LE; **Chiracel** OJ col.: iPrOH/hexane; GC-MS; sens. <0.5 ng/ml. Dru, J.D-Y. et al. (95) J Chr **B 666**, 259-267.

Higenamine [a catechol] (rabbit; incl. urine): adsorb onto acid alumina; C-18: MeCN/pH 2.5 phos.; EC; sens. 3/(urine 11) ng/ml. Lo, C-F. & Chen, C-M. (94) J Chr **B 655**, 33-39.

Idrapril (rat & man, incl. urine; & phos. homogenates of rat lung/ kidney/ heart): pptn. (MeCN) or (urine) SPE; C-18: pH 3 phos./MeCN/MeOH; EC; sens. 12 ng/ml plasma. Lippi, A. et al. (94) J Chr **B 660**, 127-134.

Labetalol [diastereoisomers]: LE; C-18: pH 11.5 DEA/MeCN/aq. NaCl; fluor; sens. 3 ng/ml (each). Grellet, J. et al. (94) J Chr **B 652**, 59-66.

* **Lacidipine**: SPE (auto); C-18: MeOH/MeCN; RIA; sens. 40 pg/ml. [Cf. LC-MS: Vol. 22, p. 31.] Braggio, S. et al. (95) J Chr **B 669**, 383-389.

Metoprolol & oxid. [1]**mtbs** (urine): LE (or, for mtb, form Me ester, & SPE); **chiral** deriv; C-18, grad: pH 3.2 phos./MeOH; fluor; sens. ~1 μg/ml. Li, F. et al. (95) J Chr **B 668**, 67-75.

Mibefendil (dog, incl. urine): LE; C-18: pH 3 phos./MeCN/pentane-sulphonate; fluor; 0.5 ng/ml. Skerjanec, A. & Tam, Y.K. (95) J Chr **B 669**, 377-382.

†**Moexipril & Moexiprilat (mtb)**: SPE; deriv; GC-MS; sens. 0.5 ng/ml. Hammes, W. et al. (95) J Chr **B 670**, 81-89.

Nadolol.- (1): SPE (non-end-capped CN); **chiral** deriv; LE; SPE (phenyl-boronic); C-18: pH 3 TEMED/MeOH/THF; fluor; sens. 2 ng/ml. Belas, F.J. et al. (95) BmChr **9**, 140-145. (2; dog): largely similar, but C-18 directly after **chiral** deriv. Hoshino, M. et al. (94) J Chr **B 661**, 281-289. (3): LE; **chiral** deriv; C-18: MeOH/THF/phos. acid pH 3; 230 nm; sens. <2 ng/ml. Srinivas, N R. et al. (95) BmChr **9**, 226-228.

Nifedipine.- (1): LE; C-18: MeOH/pH 4 acet./TEA; 350 nm; sens. 5 ng/ml. Grundy, J.S. et al. (94) J Chr **B 654**, 146-151. (2): similar, but 247 nm: artefacts discussed. Thongndpnua, P. et al. (94) J PhBmAn **12**, 119-125. (3): LE; GC-MS(EI); sens. 2 ng/ml. Martens, J. et al. (94) J Chr **B 660**, 297-302.

Nimodipine, chiral OJ col.; MS. Mueck, W.J. (95) J Chr **A 712**, 45-53.

* **Nisoldipine** (dog/rat/mouse): LE; **chiral** cellulose-type col.: heptane/ iPrOH/TFA; GC-MS on enantiomer fractions; sens. 0.5 ng enantiomer/ml. Zimmer, D. & Muschalek, V. (94) J Chr **A 666**, 241-248.

Oxprenolol (rat): LE; **chiral** deriv; C-18: MeOH/THF/pH 3.6 acet.; fluor; sens. 2.5 ng/ml. Laetyahm, M.E. et al. (93) J Chr **B 621**, 225-229.

Prazosin.- (1): phenyl col.: MeCN/pH 3.3 phos.; 394 nm; sens. 0.5 ng/ml. Niazy, E.M. et al. (95) J LqChr **18**, 977-987. (2): pptn. (MeCN); C-18: pH 8 phos./MeCN; EC; sens. 2 ng/ml. Rathinavelu, A. & Malave, A. (95) J Chr **B 670**, 177-182.

Propranolol (incl. urine): LE; **chiral** deriv; C-18: aq. MeOH; fluor; sens. 1 ng enantiomer/ml. Pham-Huy, C. et al. (94) J PhBmAn **12**, 1189-1198; also 'up-date' (95) J Chr **B 665**, 125-132: **chiral** β-cyclodextrin col.

†**Salmeterol xinafoate mtb**, 1-OH-2-naphthoic acid: SPE (auto); C-18: pH 5 phos./MeOH; fluor; <10 ng/ml. Chilton, A.S. et al. (95) J PhBmAn **13**, 165-169.

Sotalol.- (1; auto; PK): C-18 cols. (swtch): pH 6 phos./MeCN/SDS; 227 nm; sens. 10 ng/ml. Hermann, R. (95) J PhBmAn **13**, 329-333. (2): LE; **chiral** deriv; C-18: aq. MeCN; fluor; sens. <10 ng/ml. Hooper, W.D. & Baker, P.V. (95) J Chr **B 672**, 89-96.

SCH 51866 [cGMP PDase inhibitor]: pptn. (MeCN); CE (pH 2.5); 210/280 nm; sens. 0.5 μg/ml. Wang, C.J. et al. (95) J Chr **B 669**, 141-148.

UR 8225 [pyridine ring, etc.; K^+ channel blocker] (rat; auto): SPE; C-18: MeCN/pH 3 Bu_2NH-phos.; 236 nm; sens. 2.5 ng/ml. Nieto, C. et al. (94) J Chr **B 661**, 319-325. _____________

COMPENDIUM CONSULTATION GUIDANCE

Mostly human plasma samples

- Excerpts from **KEY** on pp. 162 (full version) & 168: **mtb**, metabolite; LE, liquid-liquid extraction; MeCN, acetonitrile; EC, electrochemical detection. **Journals**: Bm, Biomed.; Chr, Chromatog.; Ph, Pharm.
- For some **chemical features**, consult Analyte Index.
- COMPENDIUM scope & therapeutic groupings: p. 161

Prefixes:
* notable interest
† pro-drug
° peptide

Note: cardiovascular agents appear in Group #3 as well as #4.

4. CARDIOVASCULAR AGENTS incl. anti-thrombotic/-cholesterol, diuretic, platelet/AII-directed, inotropic; some cytokines

Acenocoumarol: LE; GC-MS - but an LC-MS method could replace. Pommier, F. et al. (94) J Chr **B 654**, 35-41.

* **Acetazolamide** (urine): C-18 precol.; pH 3 wash; swtch; C-18: aq. MeCN, grad.; 275 nm; sens. 10 ng/ml. Campins-Falco, P. et al. (94) J Chr **B 654**, 85-90. A 'model' **medium polarity drug** unsuited to SPE; circumvented by the pH 3 (phos.) washing-out of polar interferants.

Alcuronium (incl. urine): pptn. (MeCN); CN: aq. MeCN/SO_4^{2-}; 294 nm; sens. 25 ng/ml (apt for PK). Kunzer, T. et al. (94) J Chr **B 653**, 63-68.

Amiodarone.- (1; incl. aq. EtOH homogenates of tissues): pptn. (MeCN); C-18: pH 4 phos./MeCN/MeOH; 242 nm; sens. 20 ng/ml. Al-Dhawailie, A.A. (95) Anal Lett **28**, 2391-2400. (2; [1]**mtb** also): LE; NP (silica) col.: MeOH/amm.sulphate; 254 nm; sens. ~0.1 µg/ml. Manfredi, C. et al. (95) Int J Clin Pharmacol Res **15**, 87-93.

Amrinone.- (1; [1]**mtb** also): SPE; C-18: MeCN/THF/$BuNH_2$/pH 6 phos.; 254nm; sens. 1 µg/ml. Laganière, S. & Goernet, L. (94) J PhBmAn **12**, 407-411. (2): LE; CN: pH 3.2 phos./MeCN; 210 nm; sens. 0.25 µg/ml. Bansal, R. et al. (94) J LqChr **17**, 3531-3539.

Artilide: SEE Ibutilide entry

A **bicyclononane, thia- & aza-substituted** [novel antiarrhythmic] & oxid. [1]**mtb** (incl. hydrol. urine): LE; C-6: MeCN/MeOH/pH 6.8 phos.; 261 nm; sens. 0.1 µg/ml. Chen, C.L. et al. (94) J LqChr **17**, 3681-3694.

Bosentan [Ro 47-0203; endothelin receptor antagonist]: pptn. (MeCN); LE (lipid removal); C-18 (35°): MeCN/acet.; 270 nm & MS(ISP); sens. 0.5 ng/ml. Lausecker, B. et al. (95) J Chr **A 712**, 75-83.

* **Brimonidine**: LE (3-step); deriv; GC-MS(CI); sens. 2 pg/ml (notable). Acheampong, A. et al. (95) J PhBmAn **13**, 995-1002.

Clofibric acid: SEE Ethyl clofibrate entry

8-**Cyclopentyltheophylline** [adenosine A_1 agonist] (rat): LE; C-18: pH 4 acet./MeOH/MeCN; 269 nm; sens. 3 ng/ml. Mathot, R.A.A. et al. (93) J Chr **B 620**, 113-120.

Difenacoum (incl. liver): pptn. (MeCN); C-18, grad.: aq. MeCN/AcOH; make alkaline [cf. Vol. 23, p. 229]; fluor; sens. 5 ng/ml. Kelly, M.J. et al. (93) J Chr **B 620**, 105-112.

* **Digoxin & [1]mtbs**: SPE (β-cyclodextrin); deriv; SPE steps (β-cyclodextrin & C-1); NP col.: solvent mixture; fluor; sens. <0.2 ng/ml. Tzou, M-C. et al. (95) J PhBmAn **13**, 1531-1540.

Disopyramide: C-18: MeCN/MeOH/pH 3.2 phos.; 269 nm; sens. <1 µg/ml. Witek, A. et al. (94) J PhBmAm **12**, 425-427. SEE ALSO Masuhara, K. et al. (95) Int J Clin Pharmcol Res **15**, 103-113: sens. 0.1 µg/ml.

* **Dopexamine** (blood; swift addition of metabisulphite): pptn. (PCA); silica pre-col.; C-18: MeOH/phos. acid/EDTA/heptanesulphonate/diisopropylamine; EC; sens. 5 ng/ml. Baker, P.R. et al. (95) J Chr **B 667**, 283-290.

Ethmozine: SEE Moricizine entry

Ethyl clofibrate & clofibric acid (rat & man; stability of nanocapsules in plasma invstg): pptn. (TFA/MeCN); C-18: aq. MeCN; fluor; sens. 1 µg/ml. Barras, J. et al. (95) J Chr **B 665**, 178-182.

For **KEY** see p. 162 or 168

* †**Fenoldopam N-ethyltricarbamate** [SK&F 105058]: SPE; C-18: MeOH/AcOH; dried fraction onto C-8, 35°: pH 6 acet./MeOH; fluor after hydrol./OPA deriv; sens. 5 ng/ml. A general approach for **carbamate-type prodrugs**. Miller-Stein, C. et al. (94) J Chr B **661**, 291-297.

Furosemide & [2]**mtb** (incl. urine): pptn. (MeCN); C-18, grad.: MeCN/pH 2.1 phos.; fluor; sens. 5 µg (glucuronide: 10 ng)/ml plasma. Vree, T.B. et al. (94) J Chr B **655**, 53-62.

Ibutilide & Artilide: SPE; CN cols. to concentrate (swtch); C-8 & Pirkle (**chiral**) cols.: appropriate eluents; fluor; sens. 0.1 ng/ml. Hsu, C-Y.L. & Walters, R.R. (95) J Chr B **667**, 115-128.

Losartan & active acid [1]**mtb** [tetrazole/imidazole rings] (rat): C-8 (swtch) & C-18: MeCN/pH 3.5 acet.; 254 nm; sens. 20 ng/ml. Lee, H. et al. (96) Chr'ia **42**, 39-42.

Methocarbamol: LE; C-18: MeOH/phos.; 272 nm; sens. 1 µg/ml (apt for PK). Naidong, W. et al. (94) J Chr B **654**, 287-292.

Metolazone (blood or plasma): SPE (ASPEC); C-18: pH 3 phos./MeCN; fluor; sens. 1 ng/ml. Farthing, D. et al. (94) J Chr B **653**, 171-176.

* **Mexiletine**.- (1): LE; deriv (OPA); **Chiralpak** AD col.: hexane/iPrOH; fluor; sens. 20 ng/ml. Fieger, H. & Wainer, I.W. (93) J PhBmAn **11**, 1173-1179. (2): pptn./LE ($Ba^{2+}/Zn^{2+}/Et_2O$); deriv; LE; Phe-Gly **chiral** col.: solvent mixture; fluor; 2.5 ng/ml (each enantiomer). For $-NH_2$ deriv step: 2-anthroyl chloride. Kwok, D.K.W. et al. (94) J Chr B **661**, 271-280. (3): LE; deriv; GC-MS(SIM); sens. 0.02 ng/ml. Minnigh, M.B. et al. (94) J Chr B **662**, 118-122.

Moricizine (Ethmozine) & oxid. [1]**mtbs** (incl. bile & urine): LE; C-18: aq. MeOH/TEA pH 6.7; 254 nm; sens. 20 ng/ml. Xang, J.M. & Chan, K. (95) J Chr B **663**, 172-176.

Omeprazole: SPE; **chiral** (BSA) col.: pH 7 phos./PrOH; 302 nm; sens. <15 ng/ml. Cairns, A.M. et al. (95) J Chr B **666**, 323-328.

Phenprocoumon & acid [1]**mtbs** (incl. milk & hydrol. urine): SPE; C-18: aq. AcOH/MeCN; 312 nm; sens. ~0.1 µg/ml. De Vries, J.X. & Schmitz-Kummer, E. (94) J Chr B **655**, 63-71.

* **Pravastatin**: SPE (immobilized Ab); deriv; C-18, 2 cols. (swtch): MeCN/citric ac. pH 2.4 then 2.6; fluor (laser); sens. <0.1 ng/ml. Dumousseaux, C. et al. (94) J PhSci **83**, 1630-1636. [Cf. Vol. 23, p. 127.]

Probucol: LE; C-18: aq. MeCN; 241 nm; sens. 0.5 µg/ml (apt for PK). Nourooz-Zadeh, J. et al. (94) J Chr B **654**, 55-60.

Quercetin: listed in #6b class

†**Quinapril/Quinaprilat** & other [1]**mtbs** (kidney perfusates & dog urine): pptn./ultrafilter; C-8; UV/radio. Kugler, A.R. et al. (95) J Chr B **666**, 360-367. [Useful pointers to earlier lit.; cf. kindred drugs in Vol. 18.]

Rifamycin-related drug, CGP 43371 (incl. urine & faeces): LE; NP col.: isooctane etc.; 354 nm; sens. <20 ng/ml. Maniara, W.M. et al. (94) J Chr B **660**, 135-142. Cf. Vol. 23 [this series], p. 296.

Ticlodipine: LE; C-8: MeCN/MeOH/pH 3 phos./TEA; 235 nm; sens. 5 ng/ml. Dal Bo, L. et al. (95) J Chr B **665**, 404-409.

Tocainide (rabbit): SPE; **chiral** deriv; C-18: MeOH/pH 5 phos.; 262 nm; sens. <1 µg/ml. Christensen, E.B. et al. (95) J Chr B **670**, 243-249.

Triamterene (urine): C-18 cols. (swtch): MeCN/pH 3 phos., grad.; fluor; sens. 5 ng/ml. Campins-Falco, P. et al. (94) Chr'ia **38**, 29-34.

* **Verapamil &** [1]**mtb**.- (1): LE; **chiral CE**: pH 2.5 phos./Me_3-β-cyclodextrin; 200 nm; sens. 2.5 ng/ml. Dethy, J-M. et al. (94) J Chr B **654**, 121-127. (2): LE; deriv; AGP **chiral**: pH 4.8 phos./MeCN; fluor; sens. <2 ng enantiomer/ml. Stagni, G. & Gillespie, W.R. (95) J Chr B **667**, 349-354. (3; **mtbs**, invstg; dog urine): LE; EITHER: C-18, grad.: pH 4.5 acet./MeCN; OR **chiral** col. (OVM): EtOH/pH 7 phos; fluor. LC-MS(TSP) & (to identify) FAB-MS. Lankford, S.M. & Bai, S.A. (95) J Chr B **663**, 91-101.

* **Zatebradine.**- (1; PK, invstg): RP pre-col.; swtch; NP col.: aq. piperidine; fluor & radio. Roth, W. et al. (93) J PhSci **82**, 99-106. (2; mouse & man; auto): SPE (C-2; robotics); GC-MS(CI); sens. 0.2 ng/ml. Schmid, J. et al. (94) J Chr **B 658**, 93-101.

CGS 21680 [adenosine receptor agonist] (blood; auto): LE; SPE (bonded phenylboronic acid; swtch); C-18: MeCN/pH 4 acet.; fluor; sens. 5 ng/ml. Irth, H. et al. (94) J Chr **B 658**, 207-212.

CI-976 [ACAT inhibitor; a substituted dodecanamide] (rat): LE; C-18: MeCN/ AcOH; 250 nm; sens. 0.06 µg/ml. Shum, Y.Y. et al. (94) J Chr **B 653**, 205-209.

°* **DMP 728** [a platelet glycoprotein receptor antagonist; peptide core] (rat & dog): SPE (C-2); deriv; C-8 **SPS**: MeCN/pH 7.4 phos.; fluor; sens. 5 ng/ml. Wu, S.T. et al. (94) J Chr **B 657**, 254-260.

DMP 871 [platelet receptor blocker; tetrazole & imidazole moieties] (dog & rat): LE; C-8: pH 4 acet./MeCN; fluor; sens. 2 ng/ml. Agra, A.M. & Huang, S-M. (93) J Chr **B 620**, 205-209.

ER 017996 [a quinazoline; cGMP phosphodiesterase blocker] (dog): LE; C-18: MeCN/ClO_4^-; 330 nm; sens. 1 ng/ml. Tokumura, T. et al. (94) J Chr **B 658**, 202-206.

* **GR 117289X** [N-heterocyclic/-COOH; angiotensin receptor antagonist] (auto, robotics): SPE (mixed mode); C-18: MeCN/pH 4.1 acet.; 300 nm; sens. 5 ng/ml. Murkitt, G.C. et al. (94) J PhBmAn **12**, 1181-1187.

MK 0499 [antiarrhythmic] (incl. urine): RIA or (compared) LE, then LC-MS-MS (API); sens. <0.05 ng/ml. Gilbert, J.D. et al. (95) J PhBmAn **13**, 937-950.

OPC-18790 [inotropic; a quinolinone] (incl. urine): LE; chiral deriv; C-18: aq. MeCN; fluor; sens. 0.01(urine 0.1) µg/ml. Kitani, M. et al. (93) J Chr **B 620**, 97-107.

* **Ro 42-5892** [renin inhibitor] (auto): LE; deriv (ASPEC); C-18: pH 7 acet./MeCN; irradiate; EC; sens. 0.3 ng/ml. Leube, J. et al. (95) J Chr **B 665**, 373-381.

RP 73163 [ACAT inhibitor; pyrazole & imidazole moieties] (rat; PK): SPE (C-2); C-18, 40°: pH 6.5 phos./TBA/MeCN; fluor; sens. 7 ng/ml. Jones. G.D. et al. (94) J Chr **B 661**, 119-131.

TCV 116 & ¹mtbs [AII receptor antagonist] (incl. urine): LE; C-18: MeCN/pH 4 phos.; 229 nm; sens. <2 ng/ml. Different LE & C-18 conditions for mtbs. Lee, J.W. et al. (95) J Chr **B 670**, 287-298.

Prefixes: **KEY** (fuller version at start of #ABC) *Mostly human plasma samples*

* notable interest
† pro-drug
° peptide

JOURNALS:
Agents, Ag; Anti, At; Biomed., Bm; Chemother., Ch; Chromatog., Chr; Microbial, Mc Pharm., Ph

Metabolites, **mtb(s)**; **invstg** if investigated (radio denotes isotope used); Phase I: ¹ or, if including N-des, **¹ (bold)**; ² denotes Phase II if conjugates studied intact, not routinely **hydrol.** = hydrolyzed. Enantiomer studies/agents: **chiral**.
LE, liquid-liquid extraction; **pptn.** = protein precipitation. (Commonly - in SPE too - dry down and redissolve; not listed.)
Automation, **auto; swtch**, switching. Packings: e.g. C-18, CN, AGP (α_1-acid glycoprot.); cIEx = cation-exchange, aIEx, anion-. Gradient, gradient. AcOH, acetic acid; acet., acetate; iPrOH, isopropanol; MeCN, acetonitrile; phos., phosphate; TEA, Et_3N.
Derivatization/conversion: **deriv** (post-column if so listed).
Where nm value given, HPLC detection was by UV (or visible).
Approx. sensitivity (LLOQ or LOD): sens.; PK, pharmacokinetic use.

5. NEUROACTIVE AGENTS [& see #1 for some analgesics, #3 for (anti)adrenergics, #7 for smooth-muscle relaxants]

5a: 'damping-down' action, esp. anaesthetic (incl. local), narcotic/analgesic, hypnotic/sedative; anxiolytic/nootropic; **anti**-convulsant/-emetic/-insomnia/-migraine/-spasmodic

* **Amantadine (Adamantanamine** incl. urine; auto): 'SPE' & simultaneous deriv (immobilized reagent, in autosampler loop; SDS present; proteins unretained); RP col. (vague); fluor; sens. 15 ng/ml. Zhou, F-X. et al. (93) J Chr **B 619**, 93-101.

Alprazolam & oxid. **[1]mtbs** (rat; 100 μl only): LE; C-18: MeOH/MeCN/pH 2.4 acet.; 230 nm; sens. 5 ng/ml. Jin, L. & Lau, C.E. (94) J Chr **B 654**, 77-83.

Bromazepam: LE; C-18: MeCN/TEA; 254 nm; sens. 50 ng/ml. Le Solleu, H. et al. (93) J PhBmAn **11**, 771-775.

Bupivacaine & [1]mtbs incl. oxid.: SPE (precautions); C-8: pH 5.2 phos./MeCN; 205 nm; sens. 10 ng/ml. Gupta, R.N. & Dauphin, A. (94) J Chr **B 658**, 113-119. SEE ALSO **Ropivacaine** entry.

Butorphanol & [1]mtbs (urine): SPE (CN; cartridge re-usable); C-8: MeCN/MeOH/acet. pH 5.5/Me_4N^+; fluor (problems if EC); sens. 1 ng/ml. Willey, T.A. et al. (94) J Chr **B 652**, 171-178,

* **Carbamazepine** & oxid. **[1]mtb**: LE; C-18: aq. MeCN; 210 nm; sens. <0.1 μg/ml. Simple, and good accord with immunoassay (EMIT); GC more variable. Martens, J. & Banditt, P. (93) J Chr **B 620**, 169-173.

*° **Ceruletide** [a frog-skin decapeptide; CCK analogue] & shorter-chain **[1]mtbs** (dog, incl. urine): pptn (EtOH); C-18: MeCN/TFA trace, grad.; 220 nm & RIA; sens, 0.02/(urine 0.07) ng/ml. Kominami, G. et al. (94) J PhBmAn **12**, 413-418.

Chlorprothixene/Levomepromazine (Methotrimeprazine)/Promethazine: LE (elaborate); CN col.: MeCN/pyridine/pH 3.1 acet.; EC; sens. 0.5/0.2/0.1 ng/ml (apt for PK). Bagli, M. et al. (94) J Chr **B 657**, 141-148.

Citalopram & **[1]mtbs** (dog, rat; auto): 'SPE' (NP; in valve loop; swtch); C-18: pH 4.6 phos./MeCN/DEA; fluor; sens. 2 ng/ml. Matsui, E. et al. (95) J Chr **B 668**, 299-307.

Clonazepam/other nitrobenzodiazepines & [1]mtbs by HPLC: for LE, BuCl replaces Et_2O. Robertson, M.D. et al. (95) J Chr **B 667**, 179-184.

Clopenthixol: cf. Zuclopenthixol entry

Cyclizine & [1]mtb (incl. urine): pptn. (MeCN); SPE; C-18: pH 3 phos./MeCN; EC; sens. 1 ng/ml. Walker, R.B. & Kanfer, I. (95) J Chr **B 672**, 172-177.

° **'DADLE'** [peptide; a synthetic enkephalin] (dog; CSF): ultrafilter; C-18: pH 6 phos./MeCN, grad.; 210 nm; sens. 1 μg/ml. Cox, S. et al. (94) J Chr **B 657**, 243-247.

Diazepam [1]mtbs (hydrol. urine): subtle LE steps; C-18: aq. MeOH/TEA; 240 nm; sens. 2 ng/ml. Chiba, K. et al. (95) J Chr **B 668**, 77-84.

Dihydrocodeine & [12]mtbs (dog): SPE; C-18: phos./MeCN/SDS; EC (UV for glucuronide); sens. 2 ng/ml. Ohno, M. et al. (94) J Chr **B 654**, 213-219.

* **Dimiracetam** (polar! auto): ultrafilter; Bondapak cols. (swtch): water; 200 nm; sens. <0.03 ng/ml. Torchio, L. et al. (95) J Chr **B 666**, 169-177.

Dolasetron & **[1]mtb**: LE; deriv & GC-MS, **or** CN col.: MeCN/acet.; 280 nm; sens. 1 ng/ml. Gillespie, T.A. et al. (93) J PhBmAn **11**, 955-962.

†**Dolasetron** mesilate **[1]mtbs** (urine): SPE; C-4, grad.: MeCN/pH 4.7 acet.; 280 nm; sens. ~0.05 μg/ml. Sanwald, P. et al. (94) J Chr **B 661**, 101-107.

Estazolam: LE; GC-ECD; sens. 1 ng/ml. Kelly, J.F. & Greenblatt, D. (93) J Chr **B 621**, 102-104.

Famprotazone & [1]mtb (urine): LE; deriv; GC-MS; sens. <1 ng/ml. Shin, H-S. et al. (94) J Chr **B 661**, 255-261.

* **Felbamate**: pptn. (MeCN; obviates LE); CE (MECC); 214 nm; sens. 5 µg/ml. Tallied with LE & HPLC. Shihabi, Z.K. et al. (94) Clin Chem **40**, 1904-1908.

* †**Fenoldopam N-ethyl tricarbamate**: SPE; C-8: MeOH/pH 6 acet.; on-line deriv (OPA)/hydrol.; fluor; sens. <5 ng/ml. Miller-Stein, C. et al. (94) J Chr **B 661**, 291-297.

Fenoverine (rat & man; liver/muscle MeOH homogenates): C-18: MeCN/phos. (pH unstated); 214 nm; sens. 0.6 µg/g. Ratiney, R. et al. (94) J Chr **B 654**, 159-163.

Flumazenil (incl. dog): LE; GC-MS(APcI); sens. <1 ng/ml. Song, D. et al. (95) J Chr **B 663**, 263-273.

cis(*Z*)-**Flupentixol**: LE steps; GC-NPD; sens. 0.5 ng/ml. Ulrich, S. (95) J Chr **B 668**, 31-40.

* **Gabapentin**: pptn. (MeCN) combined with deriv; CE; 200 nm; sens. ~1 µg/ml. CE simpler & faster than HPLC or GC. Garcia, L.L. et al. (95) J Chr **B 669**, 157-162.

Granisetron & -OH ^{1}**mtb**: SPE (C-2); deriv; C-18: pH 4.7 acet./MeOH/TBA; fluor; sens. 0.2 ng/ml. Kudoh, S. et al. (94) J Chr **B 660**, 205-210.

Iloperidone: SPE; C-18: MeCN/pH 3.5 amm. formate; MS(ESP); sens. 0.2 ng/ml. Mutlib, A.E. et al. (95) J Chr **B 669**, 237-246.

Lamotrigine: SPE; C-18; 265 nm; sens. <0.5 µg/ml. Yamashita, S. et al. (95) J Chr **B 670**, 354-357.

Lesopitron & oxid. ^{1}**mtb** (auto): NP col.: pH 3 phos./MeCN; fluor; sens. 1 ng/ml. Garcia-Encina, G. et al. (95) J Chr **B 670**, 103-110.

Meptazinol & ^{1}**mtb** (urine & liver incubates): LE; CN col.: MeCN/MeOH/pH 5.5 acet./TEA; 220 nm; sens. 0.4 µg/ml. Parallel run with **chiral** col. (AGP; to get ratios): pH 7 phos./MeCN. Rudolphi, C. & Blaschke, J. (95) J Chr **B 663**, 315-326.

Midazolam (auto, swtch): C-18 (incl. pre-col.): pH 2.4 phos./MeCN, grad.; 230 nm; sens. 2 ng/ml. Lauber, R. et al. (94) J Chr **B 654**, 69-75.

Nalbuphine: LE steps; C-18: MeOH/pH 3.4 phos.; EC; sens. 0.5 ng/ml. Nicolle, E. et al. (95) J Chr **B 663**, 111-117.

* **Nitrazepam &** †**mtbs** (urine): CE (MECC); 220 nm; sens. 0.1 µg/ml. Tomita, M. et al. (93) J Chr **B 621**, 249-255.

° **'NT-1'** (ester) & ^{1}**mtbs** [hexapeptide] (dog): ptn. (MeOH); C-18: MeCN/PCA; fluor; sens. suits for PK. Tokumura, T. et al. (93) J PhSci **82**, 725-728.

Olanzapine: SPE; YMC basic col.: pH 7 phos./MeOH/MeCN; EC; sens. 0.2 ng/ml. Accord with GC-MS(nCI) method. Catlow, J.T. et al. (95) J Chr **B 668**, 85-90.

Oxcarbazepine & ^{1}**mtbs**: SPE; **Chiracel** OD or ODH col.: hexane/EtOH; 220 nm; sens. 5 ng/ml. Pichini, S. et al. (95) J LqChr **18**, 1533-1541.

Pazinaclone [an isoindoline] & active ^{1}**mtb** (rat): LE; C-18: aq. MeCN; fractions dried down; NP & **chiral** linked cols.: EtOH/hexane; fluor; sens. <0.5 µg/ml. Kondo, T. et al. (95) J Chr **B 666**, 291-297.

Pentazocine: LE; **chiral** (OVM) col.: MeOH/MeCN/pH 5.8 phos.; fluor; sens. 5 ng/ml. Kelly, J.W. et al. (94) BmChr **8**, 255-257.

* **Piracetam** (incl. urine): LE; C-8: aq. MeOH; 206 nm; sens. 3 µg/ml. Interferences minimized. Louchahi, K. et al. (95) J Chr **B 663**, 385-389.

Primidone & active ^{1}**mtbs** (rat): SPE; C-18: MeCN/pH 3 phos.; 210 nm; sens. 0.1 µg/ml. Simple. Moriyama, M. et al. (94) J PhSci **83**, 1751-1753.

Propofol.- (1): LE; C-18: MeCN/AcOH; UV; sens. 10 µg/ml. Some loss (discussed) if pptn. (MeCN/PCA) instead of LE. Dawidowicz, A.L. & Fijałkowska, A. (95) J Chr Sci **33**, 377-382. (2; incl. ^{1}**mtbs**; blood): LE; GC-MS; sens. 10 ng/ml. Guitton, J. et al. (95) J Chr **B 669**, 358-365.

Remifentanil (incl. mouse tissues; with human blood, suppress its esterase): RIA. Lin, J.R. et al. (94) J PhBmAn **12**, 145-150.

Remifentanil [1]mtb. (carbox. acid; GR 90291): LE; deriv; GC-MS; sens. 1 ng/ml. Lessard, D. et al. (94) J PhBmAn **12**, 659-665.

Ropivacaine & Bupivacaine: ultrafilter; 2 cols., RP and cIEx (swtch): MeCN/pH 2.6 phos.; 210 nm; sens. 3 ng/ml. Arvidsson, T. et al. (95) J Chr **B 668**, 91-98.

* **Sertindole:** SPE; NP col.: solvent mixture; fluor; sens. 0.01 ng/ml. Tzeng, T.B. et al. (94) J Chr **B 661**, 299-306.

†**Sumatriptan** succinate (incl. urine): C-18: pH 7 phos./MeOH; EC; sens. 1/(urine 200) ng/ml. Andrew, P.D. et al. (93) J PhSci **82**, 73-76.

* **Thiamylal:** SPE; C-18, **chiral** eluent: β-cyclodextrin/EtOH/pH 3 phos.; 288 nm; sens. 50 ng/ml. Sueyasu, M. et al. (95) J Chr **B 665**, 133-137.

Thioridazine & oxid. **[1]mtbs:** LE, & RP col. [published]; fractions onto **chiral** col. apt for each analyte. Eap, C.B. et al. (95) J Chr **B 669**, 271-279.

Zopiclone: LE; **Chiracel** col.: EtOH/hexane; fluor; sens. 1 ng/ml. Foster, R.T. et al. (94) J Chr **B 658**, 161-166.

* **Zuclopenthixol [clopenthixol's** *cis*-(*Z*)-isomer] & **[1]mtb** (incl. urine): SPE (CN; partly a cIEx mechanism); CN col.: MeCN/pH 6.5 phos./quat. amine; irradiate; fluor; sens. 0.05 ng/ml. Hansen, B.B. et al. (94) J Chr **B 658**, 319-325.

CGP 50068/55461 [enantiomers; a benzopyran Me ester] & free-acid **[1]mtb:** SPE; **chiral** col. (AGP): pH 7 phos./TBA; 270 nm; sens. 0.6 μg/ml. Mangoni, P. et al. (95) J Chr **B 664**, 393-400. For **CGP 33101** see **B 667**, 307-313.

CGS 19480A [anxiolytic]: LE; C-18: MeCN/pH 7 phos./MeOH; 218 nm; sens. 5 ng/ml. Stelling, M.S. et al. (94) BmChr **8**, 288-290.

* **CP 88059** [antipsychotic; dihydroindolone moiety]: SPE (cIEx); C-18, narrow-bore: MeCN/TFA/TEA; 215 nm; sens. <6 ng/ml (4-fold gain over 4.6 mm bore). Janiszewski, J.S. et al. (95) **B 668**, 133-139.

DuP 734 [a substituted piperidine] (rat): LE; C-8: MeCN/amm. acet./TEA; 229 nm; sens. 0.04 μg/ml. Kapil, R. (94) J Chr **B 653**, 221-226.

LY 300164 [a dioxolobenzodiazepine; AMPA receptor antagonist] (4 animal species): SPE; YMC basic col.: MeOH/pH 5 citrate/MeCN; 310 nm; sens. 0.1 μg/ml (apt for PK). Eckstein, J.A. & Swanson, S.P. (95) J Chr **B 668**, 153-158.

----

5. NEUROACTIVE AGENTS, continued

5b: various, incl. stimulatory/antidepressant, psychotic, (anti-)cholinergic/dopaminic; anti-MAO/-5-HT; memory-aiding

* **Amphetamine** & related compounds (urine): LE; GC, then heart-cut and **chiral** GC (cyclodextrin phase); FID. Poorer resolution if a Chirasil-Val phase. Wang, X. et al. (94) J Chr **B 653**, 98-101.

Dexfenfluramine: pptn. (TCA); **chiral** col.: solvent mixture; 235 nm; sens. 20 ng/ml. Zeng, J. et al. (94) J Chr **B 654**, 231-248.

Fenfluramine: SEE Dexfenfluramine entry

Imipramine & [1]mtb (mice; simple method): LE; C-18: MeCN/pH 3 phos./TEA; 260 nm; sens. <10 ng/ml. Yoo, S.D. et al. (95) J Chr **B 668**, 338-342.

Linopirine & oxid. **[1]mtb** (rat): LE; RP col.: MeOH/acet./TEA/octane-sulphonate; 254 nm; sens. <0.02 μg/ml. Rakestraw, D.C. et al. (94) J PhBmAn **12**, 1055-1061.

Modafinil & acid **[1]mtb:** LE; C-18: MeCN/AcOH; 236 nm; sens. 0.1 μg/ml. Moachon, G. & Matinier, D. (94) J Chr **B 654**, 91-96.

* **Naprotilidine:** LE; C-18 (TSK gel): MeCN/pH 3.2 acet.; post-col. reaction; chemilumines.; sens. 0.1 ng/ml. Ishida, J. et al. (95) J Chr **B 669**, 390-396.

Nefazodone & ^{1}**mtbs**: SPE (CN); CN col.: MeCN/MeOH/pH 6.6 phos.; EC; sens. ~1 ng/ml. Franklin, M. (93) J PhBmAn **11**, 1109-1113.

Paroxetine.- (1): LE; C-18: pH 4.5 acet./MeCN/EtOH/TBA; 295 nm; sens. 6 ng/ml (fluor improves, if for PK). Knoeller, J. et al. (95) J PhBmAn **13**, 635-638. (2): SPE; C-8: pH 3.2 phos./MeCN; fluor; sens. 1 ng/ml. Gupta, R.N. (94) J Chr **B 661**, 362-365.

Reboxetine: LE; **chiral** deriv; C-8: THF/pH 7.5 phos.; fluor; sens. 1 ng/ml Frigerio, E. et al. (94) J Chr **A [not B] 660**, 351-358.

Sertaline & ^{1}**mtb**: SPE (co-polymer); C-8: MeCN/MeOH/DEA pH 5.5-6; 214 nm; sens. 10 µg/ml. Rogowsky, D. et al. (94) J Chr **B 655**, 138-141.

Tacrine & oxid. ^{1}**mtbs**: LE; CN col.: MeCN/pH 4 acet.; fluor; sens. <0.5 ng/ml. Haughey, D.B. et al. (94) J PhSci **83**, 1582-1585.

* **Tranylcypromine** (urine, to find +/- ratio): subtle LE; **chiral** col. of crown-ether type: aq. MeOH/PCA; 235 nm. Aboul-Enein, H.Y. & Serignese, V. (95) BmChr **9**, 98-101.

Trimipramine & ^{1}**mtbs.**- (1): hydrol.; LE steps; deriv; GC-MS; sens. 2 ng/ml. Eap, C.B. et al. (94) J Chr **B 652**, 97-103. (2; auto): CPS col. (proteins unretained; swtch): pH 6.2 phos./MeCN/MeOH; 214 nm; sens. 10 ng/ml. Haertter, S. et al. (95) J LqChr **18**, 3495-3505.

Xanomeline & ^{1}**mtb** (active): LE; CN col.: pH 3 phos./THF/TEA; 296 nm; sens. 1 ng/ml. Kasper, S.C. et al. (95) J Chr **B 669**, 397-403.

CGP 36742 [amino/phosphinic acid moieties; GABA antagonist] (rat, incl. HCl homogenates of brain etc.): 2-stage deriv; GC-MS(nCI); sens. 0.5 ng/ml or 10 ng/g. Steulet, A-F. et al. (96) Biochem Pharmacol **51**, 613-619.

CI 1007 [dopamine receptor agonist] & ^{1}**mtb**: LE; C-18, grad.: MeCN/phos./TEA pH 3.5; fluor; <3 ng/ml. Feng, M.R. (95) J Chr **B 665**, 193-199.

MKS 213-492 [a xanthine]: C-18 pre-col. (swtch); C-8: MeOH/pH 4 phos.; EC; sens. <0.2 µg/ml. Berliz, G. et al. (95) J Chr **B 666**, 368-373.

X 9121 & oxid. ^{1}**mtb** [heterocyclic cognitive enhancer] (dog): SPE; CN col.: MeCN/acet./TEA; 313 nm; sens. <10 ng/ml. Kapil, R.P. et al. (95) J Chr **B 665**, 226-231.

---

6. ANTI-INFECTIVE/ANTI-CANCER AGENTS

[Compendium for this group in Vol. 20 as well as 22 & 23]

6a: some structural groups, firstly:

'ceph/cef'/'cillin'/penem/β-lactamase-inhibitory

Ampicillin: pptn. (TCA); deriv; C-18: MeCN/pH 5.6 phos.; fluor; sens. 2 ng/ml. Method simple. Lai, J. (94) J Chr **B 655**, 142-146.

Ampicillin & metampicillin (rat; incl. urine & bile): C-18 incl. pre-col. (swtch); MeCN/pH 7 phos., grad.; 230 nm; sens. 0.1 µg/ml plasma. Lee, H. et al. (95) J Chr **B 664**, 335-340.

Brombactam [6-β-Bromopenicillanic acid]: SPE; C-18: phos./MeOH; 220 nm; sens. 0.2 µg/ml. Woodcock, J.M. et al. (94) J AtMcCh **33**, 157-161.

Carbenicillin [*R* + *S*] (rabbit/rat/man; incl. urine): SPE (aIEx); C-18-AR (**chiral**): acet./MeOH; 254 nm; sens. ~10 µg/ml. Ishida, M. et al. (94) J. Chr **652**, 43-49.

Cefminox: pptn. (TCA); C-18: AcOH/MeOH/MeCN; 270 nm; sens. 0.1 µg/ml. Microbiological assay also. Aguilar, L. et al. (94) J AtMcCh **33**, 91-101.

Cefodizime (auto): SPE (CN; swtch); C-18: MeCN/AcOH/heptanesulphonate; 263 nm; sens. 0.1 µg/ml. Bompadre, S. et al. (95) J LqChr **18**, 2895-2909.

Cefpodoxime (rabbit, incl. middle-ear fluid): pptn. (MeCN); C-18: pH 4.3 acet./TEA/ MeCN; 254 nm; sens. 50 ng/ml. Lovdahl, M.J. et al. (94) J Chr **653**, 227-232.

For **KEY** (including Journal abbreviations) see p. 162 or 168.

Ceftazidime: SPE (swtch); C-8: pH 5 phos./MeCN; 258 nm; sens. 0.5 μg/ml. Bompadre, S. et al. (95) J Chr **B 669**, 265-269.

Ceftibuten (incl. urine): C-18 cols.: MeCN/acet.; 254 nm; sens. 0.1 μg/ml plasma. Lim, J.M. et al. (94) J PhBmAn **12**, 699-703.

Flucloxacillin: LE; C-18 in pre-col. mode (rapid): pH 7 phos./MeCN; 220 nm; sens. 0.2 μg/ml. Charles, B.G. et al. (94) J Chr **B 660**, 186-190.

Tazobactam (blood filtrate or serum): LE; deriv; C-18: phos. (alkaline) + tr. of MeOH; 325 nm; sens. 50 ng/ml. Peyrin, E. et al. (95) J Chr **B 672**, 160-164.

Ticarcillin (incl. urine): pptn. (MeCN; omit for urine); C-18: MeCN/ phos. acid/TBA; 205 nm; sens. <0.4 μg/ml plasma. La Follette, G. et al. (95) J PhBmAn **13**, 159-161.

BO-2727 [a carbapenem] (incl. urine & saliva; auto): cIEx pre-col. (swtch); C-18: MeCN/pH 7 phos.; 300 nm; sens. 0.05 μg/ml plasma. Accord with microbial assay. Nakashima, M. et al. (94) J AtMcCh **33**, 987-998.

E-1100 [β-lactam] (rat): pptn. (MeOH/PCA); C-18: MeCN/ClO_4^-; 254 nm; sens. <0.2 μg/ml. Tokumura, T. & Horie, T. (93) J Chr **B 620**, 153-157.

L-739,428 [a penem] (rat & monkey; 75 μl): C-18 incl. pre-col. (swtch): pH 6.5 phos./hexanesulphonate, + (main col.) MeCN; 299 nm; sens. 0.1 μg/ml. Wong, B.K. et al. (94) J Chr **B 655**, 158-162.

'oxacin' type

Ciprofloxacin (incl. urine/tissues): pptn.; C-18: MeOH/pH 3 phos./ heptane-sulphonate; fluor; sens. 0.05 μg/ml. Microbiological assays gave higher values with urine (mtbs surmised). Catchpole, C. et al. (94) J AtMcCh **33**, 103-110.

Fleroxacin & ¹mtbs (incl. urine): use of a 1988 method (C-18, fluor). Schrenzel, J. et al. (93) AtMcAgCh **37**, 2133-2138.

Flosequinan & ¹mtbs: LE; **Chiracel** col.: EtOH/MeOH; 320 nm & (sulphide mtb) fluor; sens. 5 ng/ml (apt for PK). Kashiyama, E. et al. (94) J Chr **B 652**, 179-185.

Lomefloxacin.- (1): SPE; aIEx col.: MeCN/pH 7 phos.; 280 nm; sens. 0.05 μg/ml. Carlucci, G. et al. (93) J PhBmAn **11**, 1105-1108. (2): LE; **chiral** deriv; NP col.: solvent mixt.; fluor; sens. <10 ng/ml. Foster, R.T. et al. (95) J PhBmAn **13**, 1243-1248.

Norfloxacin: LE (TCA pre-added); C-8: MeOH/TFA; fluor; sens. 0.02 μg/ml. Hussain, M.S. et al. (95) J Chr **B 663**, 379-384.

Rufloxacin.- (1, **¹mtb** also; incl. urine & bile): pptn. (PCA); PRP col.: pH 5.6 phos./TEA/MeCN/THF; fluor; sens. 50 ng/ml. Privitera, G. et al. (93) AtMcAgCh **37**, 2545-2549. (2): LE; C-18: MeCN/pH 3 phos./ Bu_4N^+; 280 nm; sens. <0.05 μg/ml. Carlucci, G. et al.(95) Analyst **120**, 2493-2495.

Sparfloxacin & ²mtb (urine: glucuronide, ±hydrol.): SPE; C-18: MeOH/MeCN /AcOH; 364 nm; sens. 25/(urine) 500 μg/ml. Fillastre, J.P. et al. (94) AtMcAgCh **38**, 733-737. Also, with only 100 μl plasma and pptn. (MeCN): El-Sayed, Y.M. (95) Anal Lett **28**, 279-293.

DU 7751a [a new amino-fluoroquinolone] (incl. urine): SPE; C-18: THF/ pH 2 phos./acet.; fluor; sens. 10/(urine 100) ng/ml. Tanaka, M. et al. (95) J Chr **B 664**, 401-407. Similar for **DU 6859a** [a quinolone], but post-col. deriv and sens. better (native fluor weak). Aoki, H. et al. (94) J Chr **B 660**, 365-374.

nucleoside-related agents

Acadesine [AICA-riboside]: ultrafilter; C-18 (Brownlee pre-col.): pH 6.3 phos./MeOH; EC; sens. 6 ng/ml. Chen, L-S. et al. (95) J LqChr **18**, 1451-1457.

Acyclovir (auto): SPE; C-18: MeOH/phos./octanesulphonate; 250 nm; sens. 10 ng/ml. Swart, K.J. et al. (94) J Chr. **A [not B] 663**, 65-69.

3'-Amino-3'-deoxythymidine (mtb; from 3'-azido...): SPE (cIEx); C-8: MeOH/ pH 5 acet./dioctylsulphosuccinate; 265 nm; sens. 5 ng/ml. Burger, D.M. et al. (93) J Chr **B 622**, 234-242.

***E*-5-(2-Bromovinyl)uracil & 1-β-D-Arabinofuranosyl...** (urine): LE (can be robotic); phenyl: pH 5 amm. acet./MeCN/MeOH; 295 nm; sens. for both: 0.1 μg/ml. Whigan, D.B. & Schuster, A.E. (95) J Chr **B 664**, 357-363.

8-Chloroadenosine & †...phosphate: SPE (C-8 or, for the phos., aIEx); C-18: pH 3.5 phos./MeOH, grad.; 260 nm; sens. (both): 10 ng/ml. Cummings, J. et al. (94) J Chr **B 658**, 183-188.

Cytarabine-ocfosfate [cytosine-related] (incl. urine): SPE steps; phenyl col., 8°: pH 6 phos./MeCN; 275 nm; sens. 2 ng/ml if pre-col. enrichment (swtch). Ramsauer, B. et al. (95) J Chr **B 665**, 183-192.

2,6-Diaminopurine dioxolane & (mtb) dioxolane guanosine: pptn. (PCA); C-18: pH 2.5 phos./MeCN; 272 nm & radio; sens. 0.1 μg/ml. Rajagopalan, P. et al. (95) J Chr **B 672**, 119-124.

Dideoxyinosine (monkey; incl. amniotic fluid): SPE; C-18: MeCN/pH 4 phos., grad.; 254 nm; sens. 0.4 μg/ml. Pereira, C.M. et al. (94) AtMcAgCh **38**, 781-786

2'-β-Fluoro-2',3'-dideoxyadenosine & †mtb (rat & monkey, incl. urine): SPE; phenyl: pH 6.8 phos/MeCN; 260 nm; sens. 50 ng/ml of either. Roth, J.S. & Kelley, J.A. (95) J LqChr **18**, 441-462.

'FTC' [a substituted 5-F-cytosine] & [1,2]**mtbs**, invstg (rat; incl. urine & faeces): ultrafilter; C-18: MeCN/pH 3 phos.; 280 nm; sens. 0.06 μg/ml. Frick, L.W. et al. (93) AtMcAgCh **37**, 2285-2292.

* **Ganciclovir**, given as †**ester**; & [1]**mtb** (restrained rats; incl. PBS/MeCN homogenates of brain & lung): pptn. (MeCN); C-18: appropriate components (e.g. pH 3.2 phos.); 254 nm; sens. (drug) 50 ng/ml. Brewster, M.E. et al. (94) AtMcAgCh **38**, 817-823.

Ganciclovir.- (1): ultrafilter; C-8: MeOH/pH 3 phos./octanesulphonate; 254 nm /fluor; sens. 50/10 ng/ml. Koel, M. & Nebinger, P. (94) J PhBmAn **12**, 429-432. (2; myocardium, homogenate (PCA), ± dephosphorylation of NP/DP/TP **mtbs**): dry down; C-18: pH 3.5 phos.; 254 nm; sens. 50 μg/ml. Bleyzac, N. & Boulieu, R. (94) J Chr **B 658**, 173-176.

Gemcitabine & a uridine [1]**mtb** (incl. urine): pptn. (solvent); $-NH_2$ col.: organic mixture incl. AcOH & (not urine) TEA; 272 nm; sens. 0.05/(urine) 20 μg/ml. Freeman, K.B. et al. (95) J Chr **B 665**, 171-181.

†Esters of 9-(2-**Phosphonylmethoxyethyl)adenine, 'PMEA'** (mouse): pptn. (TCA); C-8, grad.: MeCN/pH 5 phos./TBA; fluor; sens. 0.05 μg/ml. Naesens, L. et al. (96) Chr'ia **42**, 22-28; also (92) Clin Chem **38**, 480-485.

Zidovudine & [2]mtb (incl. urine).- (1): pptn. (PCA); SPE & C-18 (auto, swtch): pH 3.1 phos./MeOH; 270 nm; 20/(mtb)10 ng/ml. Schrive, I. & Plasse, J.C. (94) J Chr **B 657**, 233-237. (2, with acid [1]**mtb**; mouse & human): SPE; deriv; C-18: pH 7 phos./MeCN; fluor; sens. 0.9 μg/ml. Nattouf, H. et al. (95) J Chr **B 664**, 365-371. (3): an RIA (accord with HPLC). Burger, D.M. et al. (93) AtMcAgCh **37**, 1426-1431.

LY 231514 [thymidylate synthase inhibitor; pyrrole/pyramidine/glutamic moieties]: SPE; 'YMC' cols. (swtch): pH 3 phos./MeCN; 250 nm; sens. 10 ng/ml. Hamilton, C.L. & Kirkwood, J.A. (94) J Chr **B 654**, 298-303.

For **KEY** see p. 162 or 168

6. ANTI-INFECTIVE/ANTI-CANCER AGENTS, continued

6b: various agents [For immunomodulators see #7]

†**Aconiazid** (I) & [1]**mtb, Isoniazid** (II).- For (I), 5 µl run direct on C-18: MeOH/ammon. formate; 313 nm. (II): pptn. (Zn^{2+}); CN col.: aq. iPrOH/amm. formate; EC; sens. (both) 0.1 µg/ml. Hansen, E.B. et al. (95) J Chr **B 670**, 259-266.

Aldophosphamide: SEE 4-Hydroxy.... entry

Altretamine [Hexaethylmelamine]: pptn. (MeCN); C-18: MeOH/pH 8.2 phos.; 230 nm; sens. 0.2 µg/ml. Barker, I.K. (94) J Chr **B 660**, 121-126.

9-Aminocampothecin lactone (hydrol. gives the inactive acid): SPE; C-18: MeOH/pH 2.5 phos.; fluor; sens. 0.1 ng/ml. Takimoto, C.H. et al. (94) J Chr **B 655**, 97-104.

Arteether: pptn. (PCA, 55°); C-18: aq. MeCN; 254 nm; sens. 50 ng/ml. Al-Angary, A.A. et al. (94) Anal Lett **27**, 2689-2702.

Artemether & Dihydroartemisinin (mtb): LE; CN col.: pH 5 acet./MeCN; EC; sens. 2 ng/ml. Navaratnam, V. et al. (95) J Chr **B 669**, 289-294.

Atovaquone.- (1): LE; C-8: MeCN/AcOH; 254 nm; sens. 0.25 µg/ml. DeAngelis, D.V. et al. (94) J Chr **B 652**, 211-219. (2; robotics): LE; C-8: MeOH/AcOH; 254 nm; sens. 0.2 µg/ml. Studenberg, S.D. et al. (95) J PhBmAn **13**, 1383-1393.

N-Benzoylstaurosporine & oxid. [1]**mtbs** (rat): LE; C-18: MeCN/pH 4 acet.; fluor; sens. <1 ng/ml. van Gijn, R. et al. (95) J Chr **B 667**, 269-276.

†**Carzelesin** [U-80,244; a pyrroloindole] & [1]**mtbs** (auto): SPE on pre-col. (ASPEC); CN & (swtch) C-18 cols.: MeCN/pH 6.5 phos.; 360 nm; sens. 1 ng/ml. Van Tellingen, O. et al. (94) J Chr **B 652**, 51-58.

° **Cecropin B** [lytic peptide]: pH 1.5 citric ac. & ultrafilter [strong protein binding; SPE suffices for culture medium]; C-8, grad.: aq. MeCN/ TFA; 206 nm; sens. high. Instability problem. Moore, A.J. et al. (95) J Chr **B 672**, 225-231.

Chlorambucil & its †**N-oxide**: C-18: MeOH/TFA pH 2.3; 258 nm. Lability problems the focus. Chandler, K.J. et al. (94) J Chr **B 652**, 195-201.

Cisplatin: ultrafilter. (1): cIEx or (monohydrate [1]**mtb**) aIEx: pH 5.2 succinate/MeOH; deriv, 115°; 344 nm; sens. <0.1 µg/ml. Andersson, A. & Ehrsson, H. (94) J Chr **B 652**, 203-210. Cf. (2): Hanada, K. et al. (95) J Chr **B 663**, 181-186 (rat kidney & liver saline homogenates; post-col. reaction for 290 nm detection). (3): 'deriv' reaction; C-18: aq. MeOH; 260 nm; sens. ~10 ng/ml. Augey, V. et al. (95) J PhBmAn **13**, 1173-1178.

Clindamycin: LE; deriv.; GC-ECD; sens. 25 ng/ml. Gatti, G. et al. (93) AtMcAgCh **37**, 1137-1143.

Clotrimazole & [1]**mtbs** (plasma or blood): SPE; [ultrafilter for free-drug value]; C-8: MeOH/pH 6.3 phos.; 210 nm; sens. 0.03 µg/ml. Rifai, N. et al. (1995) Clin Chem **41**, 387-391.

* **Dapsone** (incl. saliva): pptn. (PCA); within-valve pre-col.; C-18: MeCN/ MeOH/pH 4.6 phos.; either 295 nm, or EC with initial SPE (IEx) for high sens., 0.2 ng/ml. Moncrieff, J. (94) J Chr **B 654**, 103-110.

Delaviridine & [1]**mtb** [piperazine moiety]: pptn.; CN col.: pH 6 phos./ MeCN; fluor; sens. ~0.01 µg/ml (apt for PK). Staton, B.A. et al. (95) J Chr **B 668**, 99-106.

7,8-Dihydrocostatolide [NSC 661123; anti-HIV]: SPE; C-18: aq. MeCN; 318 nm; sens. 15 ng/ml. Newman, R.A. et al. (94) J Chr **B 658**, 129-133.

Dimethylaminoetoposide (NK 611) & [1]**mtb** : SPE; 'reduced-activity' C-18: MeCN/phos. ac.; 205 nm; sens. 20 ng/ml. Zucchetti, M. et al. (95) J Chr **B 654**, 97-102; also **664**, 409-414 (urine).

Elsamitrucin (incl. urine): LE; C-8: aq. MeCN (urine: + MeOH); fluor; sens. ~0.5 ng/ml. Kleiun, I. et al. (93) J Chr **B 622**, 249-253.

Enloplatin (rat/dog/human): ultrafilter (plasma) or PCA pptn. (blood); C-18 or CN or phenyl (depends on species): pH 7.5 phos./PrOH; 230 nm; sens. 0.5 µg/ml. Amorusi, P. (94) J PhBmAn **12**, 1023-1033.

Etoposide.- (1; *cis*-isomer too; incl. sonicated leukaemic cells): LE; phenyl col.: MeOH/AcOH; fluor; sens. <10 ng/ml. Liliemark, E. et al. (95) J Chr **B 669**, 311-317. (2; dog): LE; phenyl col.: MeCN/citrate; EC; sens. 2 ng/ml. Eisenberg, E.J. & Eickhoff, W.M. (93) J Chr **B 621**, 110-114.

Exemestane [aromatase inhibitor]: LE; C-18: MeCN/pH 3 phos.; 247 nm; sens. 10 ng/ml. Breda, M. et al. (93) J Chr **B 620**, 225-231.

Finasteride (incl. semen): SPE (CN); C-18 (70°): MeCN/formic acid; MS (APcI); sens. <0.2 ng/ml. Constanzer, M.L. et al. (94) J Chr **B 658**, 281-287.

Fluconazole.- (1): LE; C-8: pH 4 phos./MeCN; 220 nm; sens. 0.1 µg/ml. Flores Murrieta, F.J. et al. (94) J LqChr **17**, 3803-3811. (2; incl. saliva): LE; C-18: pH 5 acet./MeOH; 261 nm; sens. 0.1 µg/ml. Koks, C.H.W. et al. (95) J Chr **B 663**, 345-351.

* **Flutamide** & oxid. **[1]mtb** (dog): pptn. (MeCN); 'mid-bore' C-18 col.: AcOH/MeCN; 300 nm; sens. 11 ng/ml. Farthing, D. et al. (94) Bm Chr **8**, 251-254.

Foscarnet [phosphonoformate] (incl. CSF): ultrafilter; C-18: pH 5.8 phos./TEA/MeOH; EC; sens. ~2 µg/ml. Hengge, U.R. et al. (93) AtMcAgCh **37**, 1010-1014.

* **Genaconazole** (PK; incl. urine): LE; GC-ECD; sens. 0.2(urine 0.5) ng/ml, 8 times better than with HPLC. Kim, H. et al. (94) J Chr **B 655**, 21-26

Halofantrine & †mtb (active).- (1): pptn. (MeCN etc.); SPE; **Chiral**pak AD: non-aq. eluent; fluor; sens. <6 ng enantiomer/ml. Terefe, H. et al. (94) J Chr **B 657**, 238-242. (2): pptn. (MeCN); LE; **chiral** deriv; C-18: pH 5 phos./TEA/MeCN/SDS; 254 nm; sens. <12 ng/ml. Brocks, D.R. et al. (95) J PhBmAn **13**, 911-918. (3; incl. whole blood): SPE; C-8: aq. MeCN/TEA; fluor; sens. 6 µg/ml. Gaillard, Y. et al. (95) J Chr **B 668**, 315-321. (4): pptn. (MeCN); LE or SPE; TLC; 256 nm scan. Cenni, B. & Betschart, B. (94) J Planar Chr **7**, 294-296.

Hexaethylmelamine: SEE Altretamine

* **Hydroxychloroquine & [1]mtbs** (blood): pptn. (PrOH); LE; CN: pH 6 phos./MeOH / MeCN; then **chiral** AGP: similar but pH 7; fluor; sens. 2 ng/ml (parent-drug enantiomers). Wei, Y. et al. (94) J LqChr **17**, 3479-3490.

4-Hydroxycyclophosphamide [cf. †cyclophosphamide].- (1): pptn. (MeCN); deriv; C-18: pH 7 phos./MeCN; 357 nm; sens. ~20 ng/ml. **Chiral** assay discussed. Johansson, M. et al. (94) J Chr **B 660**, 111-120. (2; also **Aldophosphamide** [active tautomer]; whole blood): pptn. with deriv; LE; GC-MS; sens. <25 ng/ml. Anderson, L.W. et al. (95) J Chr **B 667**, 247-257. (3): deriv (to semicarbazone); LE; C-8: MeCN/pH 6 phos.; 230 nm; sens. <20 ng/ml. Belfayol, L. et al. (95) J Chr **B 663**, 359-399.

* **Ifosfamide** [akin to cyclophosphamide].- (1; incl. urine): SPE; C-1, then **chiral** (AGP) col.: pH 4 phos./MeCN; 195 nm; sens. 2.5 ng/ml. Corlett, S.A. & Chrystyn, H. (94) J Chr **B 654**, 152-158. (2; **[1]mtbs** also; incl. urine): **chiral** (β-cyclodextrin) GC col.; MS; sens. 0.2 µg/ml. Granville, C.P. et al. (93) J Chr **B 622**, 21-31. (3; incl. **[1]mtbs**): pptn. (cold acetone); C-18 col. 1.8 mm bore: MeOH/aq. glycerol; MS, ESP or FAB on-line. Siehoff, C. et al. (95) Fresenius' J Anal Chem **352**, 801-805.

Irinotecan [related to camptothecin] & active **[1]mtb.**- (1): pptn. (MeCN/MeOH); C-18: pH 6.4 acet./MeCN/TBA; fluor; sens. <10 ng/ml. Rivory, I.P. & Robert, J. (94) J Chr **B 661**, 135-141. (2): pptn. (MeOH; supernatant dried down); TSK gel C-18: MeCN/pH 3 phos./heptanesulphonate; fluor; sens. <20 ng/ml. Sumiyoshi, H. et al. (95) J Chr **B 670**, 309-316.

For **KEY** see p. 162 or 168

Isoniazid: SEE Aconiazid entry

Itraconazole: fungal bioassays may give higher values than HPLC due to hydroxy **mtb** activity. Hostetler, J.S. et al. (93) AtMcAgCh **37**, 2224-2227.

Lometrexol [an H_4-folate] (incl. urine): SPE; C-6 col., grad.: MeCN/aq. NaOH; 288 nm; sens. 0.2 µg/ml.- But 10 ng/ml if a sample-oxidation step and final fluor instead of UV. Wedge, S.R. et al. (95) J Chr **B 663**, 327-335.

Mefloquine.- (1): pptn. (Zn^{2+}, MeCN) & LE [Variant: on-paper sample, eluted]; **chiral** deriv; C-18: MeCN/AcOH; 263 nm/fluor; sens. 0.1/<0.01 µg/ml. Bergqvist, Y. et al. (94) J Chr **B 652**, 73-81. Cf. earlier version (2): **620**, 217-224: **chiral** eluant, not deriv; Hypercarb col. (3): LE; **chiral** AGP: PrOH/pH 4.9 phos.; 222 nm; sens. ~0.05 µg/ml. Wallén, L. et al. (94) J Chr **B 655**, 153-157. (4): pptn. (MeOH/MeCN); LE; deriv; GC-MS (nCI); sens. 1 ng/ml. Neal, J.M. et al. (94) J Chr **B 661**, 263-269.

* **Methotrexate**.- (1, ^{1}**mtbs** also; incl. urine & saliva): SPE; C-18: pH 6.5 phos./$MeCN/H_2O_2$; irradiate; fluor; sens. 0.05(urine 5) ng/ml. Albertione, F. et al. (95) J Chr **B 665**, 163-170. (2): pptn. (PCA); react with $KMnO_4/H_2O_2$; C-18: aq. MeOH/TBA; fluor; sens. <0.01 µg/ml. Suzuki, T. et al. (95) Fresenius' J Anal Chem **351**, 806-807.

Mirosamicin (animal tissues): SPE (cIEx) on phos. acid/MeOH extracts; C-18: pH 2.5 phos./MeOH; 230 nm; sens. 0.05 µg/ml. Moric, M. et al. (94) J Chr **B 655**, 47-52.

Mitomycin C given as †conjugate [N-succinyl-chitosan-/carboxymethyl chitin-] (rat): LE; C-18: pH 6 phos./MeOH; 364 nm. Song, Y. et al. (93) Biol Ph Bull **16**, 48-54.

* **Mitoxantrone** (add ascorbic acid): SPE; C-18: pH 3 formate/MeCN; 658 nm; sens. 1 ng/ml. Accord with ELISA. Priston, M.J. & Sewell, G.J. (94) J PhBmAn **12**, 1153-1162.

Netilmicin: pptn. (TCA); LE (to wash); deriv; LE; C-18, grad.: AcOH/ MeCN/heptanesulphonate; fluor; sens. 0.1 µg/ml. Santos, M. et al. (95) J PhBmAn **13**, 1059-1062.

Novobicin: pptn. (MeCN); C-18: MeCN/phos. acid; 340 nm; sens. 1 µg/ml. Zuhowski, E.G. et al. (94) J Chr **B 655**, 147-152.

Paclitaxel ['Taxol'; SEE below] (urine; & mouse specimens incl. faeces & water/BSA homogenates of tissues): LE and/or SPE; C-8: MeCN/MeOH/pH 5 acet.; 227 nm; sens. 0.01 µg/ml (human; better if SPE) or 0.02 µg/ml (mouse; same for oxid. ^{1}**mtbs**). Human: Huizing, M.T. et al. (95) J Chr **B 664**, 373-382; mouse: Sparreboom, A. et al., **same**, 383-391. Cf. (93) **621**, 231 (Wiley, T.A. et al.).

Penclomedine [NSC-338720; pyridine moiety] & ^{1}**mtbs** (its action invstg, mouse cells): LE; C-18: aq. MeCN; radio. Benvenuto, J.A. (95) Biochem Pharmacol **50**, 1157-1164.

Praziquantel: SPE (CN); **Chiracel** OD: hexane-iPrOH; 212 nm; sens. 5 ng/ml. Kelly, J.W. et al. (93) J PhBmAn **11**, 1141-1144.

Pyrazinamide (rat): pptn. (MeOH)/LE; C-18: pH 4 phos./MeCN; 269 nm; sens. 0.5 µg/ml. Assay rapid. Kaka, J.S. (94) J LqChr **17**, 3793-3801.

Quercetin [flavonoid]: SPE; C-18: acetone/pH 3.5 phos./DEA; 373 nm; sens. ~1 µg/ml. Liu, B. et al. (95) J Chr **B 666**, 149-155.

Sulfamethazine (sheep; auto): SPE; C-18: MeOH/pH 6.4 phos.; 262 & 292 nm; sens. 2 µg/ml. Hubert, Ph. et al. (93) J Chr **B 622**, 53-60.

Sulofenur (4 animal species): LE; C-18: MeCN/pH 7.2 acet.; 250 nm & radio; sens. ~10 µg/ml. Ehlhart, W.J. et al. (93) J PhSci **82**, 683-688.

Sulphadiazine & ^{12}**mtbs** (incl. monkey): pptn. (MeCN; N/A if urine); 'Dynamex' col., grad.: MeCN/pH 3 acet.; 275 nm; sens. 0.2 µg/ml. Vree, T.B. et al. (95) J Chr **B 670**, 111-123.

Taxol: SEE **Paclitaxel** above, and an Editorial [(95) Nat. **373**, 370] which deplores the patenting of 'Taxol' as a company trademark.

Taxol.- (1; incl. solvent homogenates of tissues; auto): LE; C-8 (swtch) & C-18 cols.: aq. MeCN; 229 nm; sens. 5 ng/ml. Song, D. & Au, J.L-S. (95) J Chr **B 663**, 337-344. (2; mouse): pptn. (MeCN); C-18: aq. MeCN; UV. Lability at alk. pH examined. Leslie, J. et al. (93) J PhBmAn **11**, 1349-1352.

Temozolomide (incl. urine): SPE; C-18: MeOH/AcOH; 330 nm; sens. <0.4 μg/ml. Shen, F. (95) J Chr **B 667**, 291-300.

Terbinafine & [1]mtbs.- (1; incl. urine): SPE; phenyl (swtch): pH 3 TEA/ MeCN; 224 nm; sens. 20 ng/ml. Zehender, H. et al. (95) J Chr **B 664**, 347-355. (2; incl. sebum etc.): LE; C-18: MeCN/phos. acid/TEA; 224 nm; sens. <2 ng/ml. Denouel, J. et al. (95) J Chr **B 663**, 353-359.

Thalidomide [leprosy context]: SPE; C-8: MeCN/pH 3 phos./heptane-sulphonate; 295 nm; sens. <60 ng/ml. Delom, A. et al. (95) J LqChr **18**, 297-309. Cf. a Sect. #E 'annotation'.

Topotecan, lactone & (lactonized for assay) acid [camptothecin analogue]: pptn. (MeOH); C-18: MeOH/pH 6 TMEDA/heptanesulphonate; fluor; sens. 0.05 ng/ml. Rosing, H. et al. (95) J Chr **B 668**, 107-115.

Vancomycin: pptn. (PCA); LE to wash; C-18: pH 3 phos./MeCN; 229 nm; sens. 0.2 μg/ml. Luksa, J. & Marusic, A. (95) J Chr **B 667**, 277-281.

Vincozolin (urine, heated with NaOH): LE on distillate; C-18: pH 7.5 phos./MeOH: EC; sens. <8 μg/ml. Will, W. (95) Fresenius' J Anal Chem **353**, 215-218.

* **Vinorelbine** (rabbit): LE; C-18: pH 3 phos./MeCN/MeOH/heptanesulphon-ate; EC; sens. 1 ng/ml. Mouchard Delmas, C. (95) J Chr **B 663**, 390-394. - But SEE **672**, 314: author's argument for EC rather than UV criticized!

AG-331 [TMP-synthase blocker; morpholinosulphonyl & benzindole moieties] (incl. urine): SPE; C-18: MeCN/pH 3.5 phos.; 457 nm; sens. 0.02/(urine 0.04) μg/ml. Qian, M.X. et al. (95) J Chr **B 666**, 307-314.

* **CGP 53437** [HIV protease inhibitor; peptidomimetic] (animals & man): pptn. (MeCN) & LE (elaborate); deriv; C-18: pH 9 borax/MeCN; fluor; sens. ~3 ng/ml. Flesch, G. et al. (94) J Chr **B 657**, 155-161. Simplified version (mouse): Alteri, E. et al. (93) AtMcAgCh **37**, 2087-2092.

* **CI 941** [an anthrapyrazole] & **[1]mtbs** (incl. urine): SPE; C-18: aq. MeCN; 491 nm; sens. 5 ng/ml. PITFALL: set at 491 the grating gives 245 nm input too! Proksch, B. et al. (94) J Chr **B 658**, 349-353.

CI 980 [an H_2-pteridine] (incl. urine): SPE; C-18: MeCN/pH 7.5 phos.; fluor; sens. <0.2 ng/ml. Bullen, W.W. et al. (95) J Chr **B 668**, 141-151.

GI 147211 lactone & (lactonize for assay) carboxylic forms [topoisom-erase I inhibitor]: SPE; C-8: MeCN/pH 4 acet.; fluor; sens. <0.05 ng/ml. Stafford, C.G. & St. Claire III, R.L. (95) J Chr **B 663**, 119-126.

GR 99060 & 99062 [triazino-indoles] (mouse, at injection site; DMSO extract): pptn. (PCA); C-18: MeCN/pH 7.5 phos./TEA; 270 nm. Kinsman, O.S. et al. (93) AtMcAgCh **37**, 1243-1246.

KRM-1648 [rifamycin moiety] & **[1]mtb** (rat; incl. blood/urine/solvent homogenates of tissues): SPE; C-18 col. pair (2 types): MeCN/pH 4.3 citr-ate/ClO_4^-; 643 nm; sens. <20 ng/ml. Hosoe, K. et al. (94) J Chr **B 653**, 177-186.

* **L-667,639** [non-nucleoside (pyridone) reverse transcriptase blocker]: pptn. (MeCN); SPE (phenyl, centrifugal [pre-remove ink from outside of cartridge!]); alumina col.: MeOH/pH 8 phos.; 326 nm; sens. 10 ng/ml. Woolf, E. et al. (92) J Chr **577**, 129-134; cf. a PK study: Van Hewcken, A. et al. (94) Int J Clin Pharm Res. **14**, 45-50.

MK-434 [blocks 5α-reductase; steroid-like]: C-18: MeCN/MeOH/acet./TFA; MS(APcI); sens. <0.5 ng/ml. Olah, T.V. et al. (94) J PhBmAn **12**, 705-712.
NK 611 [etoposide/podophyllotoxin type]: SPE; C-18: MeCN/pH 7 phos.; 215 nm; sens. 10 ng/ml. Hüttmann, A. et al. (93) J Chr **B 620**, 233-238.
* **Ro 23-9424**: TSK gel PW col.: iPrOH/pH 6 phos. (proteins unretained; swtch); C-8: MeCN/pH 3 phos; 287 nm; sens. 0.3 μg/ml. Szuna, A.J. & Brain, R.W. (93) J Chr **B 620**, 211-216.
* **S 12363** [vinca alkaloid; ref. given to assay **review**] (mouse, dog, man): pptn. (PCA); SPE (well described); C-18: pH 3 phos./MeOH/MeCN/heptane-sulphonate; post-col. deriv & fluor; sens. 0.1 ng/ml (or, with simpler method, 1 ng/ml). Gordon, B.H. et al. (94) J Chr **B 657**, 163-171.
TNP-470 [fumagillin-related] & [1]**mtb**: LE; C-18; fractions (drug & mtb) deriv; each C-18: aq. MeCN, grad.; fluor; sens. 2.5 ng/ml. Figg, W.D. et al. (94) J Chr **B 652**, 187-194.
1069C85 [microtubule inhibitor; imidazo & pyridazin rings] (incl. tissue): LE; C-18: aq. MeOH; fluor; sens. 0.25 ng/ml. Raynaud, F. et al. (93) J Chr **B 622**, 243-248. ____________

7. UNCLASSIFIED AGENTS

Berberine (rabbit; incl. urine & bile): pptn. (MeCN); LE; C-8: MeCN/pH 6 phos.; 267 nm; sens. 20 ng/ml plasma (suffices for PK). Chen, C-M. & Chang, H-C. (95) J Chr **B 665**, 117-123.
* **Busulphan** [poor UV absorbance]: LE; 3 C-18 cols.: deriv on 1st [erratic if off-line], swtch after 1st & 2nd: MeCN/pH 4.6 phos.; 278 nm; sens. <10 ng/ml. Funakoshi, K-I. et al. (94) J Chr **B 660**, 200-204.
°* **Calcitonin** (human): SPE (immuno); C-18 micro-col., grad.: aq. MeCN/TFA; MS(ESP); sens. <0.5 ng/ml. Kobayashi, N. et al. (95) J Chr **B 672**, 17-23. SEE ALSO a citation on final p. of Sect. #A.
Canrenone [& SEE Spironolactone]: SPE; C-18: aq. MeCN; 280 nm; sens. 10 ng/ml. Zhang, H. & Stewart, J.T. (93) J PhBmAn **11**, 1341-1344.
(+)-**Catechin** [catechol moiety]: pptn. (MeCN); adsorb (alumina); C-18: MeCN/pH 2.5 phos.; fluor; sens. <20 ng/ml. Ho, Y. et al. (95) J Chr **B 665**, 383-389.
Centchroman & [1]**mtb** (incl. milk): LE; CN: MeCN/pH 3 phos.; fluor; sens. 1 ng/ml. Lal, J. et al. (94) J Chr **B 658**, 193-197.
° **Cetrorelix** [peptide] (rat): LE; C-18: MeCN/MeOH; 250 nm; sens. 0.06 μg/ ml. Raffel, H.H. et al. (94) J Chr **B 653**, 102-105.
Cimetidine.- (1): pptn. (phos. ac.); SPE (ion-pair); C-18: MeCN/pH 3 phos./octanesulphonate); 228 nm; sens. 5 ng/ml. Russel, F.G.M. et al. (94) J Chr **B 661**, 173-177. (2): LE; C-18: MeCN/pH 6.2 phos./heptanesulphonate; 228 nm; sens. <0.1 μg/ml. Kelly, M.T. et al. (95) J Chr **B 668**, 117-123.
° **Cyclosporin G** & oxid. [1]**mtbs**, invstg (human liver microsomes; radio): pptn. (MeCN) & C-18, grad. Pichard, L. et al. (96) Biochem Pharmacol **51**, 591-598; [assay details] (90) Drug Metab Disp **18**, 595-606.
Deferiprone [1,2-diMe-3-OH-pyrid-4-one; chelator] & **Fe complex** (incl. urine): pptn. (PCA); C-18, grad: pH 2 phos./MeOH/octanesulphonate [different col./conditions for urinary Fe complex]; 280 nm; sens. 10 μM Fe. Dresow, B. et al. (95) Fresenius' J Anal Chem **352**, 562-564.
Delmopinol: SPE (CN); **chiral** deriv; C-18: MeCN/pH 5.7 acet.; EC; sens. 1 ng/ml. Egginger, G. et al. (94) J Chr **A 666**, 275-282.
Evodiamine (rat): Jeng, K-F. et al. (95) J Chr **B 668**, 343-345; **rutaecarpine** method [Ko et al. (95) J Chr **B 665**, 27] adapted.
Honokiol (rat; PK): pptn. (MeCN); C-18: pH 2.5 phos./MeCN; 290 nm; sens. 0.1 μg/ml. Tsai, T-H. et al. (94) J Chr **B 655**, 41-45.

'MINI-KEY' at foot of page

Lansoprazole [ATPase inhibitor] & [1]**mtbs**: LE; C-18: aq. MeCN/ octylamine/N-acetohydroxamate (role unclear); 285 nm; sens. <10 ng/ml. Karol, M.D. et al. (95) J Chr **B 668**, 182-186.

Linogliride & †**mtbs**, invstg (rat/dog/man; incl. urine, ±hydrol.): dry down & extract; either TLC (radio) or RP-LC: pH 4.5 acet./MeOH, grad; 254 nm. IR, NMR & MS to identify. Wu, W-N. & Mutter, M.S. (95) J PhBmAn **13**, 857-867.

Metoclopramide & [1]**mtbs**: deriv; LE; GC-MS; sens. <1 ng/ml (better than HPLC). Riggs, K.W. et al. (94) J Chr **B 660**, 315-325.

†**Mycophenolate** mofetil & acid **mtb** [immunosuppressant] (rat, bile & MeOH tissue homogenates; & man): LE; CN: pH 5 phos./MeCN; 305 or (rat plasma) 215 nm; sens. 0.1 µg/ml. Sugioka, N. et al. (94) J Chr **B 654**, 249-256.

†**Procysteine** [an oxothiazolidine]: pptn. (phos. acid); C-18: pH 3 phos.; 250 nm; sens. 0.5 µg/ml. Webb, L.E. et al. (94) J Chr **B 654**, 257-262.

Quercetin: listed in #6b class

Ranitidine: LE; C-18: MeCN/pH 5.1 phos.; 300 nm; sens. 20 ng/ml. Al-Khamis, K. et al. (95) J LqChr **18**, 277-286.

* **Rapamycin** [immunosuppressant] (blood): LE; C-18: aq. MeOH; 276 nm; sens. 2 ng/ml (suffices for PK; MS not needed). PROBLEMS: light-sensitivity/interferences/losses onto glass. Napoli, K.L. & Kahan, B.D. (94) J Chr **B 654**, 111-120.

Rutaecarpine (rat): pptn. (MeCN); C-18: MeCN/pH 2.5 phos.; 344 nm; sens. 10 ng/ml. Ko, H-H. et al. (94) J Chr **B 655**, 27-31.

Spironolactone & [1]**mtb**, Canrenone [SEE its entry too] (urine; swtch): C-18, grad.: aq. MeCN; 238/300 nm; sens. 20 ng/ml. Herraez Hernandez, R. et al. (94) J Chr **B 658**, 303-310.

Terguride: pptn. (MeCN); SGX CN col.: MeOH/pH 7 phos.; 227 nm; sens. 1 µg/ml. Sochor, J. et al. (95) J Chr **B 663**, 309-313.

CI 988 [CCK-B blocker] (robotics; incl. monkey): SPE; C-18: pH 3.5 citrate/MeCN; fluor; sens. 0.2 ng/ml. Hinton, J.P. et al. (95) BmChr **9**, 94-97.

DuP 128 [S-containing ACAT inhibitor] & oxid. [1]**mtbs**: pptn. (MeCN); MeCN solution washed with hexane; 'YMC' basic col., or (sulphoxide mtb) C-18: MeCN/phos. acid; fluor; sens. 1 ng/ml. Lai, C-M. et al. (94) J PhBmAn **12**, 1163-1172.

E 3810 [H^+,K^+-ATPase inhibitor; a pyridinyl-benzimid] & [1]**mtbs**: LE; C-8: MeCN/pH 7 phos.; 288 nm; 5 ng/ml. Nakai, H. et al. (94) J Chr **B 660**, 211-220.

FCE 24928 [aromatase inhibitor; steroid-related]: LE; C-18: MeCN/ pH 4.3 phos.; 238 nm; sens. 2 ng/ml. Fontana, E. et al. (94) J Chr <u>**A**</u> **660**, 293-298.

MK-507 & [1]**mtb** [topically administered carbonic anhydrase inhibitor; S-heterocyclic] (incl. blood & urine): pptn./LE (TCA; wash step); C-8 & C-18 cols.: MeCN/phos. ac./octanesulphonate; 252 nm; sens. 5 ng/ml. Matuszewski, B.K. et al. (94) J Chr **B 653**, 77-85.

ONO-5046 [ester; neutrophil elastase inhibitor] & acid [1]**mtb** (incl. urine): C-18 incl. pre-col. (swtch): MeCN/pH 3.8 phos., grad.; 240 nm; sens. 0.1(urine 1) µg/ml. Shintani, T. et al. (94) J PhBmAn **12**, 397-405.

* **YM529** [a bisphosphonate with imidazole moiety] (incl. urine & bone): pptn. of protein (PCA) & then drug (by Ca^{2+}); C-18: MeOH/pH 7 phos./TBA; fluor; sens. 0.2 ng/ml plasma. Usui, T. et al. (94) J Chr **B 652**, 67-72.

===============

ABBREVIATIONS (KEY: p. 162) include: LE, liquid-liquid extraction; MeCN, acetonitrile; mtb, metabolite. Bm, Biomedical; Chr, Chromatography [J Chr: **A** has mis-housed arts., and vol. nos. confusable with B nos.!]
* denotes noteworthy material; †, pro-drug; °, peptide.

Section #D

SAMPLE TREATMENT INCLUDING PROTEIN REMOVAL AND SPE

Other pertinent arts. include #A-6, #C-6 and #C-13n.

The suffix **'n'** as in #D-4n distinguishes 'NOTES' from main articles. COMMENTS (starting on p. 225) comprise Forum discussion remarks and Editor's 'annotations'.

ANALYTES IN BIOFLUIDS AND TISSUES: ACHIEVING CERTAIN 'GOALS'

Ordinary 'repertoire' and, prefixed [, **occasionally used options**

- from art. by H. Lingeman, E. Reid & U.A.Th. Brinkman: p. 17 in Vol. 23, this series [1994; eds. & publisher as for this vol.]

Table 1. Ways to attain particular goals in sample preparation. RECOMMENDED FOLLOW-UP for strategies and informative applications.- Consult the General Index in this and earlier vols.; use 'keywords' in and below the Table, also (e.g.) 'Lability', 'Protein removal'.

Numbered§ GOAL: Technique	[1]Concentration	INTERFERANT REMOVAL [2]high M.W.	[3]low M.W.	[4]particulate	[5]Solubilization	[6]Stabilization
Liquid-liquid extraction (LE)	x	x	x			
Solid-phase extraction (SPE)*	x	x	x			
Precipitation▫		x(a)		(b)		
[Electrodialysis	x	x(c)				
Ultrafiltration	x	x	x	(d)		
Evaporation	x					
[Freeze-drying (lyophilization)	x					x
[Freezing	x(e)	x(f)		x		
[Hydrolysis†		x			x	x
[Detergent addition		x(g)			x	
[Chemical modification/Derivatization⁺					x	x

§Numerals below refer to the numbered columns, e.g. [3] = low M.W.

*Automated variant: 2-column HPLC, switching before analytical col.

▫A protein precipitant can serve for extraction (e.g. tissues).

†Hydrolysis is not a common approach for [2], [5] or [6], but is often applied to urine to split conjugates.

⁺**Main use** [not shown] is in chiral or GC-volatility context, but used on occasion for [5] (maybe complexation) or [6], OR [not shown] to aid extraction, separation or detection.

(a) May entail protein denaturation. | (b) Centrifugation (no pptn.).
(c) Can be conventional dialysis. | (d) Conventional filtration.
(e) Semi-freeze ('zone refining'); also can aid phase separation. | (f) Extract frozen-ground tissue. (g) Releases protein-bound drug.

'Special' treatments not shown include -
#microwave irradiation, for [2] or [5] or to thaw frozen samples [cf. γ-irradiation (or heating or strong acid) to inactivate viruses];
#enzyme inhibition, e.g. of a plasma esterase (for [5] or [6]);
#pH adjustment, for [5] (and often done to optimize extraction).

....

'Reinforcing' comments now added.- #Removing protein ('high M.W.') is circumventable (art. #D-1). #An ion-pairing agent can assist 'LE'. #SPE variant: immunoaffinity (usually an immobilized Ab).

#D-1

DIRECT-INJECTION HPLC: CHROMATOGRAPHY OF BIOFLUIDS WITHOUT PROTEIN REMOVAL

David J. Anderson

Department of Chemistry, Cleveland State University, Cleveland, OH 44115, U.S.A.

Several investigators have devised ways to allow direct injection of protein-containing samples in the RP-HPLC analysis of low mol. wt. analytes. The two approaches used, pre-column and raM techniques, are now reviewed, with critical assessment of the advantages and disadvantages of each.*

A major consideration in HPLC is sample preparation, as needed both for analytical reasons (to increase sensitivity and remove interferences, aiding specificity) and for operational reasons (to minimize detriment from the sample matrix). Sample preparation makes the procedure more lengthy and laborious, and subjects the sample to conditions that may adversely affect the analyte. Its elimination through the ability to inject samples having a complex matrix directly onto the column is accordingly a long-sought goal in bioanalytical HPLC, towards which much progress has been made in the last 5-10 years†. The present appraisal is based on a thorough review of the area [1].

Detrimental effects of proteins in HPLC

Detriment is due to the irreversible adsorption and/or precipitation of proteins that occurs in organic-containing mobile phases, such that protein-containing samples cannot be directly injected in conventional RP-HPLC techniques. Adverse effects of precipitated/irreversibly adsorbed proteins include: 1) back-pressure build-up; 2) decreased efficiency; 3) changes in retention time; and/or 4) decreased column capacity.

Limits for how much organic modifier a solution can contain without precipitation of proteins have been determined using light-scattering [2]. Expressed as % (v/v) modifier which, after

* *Abbreviations* (others in Preface, and see Table 3 for types of raM = restricted access media [was 'RAM' in the MS.- *Ed.*]; ISRP, internal surface reversed phase; MF(P), mixed-functional (phase) MeCN, acetonitrile [was 'AN' in MS.- *Ed.*]; TEA, triethylamine; THF, tetrahydrofuran. N = no. of theoretical plates.

† *Note by Eds.*- Sample preparation may also be obviated by ASTED® (Automated Sequential Trace Enrichment of Dialysates; used more in Europe than in America): particularly see Krogh & Rasmussen's art. and Index entry 'Automation' in Vol. 23 [same eds. as now].

adding serum to pH 6.8 buffer containing the modifier and then incubating (10 min), results in precipitation, the thresholds were 25% for MeCN, 20% for isopropanol and 10% for THF. These values may, however, be pH-dependent, as precipitation has been shown to occur more rapidly in organically modified buffers at pH <5 (particularly pronounced at pH <4) [3].

Taking precautions to minimize precipitation of proteins is only part of the issue, however, as even organic-modifier concentrations below the threshold lead to adverse chromatographic effects. An RP-HPLC procedure with direct injection has been reported with mobile phase buffer (pH 2.2) containing 10% methanol, without resorting to the special HPLC versions now to be discussed [4]. However this technique is far from rigorous, requiring a 200-fold dilution of serum, injection of a small amount of this dilution (10 µl), and frequent reversing and cleaning of the column with 100% MeCN (every 30-50 injections). Despite these precautions columns lasted only 200+ injections.

Approaches to direct-injection HPLC

The two approaches that several workers have developed are pre-column and raM [see previous p., *] techniques, critically evaluated below. Both can tolerate the injection of numerous (thousands) undiluted serum or plasma samples whilst preserving reasonable column performance. The focus has been on RP-HPLC techniques since this is the most utilized HPLC mode employing organic mobile phases.

PRE-COLUMN TECHNIQUE

The pre-column technique (diagram: Fig. 1) has a tandem column design consisting of a pre-column, a switching valve and an analytical column. It allows direct injection of protein-containing samples because the chromatographic process is two-step: 1) injection of the sample into an aqueous mobile phase lacking organic modifier (application buffer) that flows to waste through the pre-column which retains hydrophobic components but not hydrophilic components such as proteins (Fig. 1A), and 2) elution of the bound components from the pre-column for HPLC on the analytical column, the two columns being connected in series by rotating the switching valve (Fig. 1B) and pumping mobile phase containing organic modifier (elution buffer).

This separation of the pre-column and chromatographic processes exposes proteins in the sample only to an aqueous environment devoid of organic modifier; hence the proteins are removed from the sample by passing through the pre-column to waste, while the hydrophobic components of low mol. wt. are retained on the pre-column for subsequent chromatography. The pre-column step performs two important functions of sample preparation: a) removal of the analyte from the sample matrix, and b) concentration of the analyte.

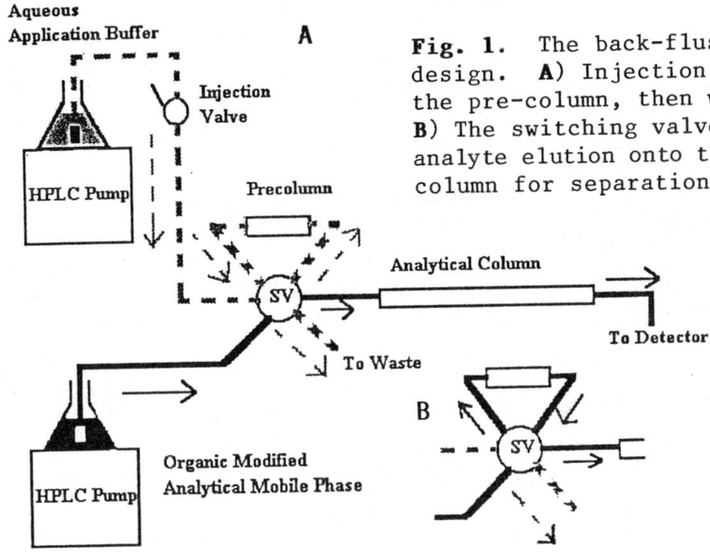

Fig. 1. The back-flush pre-column design. **A**) Injection of sample onto the pre-column, then wash-through. **B**) The switching valve, SV, allows analyte elution onto the analytical column for separation.

Typical pre-column design and operating conditions

Typically the pre-column, 10-40 × 4-4.6 mm, having either filter screens or sieves of 30 μm pore diameter replacing the frits, contains RP packing material preferably at least 30 μm in diameter (up to 50 μm). With a diameter of only 5 or 10 μm, there is retention of proteins which eventually elute onto the analytical column (Table 1), whose efficiency significantly deteriorates with a pressure increase in this column [5]. The use of frits in the pre-column similarly caused the analytical column to increase in pressure (by 18 bars) and decrease in efficiency (by 38%) after 75 100-μl serum injections, compared to pre-columns which had filter screens and showed only 1 bar pressure increase after 100 injections, the N value for the analytical column being undiminished [5].

For back-flush operation - the most widely used pre-column mode - the sample is loaded onto the pre-column in one direction and eluted off it in the opposite direction (Fig. 1), resulting in minimal increase in analyte band width. Placement of an on-line, periodically replaced, 0.5 μm filter and/or guard column immediately prior to the analytical column is also recommended. The pre-column application buffer is usually an aqueous buffer at a pH appropriate for retention of the analyte. Thus, for retention of acidic analytes, acidic application buffers have been employed (besides adjusting the sample to an acidic pH). The volume of application buffer used to wash the pre-column depends on its size: a survey of 13 published procedures where C-18 pre-columns were employed showed a range of 0.13-1.20 ml per mm pre-column length (i.d. normalized to 4 mm), with a

Table 1. Effect of pre-column particle diameter on system performance. (From ref. [5], *courtesy of authors and Elsevier.*) CONDITIONS.- Loading medium: water; eluent: 2.5 mM NaH_2PO_4, 45% (v/v) methanol, containing TEA 65 µl/L; operation: back-flush; end fittings: filter screens; pre-column: 25 × 3.9 mm; analytical column: 5 µm C-8, 150 × 4.6 mm; 100 injections of 100 µl of serum.

Pre-column		Pressure increase, bars		N (toluene), anal. col.	
Particle diam., µm	Packing material	Pre-col.	Anal. col.	Initial	Final
5	Supelcosil C-8	13.6	27.2	7080	2600
10	LiChrosorb C-18	3.4	13.6	6305	3560
37-50	Corasil C-18	0	1.36	7400	7450
30	LiChrosorb C-18	6.8	0	8325	7942

median value of 0.35 ml/min, which results in a median wash time of 7 min for a 20 × 4 mm pre-column at 1 ml/min flow rate. One report [6], however, states that the wash volume also depends on the volume of sample injected; serum samples of 100 µl and 1 ml required 2 ml and 4.6 ml of washing buffer respectively. The procedure also necessitates re-equilibrating the pre-column with application buffer prior to the next injection.

Protein binding considerations

An important issue when using pre-columns is the effect of protein binding on analyte recovery. The packing may compete with proteins for analyte binding; therefore the ability of the analyte to bind to the pre-column will depend not only on its affinity for the pre-column packing material but also on the strength of interaction between the analyte and any analyte-binding protein present in the sample. Drugs considered to be highly protein-bound, e.g. carbamazepine (82%) and phenytoin (89%), have been shown to furnish high recoveries (98% and 81% respectively) on a C-18 pre-column using a water mobile phase for application [5]. However, there are analytes which are more tightly bound to proteins, calling for modification of the procedure to ensure analyte binding to the pre-column. Examples include teniposide, for which plasma samples had to contain 38 mM SDS [7], and tefludazine and metabolites, where serum samples had to contain 10% MeCN [8].

Retinoids.- Extensive recovery experiments were done for retinoid compounds (which are >98% protein-bound) to improve recovery through variation of sample additives and mobile phase conditions [9]. As is shown in Table 2, addition of organic modifier to the sample is the single most effective way for improving recovery of most of the strongly bound retinoids,

Table 2. HPLC of retinoids (see text for identities): recovery from pre-column. Reference values (100%) were obtained by direct loading (bypassing the pre-column) of pure compounds in aqueous ethanol onto the analytical column (C-18; guard column too). Pre-column packed with Bondapak C-18 37-53 µm Corasil (Waters); 17 or 40 × 4.6 mm; 3 µm sieves in place of frits. Load vol. 500 µl. Excerpted from ref. [9]† (wherein are trials of isocratic mobile phases besides that given here). *Courtesy of R. Wyss and Elsevier.*

Plasma-sample diluent	Load/wash (10 min) medium; 'ORG' = 8:2 ammonium acetate-acetic acid (each 1%)/MeCN	% recovery of various retinoids (100 ng/ml)							
		1	2	3	4	5	6	7	8
None	Water	88	86	50	40	22	57	2	0
	ORG	93	92	56	38	75	65	11	6
9 mM NaOH/MeCN (8:2), 2:3	Water	91	89	82	84	76	91	11	0
	ORG	86	89	86	78	89	90	25	13
Ethanol, 1:2	Water	26	24	50	53	40	41	57	58
	ORG	32	31	76	81	62	68	71	63

being more effective for a greater number of retinoids than was merely increasing % organic modifier in the application buffer without modifying the sample. Evidently inclusion of MeCN in both sample and application buffer raised the recovery of one retinoid from 76% to 89% (13-*cis*-acitretin, #5 in Table), while the others showed similar recoveries compared with runs wherein only the sample had MeCN modifier. Compounds #7 (etretinate) and #8 (Ro 12-7554) are not adequately recovered regardless of the treatment employed. Ethanol addition to the sample (precipitating the proteins) increased their recovery to only ~60-70% - and severely reduced the recovery of the more polar retinoids (#1-#4; 4-oxo-iso-/4-oxo-/iso-/tretinoin, and tretinoin) which eluted off the pre-column during the wash step because of the sample's high organic content. Compound #6 is acitretin.

In a design variation conducive to disruption of analyte-protein binding without sample pre-treatment, on-line addition of MeCN at the injection stage to a retinoid-containing plasma sample is accomplished *via* a T-piece, which not only eliminates the sample-preparation step but, more importantly, minimizes contact time with organic modifier and hence retinoid isomerization such as can occur on adding MeCN prior to analysis [10].

---

†*Note by Ed.*- For a pertinent art. by the authors of ref. [9], see p. 79 in Vol. 22, this series [ed. Reid & Wilson; publisher as now]; the wash fluid composition is misstated (not an authors' error; at foot of p. 82 **delete** then) - but correct in present Table. This art. gives the gradient method developed; to get stability in the autosampler (N/A to pilot work, above) NaOH rather than water was used with MeCN for the plasma dilution.

Sample volume considerations

Two factors determine what sample volume can be injected onto a pre-column system:- 1) the breakthrough volume of the analyte from the pre-column, incorporating both injection volume and wash volume [see eqn. 4 below]; 2) loss of analytical column resolution if an increased sample volume enters it, as discussed below. Assuming that a 20% loss of resolution is acceptable, eqn. 1 gives the limit of sample volume that can be injected onto an analytical column (ac) with the parameters V_o (void vol.) and k' (capacity factor) [11]:

$$[V_{max}]_{ac} \approx 2.60[V_o(1+k')/N^{0.5}]_{ac} \quad (1)$$

In the pre-column technique, analyte enters the analytical column from the pre-column. Since the analyte is concentrated on the pre-column (pc), the volume of sample injected onto it will become concentrated by the following factor†:

$$F_{conc.} = [(k'_{inj.med.}+1)^2/(k'_{elut.}+1)^2]_{pc} \quad (2)$$

where the pre-column capacity factors (k') are for the analyte in injection medium and elution buffer respectively. Eqn. 2 differs from a previously proposed expression [11], but has theoretical and experimental support†. Hence for the maximum volume that can be injected onto a pre-column system an expression as in eqn. 3 can be derived [11 & †]:

$$[V_{max}]_{pcs} \approx \{2.60[V_o(1+k')/N^{0.5}]\}_{ac}\ [(k'_{inj.med.}+1)^2/(k'_{elut.}+1)^2]_{pc} \quad (3)$$

Knowing the maximum injection volume, one can calculate the pre-column size needed. The breakthrough volume for the emergence of the analyte from the pre-column will depend on $(k'_{inj.med.})_{pc}$. A previously derived expression [11] is modified here to include wash volume (V_{wash}), and gives the pre-column size needed, expressed as its void volume:

$$[V_o]_{pc} > 1.2\ ([V_{max}]_{pcs}+V_{wash})/[1+(k'_{inj.med.})_{pc}] \quad (4)$$

Eqn. 4 assumes that k' for the analyte on the pre-column is the same for the sample injection medium as for the wash buffer. It should be noted also that V_{wash} is a function of $[V_o]_{pc}$, ascertainable experimentally for each particular packing.

Variation of pre-column designs

Besides the basic back-flush design depicted in Fig. 1, other designs have been used, some incorporating sophisticated switching systems. A forward-flush mode has been recommended to counteract the tendency of the back-flush design to bring particulate matter onto the analytical column [12, 13]. The forward-flush mode is also helpful for removing interference peaks, by providing an additional separation of the analyte from the impurity on the pre-column [14, 15]. The back-flush

†W.J. Lough (& M.J. Mills) - talk/pers. comm. at 1995 Forum, & (at press date) art. #C-9n. [Eqn. (4), 'independent' of eqns. (2) & (3), implies that a wash step calls for a larger pre-column.]

mode, however, is the most popular because it minimizes band broadening, with benefit to sensitivity. This is underscored by published data [16] comparing, with the same analytical column (N = 3101 if no pre-column), back- and forward-flush operation of the pre-column (which contained 37-44 µm packing material): the N values were respectively 3010 and 1803 for the two pre-column modes.

Other switching design modifications have been introduced to increase throughput, e.g. multiple pre-columns in a parallel arrangement [17]. A 'boxcar' chromatographic approach using one pre-column gave an impressive throughput, 40 samples/h, in the determination of primidone and phenobarbital present in serum along with phenytoin and carbamazepine [5].

A simple pre-column arrangement, the 'loop column method', does not require an extra pump or a switching valve [8, 18, 19]: the pre-column replaces the injection loop, and conventional manual HPLC is performed. After sample loading onto the pre-column and washing with application buffer, both done by hand through the injector, the pre-column is switched on-line to the analytical column by moving the injection valve to the inject position. Automation is feasible using an autosampler [20].

Analytical column lifetime and detection limits

Reported column lifetimes range from 16 to 150 ml total serum/plasma [1]. In an assessment of analytical column performance after 2112 injections of 20 µl serum samples, N decreased from 8400 to 6500 and pressure rose from 124 to 144 bars [5]. Here a back-flush mode was used, the pre-column (as in Table 1, final line; screens in place of frits) and in-line filter were replaced every 200-300 injections, and the analytical column underwent some THF rinses and, after many runs, flow reversal.

Detection limits are enhanced insofar as large volumes of sample can be injected onto the pre-column, which then concentrates the analyte. LLOQ's of 1 ng/ml are achievable (with UV detection); 0.3-0.5 ng/ml was found for retinoids (360 nm) [21].

RESTRICTED ACCESS MEDIA (raM*)

While pre-column techniques divert protein by a switching strategy, with raM the protein problem is addressed through an RP packing material designed so that protein but not analyte is excluded from its hydrophobic surface. This is accomplished by surrounding the surface with a hydrophilic barrier that sterically prevents macromolecules from interacting with it, while allowing entry and thence chromatography of small molecules. Sample injection is done directly in a mobile phase containing organic modifier, at a level below the precipitation threshold.

*'RAM' (author's term) disfavoured (cf. data-store usage) notwithstanding 'hallowed' status.- *Ed.'s advocacy; convincing?*

Table 3. Restricted access media (raM)† commercially available (all with silica support), and pioneer refs., []; denoted ISRP, internal-surface reversed phase; SPS, semi-permeable surface; SHP, shielded hydrophobic phase; MF(P), mixed-functional (phase). GFF = glycine-L-phenylalanine-L-phenylalanine.

Type	Source/Product name	Surface chemistry: external; internal
ISRP [22]	Regis Chemical/ GFF & GFF II¤	Glycine; GFF
SPS [23]	same/SPS	Poly(oxyethylene) polymer; C-18/C-8/nitrile/phenyl options
SHP [24]	Supelco/Hisep	Hydrophilic network of polyethylene oxide with embedded hydrophobic phenyl groups
MF [25]	Shiseido/Capcell Pak MF Ph-1	Polyoxyethylene; phenyl, on polymer-covered silica
Other	Chrompack/Chrom-Spher Biomatrix	Proprietary; alkyl & phenyl
	Shandon/Ultrabiosep	ditto, 'biological polymer'; C-18/C-8

†'RAM' in lit.; see *, previous p. ¤For difference from GFF see text.

Types (listed in Table 3 if commercially available)

ISRP, the very first raM, was developed by Pinkerton in 1985 [22]. The properties of ISRP's, as reviewed [1, 2, 26], are now summarized. The peptide GFF (Table 3) is immobilized onto a diol silica support, and the hydrophobic phenylalanine groups are cleaved off from only those GFF's that are exterior to the pores, by use of carboxypeptidase A which is too large to penetrate the pores (median diam. 5.2 nm) and so cleave internal GFF's. The pores exclude the sample's proteins (size exclusion limit: mol. wt. 15000-25000) because of two characteristics of the support: 1) small pore diam.; 2) negatively charged glycine groups on the pore exterior whereby negatively charged proteins are repelled. The packing material retains analytes primarily through π interaction with the phenyl groups of the phenylalanine and secondarily through cation-exchange involving the negatively charged glycine groups. Limitations of GFF-ISRP include: a limited mobile-phase pH range (6-7.5, the lower limit reflecting the need to ensure charge repulsion of proteins from the pores, and the upper one the need for support stability); low retention capability, as the phenylalanine groups have weak hydrophobicity compared to C-8/C-18; and limited flexibility (only one type of RP ligand).

A second generation ISRP, GFF II [27], has been developed that has the same interior and exterior chemistry as GFF but differs in how the diol is chemically formed (monofunctional immobilization for GFF but multifunctional for GFF II). The GFF-II columns are more retentive and efficient (N = 60,000/m).

Although acceptable for many analytes that are sufficiently hydrophobic, GFF and GFF II columns may be unsuitable for less hydrophobic analytes, which elute along with or too near the large sample-matrix peak. The latter's tail appeared at 4 min when monitored at 240-300 nm and even as late as 18 min when monitored at <230 nm (1 ml/min, 15 cm column; T.C. Pinkerton*).

SPS, SHP, MF (Table 3).- Other marketed raM packings aim at retention better than for ISRP's. SPS and MF are based on a silica support with covalently linked RP and an outer covering of hydrophilic polyoxyethylene which blocks protein interaction with the RP. SHP is a variation of this theme, in which the phenyl groups are incorporated into a hydrophilic polyethylene oxide that is bonded to the silica and provides enough steric protection to isolate the phenyls from components of high mol. wt. such as proteins. For other raM entries in Table 3 the manufacturer has not fully disclosed the chemical make-up.

MFP is another raM packing material [28-30], fundamentally different from the MF raM given in Table 3. It consists of hydrophobic groups (e.g. phenyl, C-4, C-8, β-cyclodextrin) and diol groups on silica. Proteins pass through the column unretained because the diol groups give the support sufficient hydrophilic character to prevent proteins from adsorbing to the hydrophobic functionalities of the packing material. MFP is advantageous in that both hydrophilic and hydrophobic analytes can be determined. MFP is not yet commercially available.

Performance features

Column lifetime with raM is limited because adsorbed proteins accumulate with successive injections of serum/plasma, adversely affecting column performance. Lifetime assessments for raM have not been as quantitative as for the pre-column technique, rigorous N and pressure-increase data being scanty. However, GFF-ISRP columns were stated to survive >2000 injections of serum (20 µl; guard column replaced every 400 injections) [31], and SHP columns >1000 injections of 100 µl serum (guard column present; frequency of replacement unspecified) [32].

Retention.- It has been attempted to maximize retention of 6 sulphonamides on different raM materials [33]. Only SPS could resolve all 6 away from the matrix peak, the retention being less with SHP and even lower with GFF II. However, the retention of the sulphonamides on SPS was considerably less than with conventional C-18 columns. This is a persistent problem with commercially available raM supports, which give lower capacity factors, thus affecting the determination of some analytes that cannot be separated from the matrix peak.

---

*1988 review in the 'magazine' *Am. Lab.* (*20,* 70-76)

Limits of detection for raM are at least several orders of magnitude poorer than with pre-columns, mainly because these can concentrate large sample volumes, whereas raM is limited to small volumes (which are not concentrated) to maintain resolution. Another reason is that the tailing sample-matrix peak impairs detectability. A UV detection limit of 25 ng/ml has been reported for SPS [33]; but most investigators reckon the lower operating limit for raM to be ~1 μg/ml.

Other features of raM operation.- The organic content of the mobile phase must not exceed the applicable threshold at which precipitation occurs. Column efficiency (N) has been reported as 60,000 plates/m for both GFF II [27] and SPS [23].

CONCLUDING COMMENTS

Direct-injection HPLC - by pre-column techniques or raM - is highly advantageous because of its time-saving attributes. It may be the unique remedy for achieving HPLC determination of analytes that are not stable with the passage of time or in the conditions used for sample preparation. Pre-column techniques substantially surpass raM in detection limits, capacity factors and versatility that embraces a wider analyte-polarity range; moreover they are not limited in mobile-phase conditions, and the analytical column's lifetime appears to be better although more in-depth characterization of raM packings is awaited. Yet there are many applications apt for raM, which has the distinct advantage of not needing extra HPLC hardware such as switching valve(s) and an extra HPLC pump.

Acknowledgements

The author gratefully acknowledges the work of Dr W.J. Lough (Univ. of Sunderland, U.K.) and his helpful consultation on updating equations that quantify the maximum injection volume on pre-column systems.* The author also thanks Chemical Abstracts Service for providing access to STN International.

References

[*Chr.* as in *J. Chr.* denotes *Chromatography*]

1. Anderson, D.J. (1993) *Anal. Chem.* *65*, 434R-443R.
2. Pinkerton, T.C., Miller, T.D., Cook, S.E., Perry, J.A., Rateike, J.D. & Szczerba, T.J. (1986) *BioChr.* *1*, 96-105.
3. Kudoh, S. & Nakamura, H. (1991) *Anal. Sci.* *7* (Suppl.), 909-912.
4. Shihabi, Z.K., Thompson, E.N. & Constantinescu, M.S. (1993) *J. Liq. Chr.* *16*, 1289-1296.
5. Nazareth, A., Jaramillo, L., Karger, B.L., Giese, R.W. & Snyder, L.R. (1984) *J. Chr.* *309*, 357-368.
6. Altmayer, P., Büch, U., Büch, H.P. & Larsen, R. (1993) *J. Chr.* *612*, 326-330.

For stop-press corrections/comments see p. 188 (maybe 193/228 too); W.J.L.'s MS., art. #C-9n, and response to an "O.K.?" query were belated - Ed.

7. Van Opstal, M.A.J., Van der Horst, F.A.L., Holthuis, J.J.M., Van Bennekom, W.P. & Bult, A. (1989) *J. Chr. 495,* 139-151.
8. Nielsen, H. (1986) *J. Chr. 381,* 63-74.
9. Wyss, R. & Bucheli, F. (1988) *J. Chr. 456,* 33-43.
10. Wyss, R. & Bucheli, F. (1992) *J. Chr. 593,* 55-62.
[For a Vol. 22 art. by these authors, see footnote †, Table 2]
11. Lankelma, J. & Poppe, H. (1978) *J. Chr. 149,* 587-598.
12. Van der Horst, A., Goede, P.N.F.C., Willems, H.J.J. & Van Loenen, A.C. (1986) *J. Chr. 381,* 185-191.
13. Juergens, U. (1984) *J. Chr. 310,* 97-106.
14. Matsuura, A., Nagayama, T. & Kitagawa, T. (1989) *J. Chr. 494,* 231-245.
15. Posluszny, J.V. & Weinberger, R. (1988) *Anal.Chem. 60,* 1953-1958.
16. Hux, R.A., Mohammed, H.Y. & Cantwell, F.F. (1982) *Anal. Chem. 54,* 113-117.
17. Roth, W., Beschke, K., Jauch, R., Zimmer, A. & Koss, F.W. (1981) *J. Chr. 222,* 13-22.
18. Bechgaard, E. (1989) *J. Chr. 490,* 219-223.
19. Voelter, W., Kronbach, T., Zech, K. & Huber, R. (1982) *J. Chr. 239,* 475-482.
20. de Jong, J., Munniksma, C.N., Guerand, W.S., Bast, A. & Van der Vijgh, W.J.F. (1992) *J. Chr. 574,* 273-281.
21. Wyss, R. & Bucheli, F. (1990) *J. Pharm. Biomed. Anal. 8,* 1033-1037.
22. Hagestam, I.H & Pinkerton, T.C. (1985) *Anal. Chem. 57,* 1757-1763.
23. Glunz, L.J., Perry, J.A., Invergo, B., Wagner, H., Szczerba, T.J., Rateike, J.D. & Glunz, P.W. (1992) *J. Liq. Chr. 15,* 1361-1379.
24. Gisch, D.J., Hunter, B.T. & Feibush, B. (1988) *J. Chr. 433,* 264-268.
25. Kanda, T., Kutsuna, H., Ohtsu, Y. & Yamaguchi, M. (1994) *J. Chr. A 672,* 51-57.
26. Perry, J.A. (1990) *J. Liq. Chr. 13,* 1047-1074.
27. Perry, J.A., Invergo B., Wagner, H., Szczerba, T.J. & Rateike, J.D. (1992) *J. Liq. Chr. 15,* 3343-3352.
28. Haginaka, J. & Wakai, J. (1990) *Chromatographia 29,* 223-227.
29. Haginaka, J. & Wakai, J. (1990) *Anal. Chem. 62,* 997-1000.
30. Haginaka, J. & Wakai, J. (1992) *J. Chr. 596,* 151-156.
31. Rainbow, S.J., Dawson, C.M. & Tickner, T.R. (1990) *J. Chr. 527,* 389-396.
32. Ueno, R., Uno, K. & Aoki, T. (1992) *J. Chr. 573,* 333-335.
33. Brewster, J.D., Lightfield, A.R. & Barford, R.A. (1992) *J. Chr. 598,* 23-31.

[*Chr.* denotes *Chromatography*]

Note by Editor.- In eqns. (2) & (3) [credited to Lough] the final term has a ^{2} which is lacking in the similar-looking equation in art. #C-9n but is correct in the present context as Dr Lough has explained.- "Our paper [#C-9n] does *not* contain a $(k' + 1)^2$ term but this [Anderson's (2) & (3)] does need a square term because he is using a different form of the equation."

#D-2

NEW APPROACHES TO SOLID-PHASE EXTRACTION

[1]Colin F. Poole†, [1]Salwa K. Poole, [2]Donna S. Siebert & [2]Kenneth G. Miller

[1]Zeneca/SmithKline Beecham Centre for Analytical Chemistry, Imperial College of Science, Technology & Medicine, South Kensington, London SW7 2AY

[2]Department of Chemistry, Wayne State University, Detroit, MI 48202, U.S.A.

'Disk technology' devices that complement the traditional cartridge design for SPE have lately appeared: particle-loaded membranes and particle-embedded glass fibre disks. Their reduced height and larger cross-section besides improved packing density compared to cartridge devices is claimed to shorten processing times, decrease plugging by particles, reduce channelling and give cleaner extracts due to lower non-specific matrix adsorption.

Forced-flow planar chromatographic methods allow study of the complete set of kinetic and thermodynamic properties that affect the sampling performance of these disk devices. Also, a solvation parameter model has been developed which can predict breakthrough and elution volumes from sorbent characteristics. We can now offer a quantitative model for sampling performance (cartridge vs. *disk devices), selection of the best sorbent for a particular extraction, full experimental design, and prediction of safe sampling volumes. The model can explain the large changes in retention for porous polymer sorbents (on membranes) when different conditioning solvents are used, and predicts that polar sorbents, e.g. cyanopropylsiloxane-bonded silica, have smaller breakthrough volumes for all solutes compared with typical octadecylsiloxane-bonded sorbents. The model attributes the often observed poor performance of cartridge devices to inadequate packing density, which could be simply overcome.*

For pre-chromatographic isolation and concentration, SPE is widely used [1-4]. It has largely replaced liquid-liquid extraction, viewed as too labour-intensive, difficult to automate, and needing rather large volumes of expensive high-purity solvents. Moreover, extraction selectivity was limited by the narrow choice of immiscible solvents, and emulsion formation could add to the difficulty of obtaining two distinct phases. SPE using cartridge or disk devices may not have solved all these problems, and does have its own characteristic problems; but most have been satisfactorily addressed, as evidenced by the wide acceptance of SPE.

†addressee for any correspondence

The basic design of SPE cartridges has changed little since their introduction in the late 1970's. Typically a cartridge consists of a short column (generally an open syringe barrel) containing a sorbent of an average nominal particle size ~40 µm packed between porous metal or plastic frits. This design has certain attendant disadvantages: (1) the small cross-sectional area of the cartridges results in slow sample-processing rates and low tolerance to blockage by particles and adsorbed matrix components; (2) channelling reduces the cartridge's capacity to retain analytes; (3) sorbent properties can vary from lot-to-lot and between manufacturers; (4) incomplete reversibility of the sorption of some analytes at active sorbent sites impairs recovery; and (5) the isolated analytes may be contaminated by impurities originating from the manufacturing and packaging process. Maybe some of these problems could be addressed by improved quality-control procedures and design modification; but perhaps more radical solutions should now be considered.

In the last few years new products for SPE have appeared in the form of particle-loaded membranes [5-7] and particle-embedded glass fibre disks [8] referred to generically as disk technology. The former, in various diameters and 0.5 mm thick, consist of sorbent particles (~8 µm diam.; 90% w/w) immobilized in a web of PTFE microfibrils; the membranes are flexible and generally supported on a sintered glass disk (or other support) in a standard filtration apparatus, using suction to generate the desired flow through the membrane. The latter type of disk contains sorbent particles embedded in a glass fibre supporting matrix. Small-diameter glass fibre disks are rigid and self-supporting; for wider ones a support is usually employed.

The specific claims made for the disk rather than cartridge format are: (a) shorter sample processing time due to the larger cross-sectional area and reduced pressure drop enabling higher sample flow rates to be used; (b) decreased plugging by particles due to the larger cross-sectional area; (c) reduced channelling and improved kinetic performance resulting from the use of smaller sorbent particles and the greater mechanical stability of the sorbent bed; and (d) cleaner background and lower interferences by optimizing the bed mass to reduce non-specific matrix adsorption. From this perspective, the disk technology can be considered a further evolution in the SPE concept, addressing some of the shortcomings of cartridges.

Investigation of SPE devices.- Although SPE cartridges have been in use for >20 years, only scant information exists concerning their operating characteristics. This led us to study the kinetic and retention properties of SPE devices as a prelude to developing a model for computer-aided development of methods* and to compare devices of different formats and brands so as to delineate particular strengths and weaknesses.

*Cf. D.L. Massart's 'system': p. 335, Vol. 23, this series.-*Ed.*

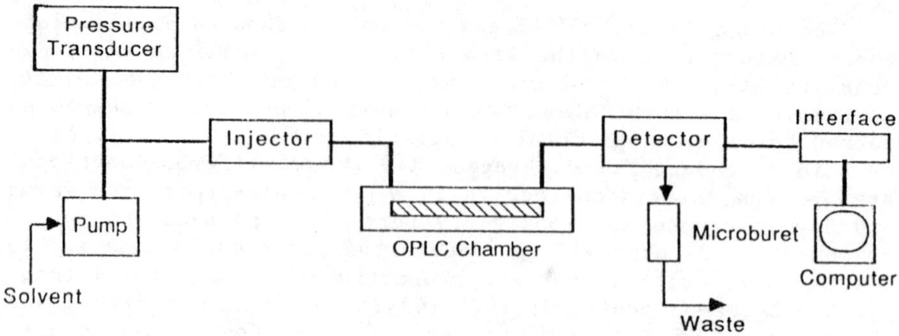

Fig. 1. Block diagram of the apparatus used to determine the kinetic properties of particle-loaded membranes and particle-embedded glass fibre disks for SPE. *Adapted from a diagram in ref. [10].*

The characteristic properties of cartridge sorbents can be studied by HPLC [9]; but disk devices need a special apparatus (Fig. 1) devised originally for forced-flow TLC [10] (and used by us to establish the kinetic properties of pre-coated TLC plates [11]). In the present context the disk is supported on an aluminium foil and sealed along its edges with a latex glue. The supported disk is then placed in the over-pressure development chamber (OPLC) and the top surface is sealed hydraulically to the foil support by intimate contact with a pressurized plastic membrane that withstands pressures up to 25 bar. Cut into the disc material are a trough that accepts solvent flow from a pump and, opposite, another for solvent egress with on-line connection to a UV detector. This apparatus serves to set and control the mobile phase velocity and to accurately determine disk properties through peak-width and capacity-factor measurements. To carry out such studies using standard experimental procedures with a suction apparatus would be tedious or even impossible.

KINETIC PROPERTIES OF CARTRIDGE DEVICES STUDIED BY HPLC

Table 1 summarizes the characteristic kinetic properties of three different SPE cartridge sorbents, all prepared from silica substrates nominally small-pore (6 µm) and high surface area, having a high ligand-bonding density as indicated by the hydrophobicity index [9, 12]. The intra-particle porosity was very small in the case of the C-18 material (Table 1), which indicates considerable filling and/or constriction of the pores by the heavy loading of bonded ligands. The C-18 material is near-pellicular in character while the more polar bonded-phase sorbents with their shorter chain length are significantly more porous; these differences should influence their characteristic mass transfer properties (see later).

Table 1. HPLC determination of the characteristic kinetic properties of 3 cartridge sorbents manufactured by J.T. Baker. Average particle size was determined by the Kozeny-Carman equation from the column permeability.
For specific permeability (unit: m^2; the exponential factor 10^{12} serves to aid numerical tabulation) the reference state is an open tube with the cross-sectional area indicated.
The no. of theoretical plates per cm refers to a cartridge having packing density as for the HPLC column and operated in a 0.5-30 ml/min flow-rate range.

Property		C-18	Cyanopropyl	Diol
Packing density, g/ml	Cartridge	0.667	0.599	0.557
	HPLC column	0.726	0.634	0.794
Apparent particle size, μm		55.1	30.3	92.9
Porosities	total, HPLC column	0.47	0.52	0.58
	inter-particle	0.41	0.41	0.41
	intra-particle	0.06	0.11	0.17
Specific permeability, $10^{12}m^2$		2.14	34.1	3.29
Flow resistance parameter		707	2050	
No. of theoretical plates, n/cm		15 to 42	6 to 10	9 to 19
Hydrophobicity index [9, 12]		1.53	1.04	1.14
Silanophilic index [9]		0.60	0.64	0.61

The sorbent's kinetic performance will also be influenced by the particle size, which differed significantly amongst the three materials (Table 1). The 55.1 μm value for C-18 is close to the nominal value claimed by the manufacturer (50 μm); but the cyanopropylsiloxane-bonded and spacer-bonded propanediol sorbents deviate markedly, and the former has an appreciable fraction of sub-size particles, skewing the average size to a lower value which is also reflected in the specific permeability and flow resistance parameter. This is undesirable since it leads to reduced flow capacity under suction and potential contamination of isolated extracts with sorbent particles. The spacer-bonded propanediol phase shows particle-size skewing in the opposite direction, a significant proportion being above average size. This is also undesirable; although it creates a packed bed with acceptable flow characteristics it confers a spatially heterogeneous particle density. The silanophilic index (Table 1) is fairly typical of bonded phases formed from trifunctional silane monomers and indicates a significant but unexceptional capacity for silanophilic interactions.

Further aspects of cartridge packings and performance

It is noteworthy that the packing density of the cartridge devices is significantly less than that of HPLC columns prepared from the same sorbent by the tap-and-fill method (Table 1); so the tabulated column efficiencies have to be reckoned a ceiling for a well prepared cartridge, and typical cartridges provide significantly fewer theoretical plates. The efficiency of the column (cartridges) depends significantly on flow rate in the inter-particle velocity range ($u_e = 0.5$-4.0 mm/sec) corresponding to a flow rate of ~0.5-30 ml/min through a cartridge of ~1 cm i.d. Plate height, H (mm), at the column packing density for the three cartridge sorbents varied linearly with mobile phase velocity over the typical sampling flow-rate range:
C-18, $0.227 + 0.052 u_e$; CN, $0.844 + 0.202 u_e$; diol, $0.471 + 0.172 u_e$

The main contributions to the plate height arise from flow anisotropy in the packed bed and resistance to mass transfer. The near-pellicular nature of the C-18 material may contribute favourably, by its reduced resistance to mass transfer, to this influence on plate height compared with the other bonded phases which have significantly greater porosity. Typical cartridges, however, due to their low packing density contain substantial free space that will result in heterogeneous flow through the sorbent bed as the sample-flow streams encounter regions of different packing density. Thus, the C-18 sorbent cartridges contained on average ~19% additional unpacked volume compared with the stable HPLC column bed. It is, then, unlikely that typical SPE cartridges will provide more than ~5-15 plates per cm of bed height under typical sample-processing conditions.

We can deduce from these studies that typical cartridges show variations in sampling performance that result from an inadequate packing density and a heterogeneous particle-size distribution. A simple remedy to reduce performance variations is to consolidate each cartridge bed prior to sampling, by vibration and re-positioning of the upper frit if the design allows this. From the manufacturing perspective the end-user has the right to expect tighter control of the particle size range, a higher packing density, and a tighter range of packing densities. There is no evidence that acceptable sampling performance, with cartridges of a conventional bed height, calls for a smaller average particle diameter. On the other hand, the typical bed heights used in cartridge devices are probably greater than required to compensate for the bed's heterogeneous packing structure, implying use of more sorbent than would be required if the bed characteristics were optimized, which in turn contributes to increased contamination of sample extracts by elevating the concentration of non-specifically adsorbed sample matrix components.

Table 2. Determination by forced-flow TLC of the characteristic kinetic properties of two C-18 particle-loaded membranes and a C-18 particle-embedded glass fibre disk.

Property	Membranes		Disk, Ansys
	Baker	Varian	
Apparent particle size, µm	7.7	5.8	15.3
Porosities: total	0.52	0.54	0.51
Porosities: inter-particle	0.37	0.48	0.47
Porosities: intra-particle	0.15	0.06	0.04
Specific permeability, $10^{14}m^2$	2.50	2.21	8.38
Flow resistance parameter	1100	1040	970
Plate height minimum, mm	0.056	<0.12	
Optimum mobile phase velocity, mm/sec	0.13	<0.05	
Hydrophobicity index	1.55	1.51	1.52
Silanophilic index	1.34	1.14	1.74
Nominal pore diam. before bonding, nm	6	6	8

KINETIC PROPERTIES OF DISK DEVICES STUDIED BY FORCED-FLOW TLC

Table 2 summarizes the characteristic kinetic properties of C-18 products - two particle-loaded membranes [10, 13] and a particle-embedded disk [8], all silica-based with a small pore diameter and a high ligand bonding density as indicated by the hydrophobicity index and intra-particle porosity. The high silanophilic index indicates a significant concentration of accessible silanol groups. Taken together, the high hydrophobicity and silanophilic indexes are fairly typical of polymeric bonded phases formed from multifunctional silanes with water present. The low intra-particle porosity results from filling and/or constriction of the pores, a further general property of this type of sorbent.

The devices contain both sorbent particles and a support matrix; so the interpretation of their properties based on models for a homogeneous particle bed have to be treated carefully. The assigned values of average particle size (Table 2) seem in fair accord with the manufacturers' specifications for the particle-loaded membranes; those for the particle-embedded glass fibre disks are somewhat larger but plausible given the unusual microscopic structure of these materials. The specific permeability of the disk materials is similar to that of HPLC columns packed with a similar sized sorbent, indicating that it is the thinness of the disks (~0.5 mm) as well as their large diameter that is important in generating their high sample flow rates. The flow resistance parameters indicate a homogeneous structure for the materials and absence of channels or holes.

Plate heights with disks

The two particle-loaded materials differ in the relationship between plate height and flow rate: it was linear for the Varian product (see below), indicating that flow anisotropy and resistance to mass transfer govern the material's efficiency:

$$H = 0.081 + 0.905 u_e$$

The Varian product's optimum mobile phase velocity is too slow (<0.05 mm/sec) to be practically useful. Likewise in the above expression the large coefficient associated with the velocity term suggests poor mass transfer. It should be relatively easy to make improvements in the kinetic properties of this material within its normal sampling flow-rate range.

Fig. 2 shows, for the J.T. Baker particle-loaded membrane, the change in plate height as a function of the inter-particle mobile phase velocity. The minimum plate height at ~7 particle diameters is somewhat higher than for a typical well-packed column, and occurs at an optimum inter-particle mobile phase velocity of 0.19 mm/sec. This is ~1 order of magnitude slower than for a well packed column. Curve-fitting of the data in dimensionless form (reduced parameters) gives **A** = 3.75 (flow anisotropy), **B** = 1.72 (longitudinal diffusion) and **C** = 1.54 (resistance to mass transfer). Compared to a well-packed column the packing density of the particle-loaded membrane is somewhat lower. The heterogeneous nature of the membrane structure may contribute to the unfavourable flow anisotropy term, whereas the mass transfer term seems to be influenced by this heterogeneity and/or the bonding density of the organic ligands to the silica particles. The microfibrils that hold the sorbent particles in place in the particle-loaded membranes conceivably contribute adversely to the flow-anisotropy and resistance-to-mass-transfer terms (**A, C**) by disruption of the flow streams

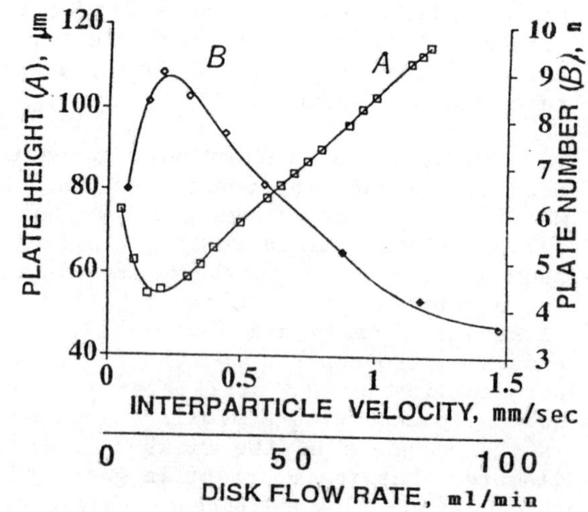

Fig. 2. Efficiency of a particle-loaded membrane (J.T. Baker) *vs.* inter-particle mobile phase velocity. *A*, a van Deemter-type plot: plate height as a function of velocity. *B*, transformation of to represent no. of theoretical plates achieved *vs.* sample flow rate through an 0.5 mm disk with a 38 mm diam. active sampling area.

through the membrane. Also, the web of microfibrils may bring about trapping, in the inter-particle space, of a portion of the mobile phase, increasing the volume that is stagnant and leading to an increase in the resistance-to-mass-transfer term.

Sample processing in SPE with a particle-loaded membrane is generally performed in the range of ~4-9 theoretical plates (Fig. 2B), comparable to that anticipated for cartridge devices containing a similar amount of sorbent (500 mg). So the number of plates is never large, and although its value may influence the breakthrough volume (see later) it is unlikely to be the dominant term for well retained analytes unless a threshold plates value is required to avoid instantaneous breakthrough.

BUILDING A MODEL TO INTERPRET BREAKTHROUGH AND ELUTION VOLUMES

Under normal operating conditions, breakthrough volume is governed by kinetic parameters and by retention. To understand how this volume depends on these parameters a suitable model is required to describe the extraction process in terms of them. The retention of an analyte by an SPE device can be described by a frontal analysis model assuming a linear distribution isotherm as proposed by Lovkvist & Jonsson [14], later shown by us to describe reasonably the sampling characteristics of SPE cartridge and disk devices [9, 10, 15]. The key equation can be written as:

$$V_B = (a_o + a_1/n + a_2/n^2)^{-\frac{1}{2}}(1 + k_S)V_m$$

where V_B is the breakthrough volume, V_m the hold-up volume for the device, k_S the analyte's capacity factor, and n the total no. of plates for the device. The coefficients a_o, a_1 and a_2, tabulated in the cited paper [14], are determined by the breakthrough level. The breakthrough volume increases proportionally with the hold-up volume, but this is not the preferred approach to its optimization since large sampling devices need large volumes of eluting solvent and would be difficult to use with a simple suction apparatus. Thus, in practice, V_m is limited to a small range of values given the intended use of the device, and for the present discussion we can reckon that V_m is fixed by the design of the device.

The influence of the kinetic performance of the sampling device on the breakthrough volume appears in the first part of the above equation's right-hand side. The homogeneity and quality of the sorbent bed and the sample processing rate will influence breakthrough volume through the dependence of n on these parameters for any defined breakthrough level. This contribution is evaluated in Table 3 for a J.T. Baker particle-loaded membrane and sorbent cartridge, assuming a 1% breakthrough level. Their kinetic properties differ: cartridges provide larger breakthrough volumes at low flow rates, while

Table 3. Kinetic contribution to the breakthrough volume for J.T. Baker C-18 devices (500 mg): particle-loaded membrane and and cartridge. High density = packed column density (Table 1; 0.726 g/ml); typical density = an estimated more likely value at the lower packing density typical of cartridges.

Sample flow rate, ml/min	Particle-loaded membrane	Cartridge of high density	Cartridge of typical density
5	0.535	0.769	0.651
10	0.592	0.732	0.608
20	0.595	0.669	0.518
30	0.573	0.618	0.477
40	0.539	0.560	0.427
60	0.485		
80	0.438		
100	0.405		

the membrane has a maximum midway in the flow-rate range. The cartridge shows greater dependence of breakthrough volume on flow rate; but for volumes determined within a modest range of flow rates - which would be considered good laboratory practice - the kinetic contribution is near-constant. Hence flow-rate variations are less important as an experimental optimization variable than is retention [6].

With V_m fixed and the kinetic contribution to the breakthrough volume varying little with typical sample-flow rates, the dominant parameter in the breakthrough volume equation is the retention term. Retention can be predicted using the solvation parameter model wherein the dependent variable, SP, is either $\log k_s$ or $\log V_B$:

$$SP = c + mV_X/100 + rR_2 + s\pi_2^H + a\alpha_2^H + b\beta_2^H$$

where V_X is the solute's characteristic volume, R_2 its excess molar refraction, π_2^H is a measure of the solute's ability to stabilize a neighbouring dipole by virtue of its capacity for orientation and induction interactions, and α_2^H and β_2^H are parameters characterizing respectively the solute's hydrogen-bond donor acidity and acceptor basicity. Solute descriptors for >2000 compounds are available, and unknown values can be determined by calculation or experiment.

The system constants m, r, s, a, b and c in the equation (also see below) are characteristic of the sampling system (sorbent and sample solvent). Evaluation of these parameters is by multiple linear regression analysis: (1) the breakthrough volume is settled for a solute series with known descriptors [15, 16] *or* by HPLC (or forced-flow TLC); and (2) the solute capacity factor either (2a) is determined with a mobile phase of identical composition to the sample solvent [12, 17] *or* (2b) - if (2a) is impracticable - is estimated by extrapolating the capacity factor-mobile phase composition relationship to the sample-composition value [8, 9, 13].

The system constants are unambiguously defined: r and s refer to the ability of the solvated sorbent to, respectively, interact with solute n- or π-electron pairs and take part in dipole-dipole and dipole-induced dipole interactions; a and b are measures of the solvated solvent's hydrogen-bond basicity and acidity respectively; and m indicates the relative ease of forming a cavity for the solute in the sample solvent and solvated sorbent. The solvation parameter model has served very well to explain other solubility-related properties including GC and HPLC retention and octanol-water partition coefficients, and to characterize coatings for sensors [18-20].

METHOD-DEVELOPMENT OPTIMIZATION USING THE MODEL

In SPE the sorbent is normally conditioned prior to sample application with a small volume of organic solvent - likewise added, for large-volume samples, to the sample (typically 1% methanol, v/v) as a processing aid to ensure efficient sample extraction and reasonable and stable sample flow rates [6]. The influence of the processing solvent on retention and elution can be gleaned from a plot of the system constants *vs.* solvent composition (Fig. 3). The left-hand side of Fig. 3 provides the system constants for the chosen sampling conditions and allows the breakthrough volume to be estimated for any analyte whose solute descriptors are ascertainable or calculable. The right-hand side indicates the range of solvent compositions at which the same analyte can be eluted with a low capacity factor from the sorbent (k_s <1 is preferred). Retention maps of the

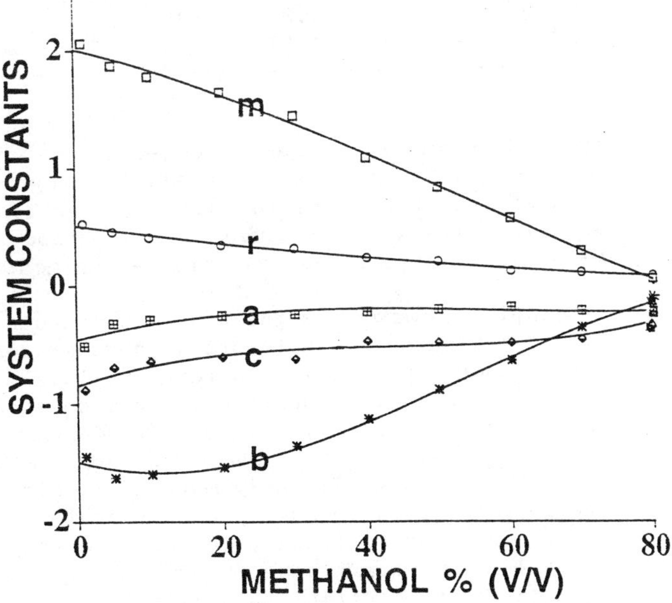

Fig. 3. Variation of the system constants with methanol concentration for a cyanopropylsiloxane-bonded silica-based sorbent. The s-constant was statistically insignificant at all compositions.

Table 4. System constants for three cartridge sorbents (1% MeOH). - *See text.*

System constant	C-18	Cyanopropyl	Diol
m	5.65	2.06	1.57
r	0.75	0.53	0.61
s	-0.76	0.0	0.0
Statistics:			
r	0.982	0.991	0.982
S.E. of the estimate	0.18	0.064	0.068
F = Fisher statistic	140	175	124
n = no. of solutes	18	18	27

(*r* = multiple correlation coefficient)

Contribution of individual intermolecular interactions to retention

Solute	Sorbent	Contribution to retention						
		$mV_X/100$	rR_2	$s\pi_2^H$	$a\alpha_2^H$	$b\beta_2^H$	c	$\log k_s$
n-propyl-benzene	C-18	6.44	0.42	-0.38		-0.49	-1.18	4.81
	CN	2.20	0.33			-0.22	-0.88	1.44
	diol	1.79	0.37			-0.12	-1.05	0.99
benzo-nitrile	C-18	4.92	0.52	-0.84		-1.08	-1.18	2.34
	CN	1.68	0.40			-0.49	-0.88	0.71
	diol	1.37	0.45			-0.26	-1.05	0.51
acetanilide	C-18	6.29	0.61	-1.06	-0.20	-2.18	-1.18	2.28
	CN	2.15	0.47		-0.26	-0.98	-0.88	0.50
	diol	1.75	0.53		-0.23	-0.54	-1.05	0.47
phenol	C-18	4.38	0.56	-0.78	-0.24	-0.98	-1.18	1.76
	CN	1.49	0.43		-0.31	-0.44	-0.88	0.29
	diol	1.22	0.49		-0.27	-0.24	-1.05	0.15
hexan-2-one	C-18	5.47	0.09	-0.52		-1.66	-1.18	2.20
	CN	1.87	0.07			-0.75	-0.88	0.31
	diol	1.52	0.08			-0.41	-1.05	0.15

type shown in Fig. 3 can be constructed for different sorbents and/or co-solvents, and the breakthrough and elution volumes predicted in each case for all analytes of interest, so as to establish the working range of the method and compare findings with requirements, thus avoiding experiments that will fail.

Table 4 summarizes the system constants for three different J.T. Baker cartridge sorbents with water containing 1% methanol as the sample processing solvent [9, 12, 17]. The main driving force for retention, positive system constants, is the difference in the work required to form a cavity in the stationary phase compared to the sample solvent, m-constant. In this respect it is significantly more difficult to form a cavity in the solvated polar bonded phases than for C-18, and this is a major source of difficulty in their retention properties. The polar bonded phases have a greater capacity for interactions with π- and n-electrons than does the C-18 sorbent, but their magnitude is too small to compensate for the difference in the cavity term. The spacer-bonded propanediol and cyanopropylsiloxane-bonded sorbents are more competitive with the sample solvent than the C-18 sorbent for interactions of a dipole type (s-constant) and are stronger hydrogen-bond bases and acids (a- and b-constants) than the C-18 sorbent.

Overall the sampling system is dominated by the characteristic properties of water - the source of the favourable cavity term and the unfavourable polar interaction terms. Sorbent selectivity is manifested by the solvated solvent's capacity to compete with, but not dominate, the sample solvent for the full concert of intermolecular interactions. What this means in terms of isolation of a varied group of analytes appears in the latter part of Table 4. Evidently, from the data presented, for all practical purposes the relative difficulty of cavity formation in the cyanopropylsiloxane-bonded and spacer-bonded propanediol sorbents is never fully offset with these sorbents by the more favourable interactions of a dipole type ($s\pi_2$) and hydrogen-bond type ($a\alpha_2^H$, $b\beta_2^H$). Hence these sorbents would never be more useful than the C-18 sorbent for the SPE isolation of organic non-electrolytes from water.

With porous polymer sorbents (on membranes) of the styrene-divinylbenzene type, the selective sorption of the organic solvent from the sample changes stationary phase volume and the system selectivity (Table 5), which can cause large changes in breakthrough volume [15, 16]. Such large changes, exceptional with silica-based bonded-phase sorbents, are most likely due to the flexibility of the porous polymer matrix, conducive to selective solvation and swelling of the sorbent particles.

The polymer's organic-solvent uptake is probably accompanied by changes in the phase ratio, which together with other statistical and chemical factors is incorporated into the c-constant. The composite expression $(mV_X/100)+c$ can be taken as a rough measure of the influence of solvent uptake on ease of cavity formation for the solvated sorbent. The wide range of m-constant and c-term values for the four solvents is a firm indication of the influence of solvent uptake by the porous polymer on the relative ease of cavity formation and hence on retention. The tabulated system constants observed for polar interactions illustrate that significant changes in sorbent selectivity accompany the uptake of the sample co-solvent. An indication of the relative contribution of solute size and polar interactions to the breakthrough volume is apparent (for selective solutes) in the middle portion of Table 5. Isopropanol and methanol are the preferred co-solvents for toluene, providing larger breakthrough volume volumes than those of acetonitrile and tetrahydrofuran. The dominant consideration

Table 5. System constants for particle-loaded membranes having a porous polymer or C-18 sorbent, with 1% supplements (iPrOH, isopropanol; MeCN, acetonitrile; THF, tetrahydrofuran).

System constants	Porous polymer				C-18,
	methanol	MeCN	THF	iPrOH	methanol
m	5.16	5.72	5.09	7.20	5.14
r	0.81	-0.26	0.0	0.41	0.0
s	-0.65	-0.35	0.0	-0.34	-0.92
a	-1.85	-1.17	-0.92	-1.50	-1.05
b	-2.93	-2.81	-4.19	-4.53	-2.24
c	-0.77	-1.19	-1.00	-2.10	-1.23
Statistics:					
r (as in Table 4)	0.999	0.999	0.999	0.999	0.996
S.E. of the estimate	0.036	0.041	0.039	0.035	0.070
F = Fisher statistic	1108	1269	2478	1313	594
n = no. of solutes	22	25	35	22	23

Contribution of individual intermolecular interactions to breakthrough vol. for the polymer sorbent with 1% v/v organic co-solvent

Solute	Solvent	Contribution to log V_B: $(mV_X/100)+c$	rR_2	$s\pi_2^H$	$a\alpha_2^H$	$b\beta_2^H$	Predicted V_B, cm^3
toluene	iPrOH	4.070	0.246	-0.177		-0.634	3199
	methanol	3.652	0.487	-0.338		-0.412	2449
	MeCN	3.712	-0.156	-0.182		-0.393	957
	THF	3.362				-0.587	596
benzo-nitrile	iPrOH	4.171	0.304	-0.377		-1.494	402
	methanol	3.724	0.601	-0.722		-0.967	433
	MeCN	3.792	-0.193	-0.389		-0.927	192
	THF	3.433				-1.383	112
heptan-1-ol	iPrOH	6.223	0.087	-0.143	-0.555	-2.174	2742
	methanol	5.195	0.171	-0.273	-0.685	-1.406	1005
	MeCN	5.422	-0.055	-0.238	-0.433	-1.349	2742
	THF	4.884			-0.340	-2.011	341

Contribution of individual intermolecular interactions to breakthrough vol. for the C-18 sorbent with 1% methanol as the sample solvent

Solute	Solvent	$(mV_X/100)+c$	rR_2	$s\pi_2^H$	$a\alpha_2^H$	$b\beta_2^H$	Predicted V_B, cm^3
toluene		3.175		-0.478		-0.314	242
benzonitrile		3.247		-1.021		-0.739	31
heptan-1-ol		4.712		-0.386	-0.389	-1.078	723

is size, since toluene is only weakly polarizable and a weak hydrogen-bond base. For larger molecules similar to toluene in capacity for polar interactions isopropanol will provide larger breakthrough volumes than methanol due to its more favourable $(mV_x/100) + c$ term. Benzonitrile is slightly larger than toluene but has a greater capacity for dipole-type interactions and is a weak hydrogen-bond base. The methanol-solvated sorbent is the least competitive for dipole-type interactions, but this is largely offset by its capacity to interact with solute n- and π-electrons. For benzonitrile methanol is the preferred co-solvent because the methanol-modified sorbent is a weaker hydrogen-bond acid than the isopropanol-modified sorbent; isopropanol would be preferred for larger solutes with similar hydrogen-bond basicity. Heptan-1-ol has a modest capacity for dipole-type and n- and π-electron interactions and is both a hydrogen-bond acid and a hydrogen-bond base. Acetonitrile and isopropanol are the preferred co-solvents; the acetonitrile-modified sorbent is the most competitive in hydrogen-bond type interactions and can offset the less favourable cavity term.

For higher homologues isopropanol is the preferred co-solvent because the favourable cavity term can offset its less favourable capacity for hydrogen-bond type interactions. On such grounds, in descending the series acetonitrile would be an increasingly better choice to maximize the breakthrough volume.

Breakthrough volumes are generally larger for the porous polymer sorbent than for the C-18 sorbent with 1% methanol as the sample solvent (Table 5). This is accounted for by a more favourable cavity term supplemented by a significant contribution from n- and π-electron interactions. Polar interactions of a dipole type favour the porous polymer sorbent over C-18, while hydrogen-bond type interactions show the opposite preference. The balance of interactions will generally favour the porous polymer sorbent when methanol is the sample co-solvent; but the differences in magnitude for the breakthrough volume are again dependent on the solute's size and its capacity for polar interactions.

CONCLUSIONS

Although kinetic and retention properties contribute to the breakthrough volume in SPE, retention is recognized as being the most important term. In fact devices with very large plate numbers are not required for SPE; rather the requirement is for a homogeneous packing structure providing a minimum threshold efficiency. The primary reason for the oversizing of cartridge devices is their low packing density, resulting in channelling, which is offset by increasing the bed mass to ensure adequate performance with the penalty of an increase in non-specific sample adsorption and more complex fractions compared to the membrane-based devices.

The solvation parameter model can predict the retention properties of SPE devices and yields an easily understood picture of the sampling process in terms of the fundamental intermolecular interactions involved. It clearly indicates for aqueous samples that the main driving force for retention is the difference in the work required to form a cavity in the solvated sorbent and sample solution. For polar bonded phases the former needs more energy than is realized by the additional analyte-sorbent polar interactions, resulting in lower breakthrough volumes compared with octadecylsiloxane-bonded sorbents. The propensity of porous polymers to swell by selectively absorbing co-solvent has a large impact on their sampling properties, as has to be taken into account in any discussion of their general properties and applications.

References

[*Chr.* as in *J. Chr.* denotes *Chromatography*]

1. Poole, C.F. & Poole, S.K. (1991) *Chromatography Today*. Elsevier, Amsterdam; see pp. 777-790.
2. Poole, S.K., Dean, T.A., Oudsema, J.W. & Poole, C.F. (1989) *Anal. Chim. Acta 236,* 3-42.
3. Liska, I. (1993) *J. Chr. A 655,* 163-176.
4. Berrueta, L.A., Gallo, B. & Vicente, F. (1995) *Chromatographia 40,* 474-483.
5. Hagan, D.R., Markell, C.G., Schmitt, G. & Blevins, D.D. (1990) *Anal. Chim. Acta 236,* 157-164.
6. Mayer, M.L. & Poole, C.F. (1994) *Anal. Chim. Acta 294,* 113-126.
7. Barcelo, D., Chiron, S., Lacorte, S., Martinez, E., Salau, J.S. & Hennion, M.C.(1994) *TRAC 13,* 352-358.
8. Mayer, M.L., Poole, C.F. & Henry, M.P. (1995) *J. Chr. A 695,* 267-277.
9. Miller, K.G. & Poole, C.F. (1994) *J. High Resolu. Chr. 17,* 125-134.
10. Fernando, W.P.N., Larrivee, M.L. & Poole, C.F. (1993) *Anal. Chem. 65,* 588-595.
11. Poole, C.F. & Fernando, W.P.N. (1993) *J. Planar Chr. 6,* 357-361.
12. Seibert, D.S. & Poole, C.F. (1995) *J. High Resolu. Chr. 18,* 226-230.
13. Mayer, M.L., Poole, S.K. & Poole, C.F. (1995) *J. Chr. A 697,* 89-99.
14. Lovkvist, P. & Jonsson, J.A. (1987) *Anal. Chem. 59,* 818-821.
15. Larrivee, M.L. & Poole, C.F. (1994) *Anal Chem. 66,* 139-146.
16. Poole, S.K. & Poole, C.F. (1995) *Analyst 120,* 1733-1738.
17. Seibert, D.S. & Poole, C.F. (1995) *Chromatographia 41,* 51-60.
18. Abraham, M.H. (1993) *Chem. Soc. Rev. 22,* 73-78.
19. Poole, C.F., Kollie, T.O. & Poole, S.K. (1992) *Chromatographia 34,* 281-302.
20. Carr, P.W. (1993) *Microchem. J. 48,* 4-24.

#D-3

SOME EXPERIENCES WITH SPE IN BIOANALYSIS

Selina C. Bridges, Mark R. Taylor, Steven A. Westwood† and Karl Woodward

Horseracing Forensic Laboratory, P.O. Box 15, Snailwell Road, Newmarket, Suffolk CB8 7DT, U.K.

Drug screening on a complex matrix such as equine urine calls for an extraction system which is both highly selective in rejecting the matrix and yet 'universal' in extracting all possible drugs and metabolites. For this application, SPE using mixed-mode cartridges has been highly successful [1-3]. Mostly when performing SPE we are using mixed retention modes anyway but are just unaware of this [4]. Thus with C-8 phases the underlying silanols allow polar and ion-exchange mechanisms in addition to the primary hydrophobic interaction.♦ In order to get good recovery of analytes, all retention mechanisms must be suppressed during the elution step. This is probably the biggest source of difficulty in developing good SPE methods.

Moreover, 'equivalent' phases from different manufacturers may differ*, as shown in Fig. 1 for two makes of C-8 cartridge. Selectivity differences can be attributed to several factors, notably the availability of underlying silanols or other retention sites and the amount of phase available to the sample.

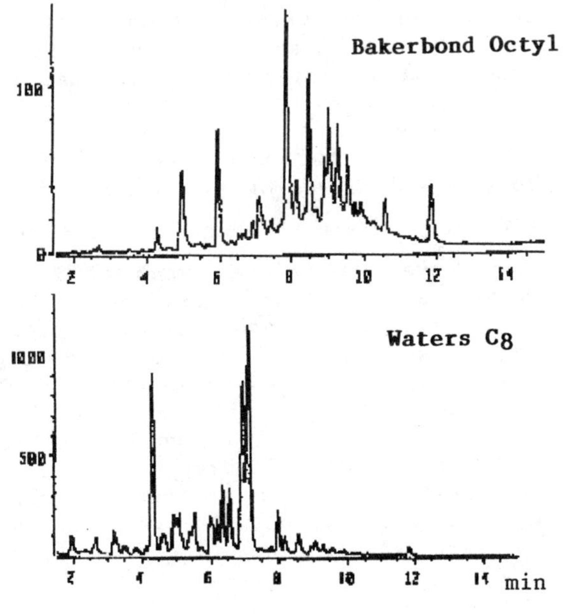

Fig. 1. Comparison of two commercial SPE C-8 cartridges, used for RP-HPLC analysis of an equine urine extract. *Elution profiles very different, indicative of different selectivity.*

†addressee for any correspondence

♦Earlier Vols. have pertinent Index entries ('Packings', 'SPE') and an art. (B. Law, Vol. 22) on ion-exchange exploitation of NP phases.- *Ed.*

*Vol. 23 has a pertinent art. (Rees & Austin; eds. and publisher as now).- - *Ed.*

Mixed mode cartridges

The term mixed mode usually refers to cartridges which in effect contain more than one phase, thereby exhibiting more than one primary retention mechanism. The most successful type to date contains a hydrophobic phase, C-8 or similar, and a strong cation-exchange (sulphonic acid) functionality. Several manufacturers produce cartridges of this type which we have been using for some years [5] in the analysis of a hydrolyzed basic-drug fraction in our general screen. However, different manufacturers produce these phases by different routes. One method is to synthesize separately the hydrophobic phase and the sulphonic acid phase and then to mechanically admix them. This approach has the advantage that better control can be achieved during chemical synthesis of the individual phases. The other approach produces a multi-functional single material by reacting the organosilanes onto the same 'base' silica.

Manufacturers are reluctant to give detailed information on their manufacturing routes, but a simple test indicative of the phase type can be applied. When packings of mechanically mixed origin are added to a 1:1 hexane/water mixture, the hydrophobic material settles at the interface and the denser hydrophilic cation-exchange material sinks to the foot of the water layer. In contrast multi-functional materials form a rather emulsified layer at the interface only or spreading into the hexane layer.

Significant selectivity differences are observed between these differently produced materials, and we have found differences in the tendency to block during sample loading, unrelated to particulate matter that may clog the frits.- There seems to be a chemical reaction between sample components and sorbent, leading to an impermeable layer at the top of the bed. Bed permeability can be regained by a solvent wash step. Formation of an impermeable layer is much more marked with the biphasic compared with the monophasic material, the latter producing a more diffuse layer of sorbed material.

SPE using membranes, and cartridge behaviour

There have been a number of new developments in SPE using membranes rather than packed beds. A major advantage of these is the absence of channelling. A packed-bed cartridge may have regions where the sorbent is not so tightly packed, maybe near the wall. This is especially likely when cartridges are transported, or stored, lying on their side. When the sample, or a reagent, is passed through the bed it will take the line of least resistance, leading to localized overload, or else non-interaction with the sorbent and hence low recoveries. The membrane approach obviates this problem since the silica particles are held in place by the membrane matrix material.

The nature of the problem is indicated by irreproducibility that may be encountered in a typical SPE method with packed-bed SAX cartridges. The results are mostly quite uniform but have a few obvious outliers. These are replicates where recovery sharply drops, such that the overall C.V. (precision criterion) may be as high as 25-50% whereas with the outliers removed the value drops to a more acceptable 5-15%. Typical behaviour is exemplified by peak-height data (mAU) for 100 mg SAX cartridges and 20 mg membranes, with 4 neutral drugs: (1) theophylline; (2) hydroxyethyl-(1); (3) trichloromethiazide; (4) sulphamethoxypyridazine.-

SAX, duplicates: (1) 6, 8; (2) 18, 50; (3) 12, 25; (4) 22, 55.
Membranes, " : (1) 20, 20; (2) 140, 140; (3) 30, 30; (4) 80, 80

Membranes in cartridge format have the great attraction of low bed volume which allows low elution volumes, pertinent to on-line automation. However, problems remain to be overcome, especially with loading samples onto cartridges. High back-pressures and proneness to blockage have seriously impeded the implementation of this technology for widespread screening; but progress is being made, and new products are in prospect.

Automation

There is a wide choice of instrumentation for automating SPE, mainly full robotic systems *vs.* autosampler-based semi-robotic systems. The former type's complexity and great cost can be justified where the process carried out involves more than just the normal SPE steps, e.g. a centrifugation. Where relatively simple procedures are in use the autosampler-type systems can be very cost-effective. Recent developments are enabling the simpler type of instrumentation to carry out increasingly complex tasks, aided by much simpler programming.

New instrumentation is beginning to appear which will allow batch rather than sequential processing of samples - a welcome idea to many people who have performed manual extractions with vacuum boxes. Yet the merits of this approach depend largely on the analytical system with which the SPE instrumentation is to be used. Schemes 1-3 show a comparison of batch and sequential processing used with batch and sequential analyses.

Another area of increasing interest is direct coupling of the SPE system to the analytical instrument. This is quite straightforward for SPE-HPLC systems where the eluent is HPLC-compatible: if not, the eluting solvent must be removed, a step which is often rate-limiting. Similarly, coupling SPE and GC is much more problematic; solvent elimination and derivatization are again often rate-limiting. Here lies the immense advantage afforded by the greatly reduced elution volumes achievable with membrane cartridges. Only 100 µl may suffice; so even a rather involatile solvent, e.g. methanol, may be removable in 1-2 min.

Schemes 1-3. Sequential *vs.* batch processing in SPE: notional analyses on 50 samples (t = min elapsing for each) with an 8-min SPE step.

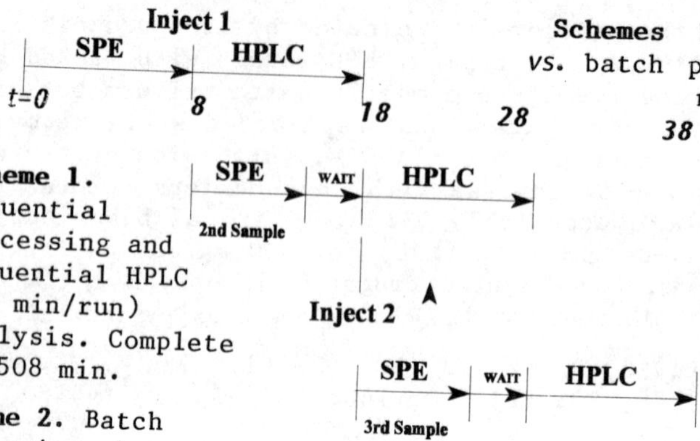

Scheme 1. Sequential processing and sequential HPLC (10 min/run) analysis. Complete in 508 min.

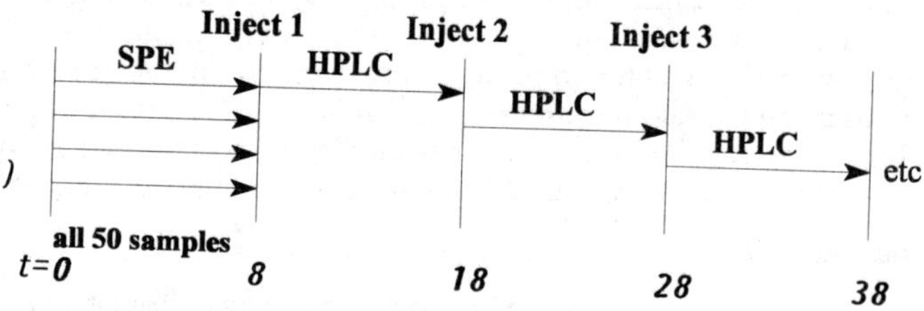

Scheme 2. Batch processing, but otherwise as for Scheme 1. Complete in 508 min.

Batch SPE (rather than sequential SPE) with sequential analysis does not save time.

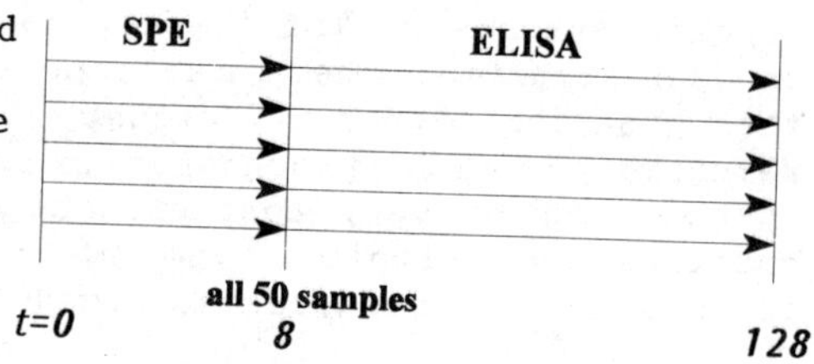

Scheme 3: batch processing and simultaneous batch analyses, e.g. ELISA (120 min). Complete in 128 min. *Batch SPE with either very fast sequential analysis or (as here) batch analysis can save time.*

A further consideration with batch *vs.* sequential SPE is the concept of just-in-time processing. Situations occur where the sample cannot be left for any length of time due to instability of one kind or another, e.g. solvent evaporation, the use of unstable reagents (which can be prepared by the robotic system and used immediately) and, with labile solutes, elution-solvent detriment maybe due to pH, e.g. hydrolytic or oxidative losses.

References

1. Singh, A.K., Ashraf, M., Granley, K., Mishra, U., Madusudana Rao, M. & Gordon, B. (1989) *J. Chromatog. 473,* 215-226.
2. Dumasia, M.C. & Houghton, E. (1991) *J. Chromatog. 564,* 503-515.
3. Chen, X-H., Franke, J-P., Wijsbeek, J. & de Zeeuw, R.A. (1991) *J. Anal. Toxicol. 16,* 351-355.
4. Ruane, R.J., Wilson, I.D. & Tomkinson, G.P. (1988) in *Bioanalysis of Drugs and Metabolites, Especially Anti-Inflammatory and Cardio-vascular* [V. 18, this series] (Reid, E. *et al.* eds.), Plenum, N.Y., pp. 295-8.
5. Westwood, S.A. & Dumasia, M.C. (1994) in *Sample Preparation for Biomedical and Environmental Analysis* (Stevenson, D. & Wilson, I.D., eds.), Plenum, New York, p. 163 onwards.

#D-4n

A Note on

RAPID SAMPLE-PREPARATION APPROACHES, NOTABLY DISC-BASED SPE, FOR HPLC-MS-MS(API) [▫]

Terence A.G. Noctor[§], Martyn Hemsley & [†]John N. Robson

Corning Hazleton, Otley Road, Harrogate HG3 1PY, U.K.

MS-MS allows HPLC to be unsubtle and brief; but benefit to sample throughput is limited by the usual slowness of sample preparation (which, if SPE, may take near-80% of total time; 4% for the analysis, and the rest for data management). One new approach may ameliorate the sluggishness of sample-preparation advances, viz. disc-based SPE. This article focuses on SPE (especially automated) and, firstly, on protein precipitation[▫].

PROTEIN PRECIPITATION METHODS

Deproteinization is a simple approach, as reviewed [e.g. 1]. It merely needs an agent such as acetonitrile (water-miscible), perchloric or trichloroacetic acid or ('old-vintage' example) $Zn(OH)_2$. The common practice of adding acetonitrile (at least 1 vol.), the supernatant being directly loadable for HPLC, often results in peak-shape distortion, ameliorated by suitable pre-dilution to provide a pseudo-mobile-phase composition.

Sensitivity improvement in API* methods by drying down the supernatant under a N_2 stream and redissolving the residue in a small volume of chosen solvent may bring a variability penalty, especially with very low analyte levels. The probable reason is that relatively large amounts of endogenous material may get carried through and variably influence analyte ionization - and maybe also cause blockage, more frequently than usual, in the capillary tubing of the mass spectrometer's ion source. There may be a need, especially where pg/ml sensitivity is wanted, for an additional or alternative step; the best choice is SPE.

[▫]To complement the authors' text (now abridged), pertinent arts. in this and earlier vols. warrant consultation, including the #ABC sections which have examples of unorthodox procedures, e.g. 'old-fashioned' means of deproteinization.- *Ed.*

[§]addressee for any correspondence

[†]now at Perkin-Elmer, Post Office Lane, Beaconsfield HP9 1QA

**Abbreviations include:* in MS (MS-MS = tandem): API, atmospheric pressure ionization. LLE, liquid-liquid extraction.

SOLID-PHASE EXTRACTION (SPE) METHODS

SPE has many advantages: it gives a cleaner extract than protein precipitation; it suits well for sample-concentration and automation, and surpasses LLE in speed and solvent economy and choice - there being no need for immiscibility or circumvention of emulsion problems. Conventional SPE, with sorbent (usually modified silica) packed in a barrel between two frits, has been around, with little change, since the mid-1970's [2].

Disadvantages.- Fast channels may develop in the bed, reflecting its fluid nature or particle irregularity, if not poor packing. Analyte then in effect becomes smeared deeply in the bed rather than staying as a sharp topmost band, and breakthrough during loading has to be guarded against. Moreover, solvent volumes may have to be larger (and more time allowed; near 10 min even normally) for pre-conditioning, washing and elution. Other troubles are now less common: batch differences and product-derived interferences, e.g. fines or rogue peaks.

Disc-based SPE embraces several formats: glass microfibre discs, and PTFE microfibril membranes; other impregnated membranes (PVC) and derivatized membranes are less often used. Glass microfibre discs with only 15 mg of silica-based sorbent may have as much capacity as a conventional cartridge and, being rigid, are not prone to channelling*. As void volume is <50 µl solvent volumes are minimal (Fig. 1) and processing is brief (4.5 *vs.* 18 min); similarly for PTFE microfibril membranes.

Automated disc operation is rather slow with commercial systems, as they are based on those for conventional SPE with rather slow liquid-transfer times. Thus it pays to scrutinize each step in an automated method, speeding up those transfers that are not flow-critical (conditioning, washing, rinsing) but leaving alone those not apt for speed-up (sample aspiration and application to the disc; elution). With large batches it saves time if rinse solvents etc. are optimally located according to frequency of use, so reducing wasted instrument time. Shrinkage from 15-20 to <5 min per sample does not impair performance and offers throughput approaching that of modern HPLC-MS-MS(API).

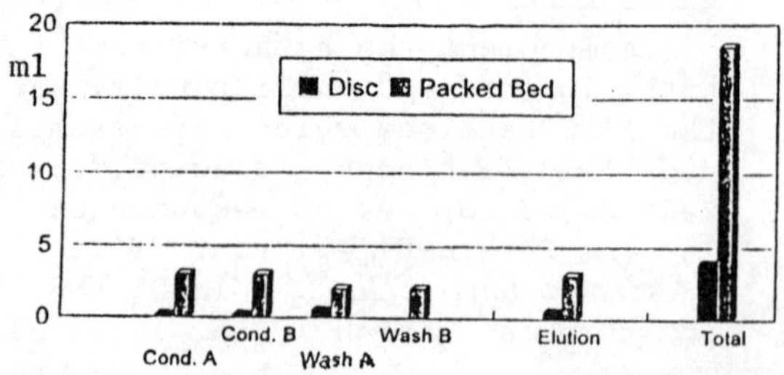

Fig. 1. Disc (glass fibre) *vs.* conventional SPE for a plasma analyte: solvent use. (Cond. = conditioning.) No difference in recovery or reproducibility.

*Dye studies by G.M. Hearne & D.O. Hall: *Am. Lab.*, Jan. 1993 (a 'magazine').

References

1. Blanchard, J. (1981) *J. Chromatog. 226,* 455-462.
2. McDowall, R.D. (1989) *J. Chromatog. 492,* 3-11.

#D-5

IMMUNOAFFINITY AS A SAMPLE CLEAN-UP STEP FOR DETECTING DRUGS AT LOW LEVELS IN EQUINE BODY FLUIDS

M.A. Seymour✦, J. Byard, [1]S. Feely, H. Marsh, K. Reeve, C.J. Roberts, P. Teale, [1]C.S. Creaser & E. Houghton

Horseracing Forensic Laboratory, P.O. Box 15, Snailwell Road, Newmarket, Suffolk CB8 7DT, U.K.

and [1]Department of Chemistry & Physics, Nottingham Trent University, Clifton Lane, Nottingham NG11 8NS, U.K.

The selectivity of Ab's* and the consequent potential of IAf to effect purifications with factors of perhaps several thousand-fold in a single step≠ make it an attractive technique for forensic applications. IAf is particularly applicable to the analysis of very potent drugs or drugs in slow-release formulations. Such analytes and/or their metabolites may be present only at extremely low concentrations in biofluids.

IAf extraction

An automated extraction system has been developed, usable for a range of analytes, e.g. dexamethasone (DX; below). To purify antisera, normally raised in sheep or rabbits, Protein-G Sepharose is used to isolate an IgG fraction. This is then covalently bound to CNBr-activated Sepharose. Any residual active groups are blocked with a small amine, e.g. Tris.

The column is equilibrated with pH 7.0 20 mM phosphate buffer/0.5 M NaCl (loading buffer). For all solvents the flow rate, using a peristaltic pump, is 1-2 ml/min. After sample application the column is washed with loading buffer and eluted with 1:1 (by vol.) 2 M propionic acid/methanol. Collected eluate fractions are, depending on the compound, concentrated for GC-MS analysis or applied directly to an LC-MS system *via* a trace-enrichment cartridge. Before applying further samples the column is flushed with a second cycle of loading buffer followed by elution solvent. Columns are stored at 4° in loading buffer containing sodium azide as a bacteriostat.

Pilot work was carried out using a column prepared from an Ab raised against a synthetic corticosteroid (DX). The loading and elution solvents were optimized using ^{3}H-DX. Fig. 1 shows

---

✦addressee for any correspondence

**Abbreviations.*- Ab, antibody; IAf, immunoaffinity; DX, dexamethasone; S/N, signal-to-noise; nCI, -ve ion chemical ionization.

≠*Note by Ed.* on 'background' to IAf: see Index entry 'Affinity' in earlier vols. and key arts. in Vols. 22 (p. 297) & 20 (p. 365; IAf on-line with LC) [eds. E. Reid *et al.*; publisher as now].

Fig. 1. IAf extraction of ^{3}H-DX from spiked equine urine (0.4 µg/ml): 5 ml applied at 1 ml/min to an anti-DX column (~6 × 1 cm); see text for elution.

Liquid scintillation counting of 1 ml fractions showed that two contained 86% of the applied activity.

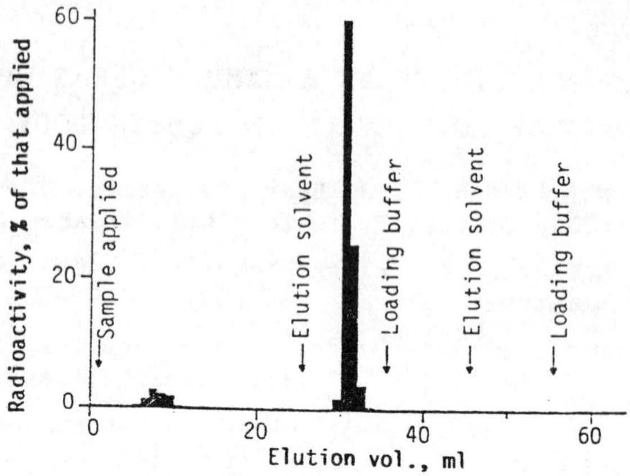

a typical profile of radioactivity eluted from the column. Its recovery was quantitative, with most of the radioactivity being eluted after switching to the elution solvent and hardly any in the second elution cycle. The column was shown to extract DX and other corticosteroids which cross-react with the Ab (betamethasone, flumethasone, 9-α-fluoroprednisolone, prednisolone) from equine urine. The column capacity was ~0.4 µg DX/ml of gel; at least 100 ml of urine could be applied with little loss of analyte or S/N deterioration on GC-MS analysis.

Coupling to LC-MS.- An on-line IAf-LC-MS system has also been developed. IAf extraction is carried out essentially as in the off-line system. Using valve switching, a portion of the column eluate containing the analyte is then trapped in a loop, diluted with water to facilitate focusing of the sample at the head of the analytical column, and transferred to the LC-MS(APcI, positive ion). The system has been used to detect DX at levels down to 0.2 ng/ml in spiked urine. Refinements that are planned include incorporation of a trace-enrichment cartridge in place of the loading loop. Work has also started on developing an on-line IAf-GC-MS system.

Illustrative applications

Following administration of DX (20 mg, i.m.) to a thoroughbred gelding, urine samples were taken and extracted using the anti-DX column. For comparison, the same samples were also extracted with dichloromethane. After forming the methoxime-trimethylsilyl derivatives, both sets of extracts were analyzed by GC-MS(nCI). The sensitivity of the IAf method was at least as good as that of solvent extraction, and the IAf extracts gave 'cleaner' chromatograms, with a better S/N ratio. The

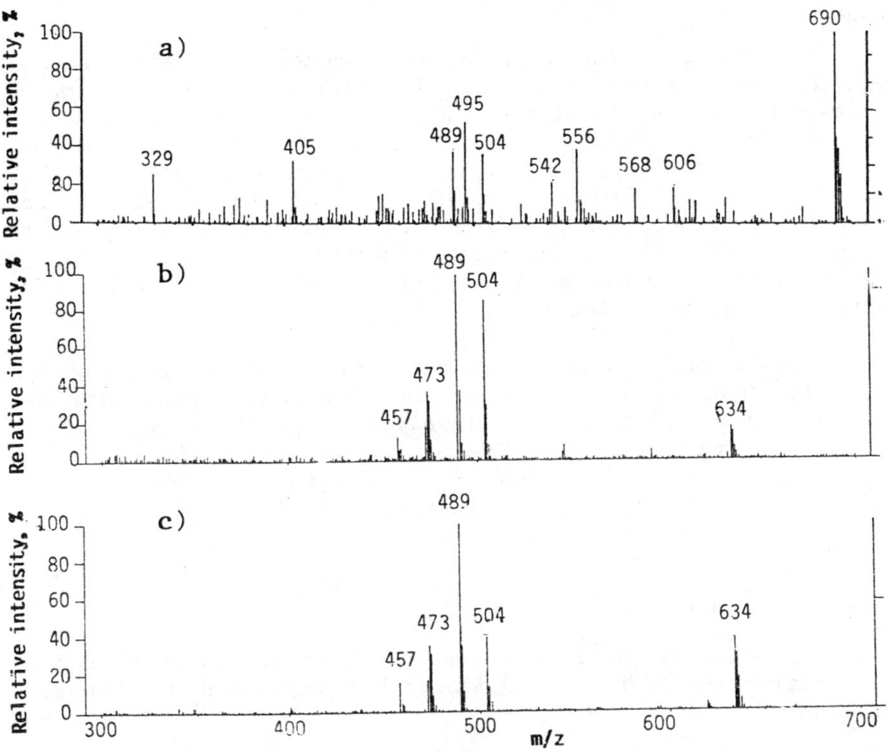

Fig. 2. GC-MS analysis of extracts of urine collected 23 h after dosing: nCI mass spectra from: **a**) solvent extract (from 10 ml urine), and **b**) IAf extract (from 5 ml urine); **c**) is a derivatized DX standard.

mass spectra from IAf extracts of low-level samples, but not those from solvent extracts, met in-house acceptance criteria for confirmatory data (Fig. 2). Thus the detection limit was lowered, and the effective detection period after dosing was longer using IAf than for solvent extraction.

Reserpine is a highly potent and long-acting tranquillizer and anti-hypertensive agent. Whilst not widely used in humans, it is still used in veterinary medicine and is reportedly abused in the equine sporting world to calm excitable horses prior to competition in events such as dressage. It is extensively metabolized, little or no unchanged drug being excreted in the urine. However, following the development in this laboratory of an ELISA to the drug, it was noted that a small proportion of the urine samples being screened were giving positive results. IAf has been used successfully to isolate a number of cross-reacting substances, which appear to be metabolites of reserpine. Further work is under way to confirm their identities and structures.

DISCUSSION

IAf evidently can be a powerful technique in the field of equine dope testing. It is particularly useful for samples containing low levels of analyte, when large volumes of matrix may be applied to the column to increase the analyte load with little or no increase in background, so greatly improving S/N. The technique is limited to those compounds for which Ab's can be raised, although in practice this is seldom a problem. Care should be taken with drugs that are extensively metabolized, to ensure that the Ab's selected for immobilization cross-react with appropriate analytes.

Development of the IAf extraction system is continuing, e.g. to investigate the use of rigid supporting media and the use of columns of immobilized analyte to purify Ab's prior to coupling to the support. The use of monoclonal Ab's is also being investigated. A number of clones that show specific binding and displacement have been identified. These are being cultured, to provide sufficient Ab for full evaluation.

Acknowledgement

The work on on-line systems has been partly supported by the Engineering & Physical Sciences Research Council, through a CASE award to S.F.

#D-6n

A Note on

DRUG EXTRACTION BY SELECTIVE SPE USING IMMOBILIZED ANTIBODIES: TRIALS WITH MORPHINE AND CLENBUTEROL

°B. Rashid, †P. Kwasowski & °D. Stevenson

°Robens Institute †ClifMar Associates, Sch. of Biol. Sciences
University of Surrey, Guildford GU2 5XH, U.K.

Whilst there have been many elegant advances in analytical instrumentation, the most time-consuming step in trace-organic analysis is sample preparation prior to instrumental analysis. Automation of injection and data handling is now commonplace; but manual sample preparation still prevails. Although sample preparation is important and is reckoned to be the most notable source of errors, new investigation has been rather scant. SPE (typically with silica, bonded silicas or carbon) is gaining favour over liquid-liquid extraction; it is easier to automate, produces less solvent waste, and allows trace enrichment.

Bonded-Ab* phases [1-8] have effected selective extraction both on-line with column-switching and off-line using typical SPE apparatus. One problem is to bind the Ab to a rigid matrix without losing Ab activity. A further problem is development of an extraction based only (or almost entirely) on the Ab-antigen interaction with no secondary interactions, which if non-selective can lead to difficulty in eluting analytes in a small volume such that pre-concentration is unnecessary.

We have developed a protocol that has proven successful for two herbicides - chlortoluron [9] and isoproturon [10] - which, with the respective Ab's, were enrichable from water, at least 1000-fold, by virtually identical methods. We also showed that analyte retention was due to Ab-antigen interactions. We now sought to extend this work by similarly immuno-extracting the drugs morphine and clenbuterol.¤ We emphasize that the aim of this preliminary work, not with a bio-matrix, was to develop, optimize and clarify the IAf extraction procedures rather than to develop at this stage new methods for these drugs - chosen as model compounds because, for each, antibodies raised in sheep were already available. HPLC was based on published methods [11,12].

EXPERIMENTAL

Morphine.- The HPLC apparatus consisted of a #110B pump (Beckman Insts., High Wycombe), a Coulochem #5610OA EC detector

**Abbreviations*.- Ab, antibody; IAf, immuno-affinity.
¤cf. Kwasowski's clenbuterol study, end of Sect. #D

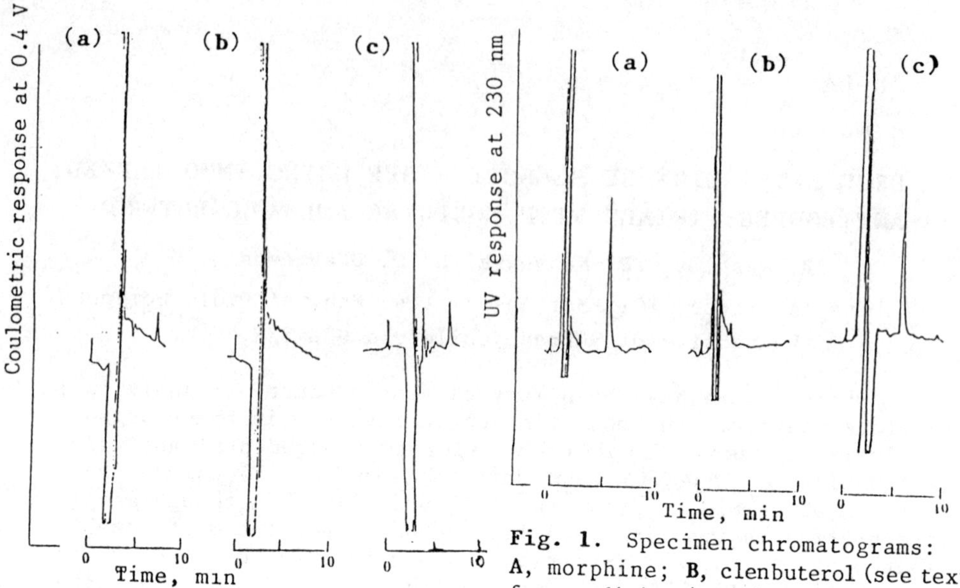

Fig. 1. Specimen chromatograms: **A**, morphine; **B**, clenbuterol (see text for conditions). (**a**) Unextracted standard; (**b**), wash fraction; (**c**) immuno-extracted standard.

(ESA, Bedford, MA, U.S.A.) operated at +0.40 V, and a Hypersil cyanopropyl 5 µm column, 150 × 5 mm (Jones Chrom'y, Hengoed). The mobile phase was 15% (v/v) acetonitrile in 65 mM NaH_2PO_4 pH 2.5 containing 1.5 mM SDS; flow rate 1 ml/min; typically 100 µl was injected, by Waters WISP.

Clenbuterol HPLC was performed generally as above but with a Spherisorb ODS2 5 µm (150 × 5 mm) column, 230 nm detection and 25% acetonitrile in pH 3 buffer containing 1% TEA.

IAf sorbents.- The solid support was aldehyde-activated controlled pore-size (10 mµ) glass beads (100 µm diam.; ~2 g used); for Ab binding (by ClifMar Associates) 1 ml of morphine antiserum or 0.5 ml clenbuterol antiserum per g beads was used.

RESULTS AND DISCUSSION

IAf columns were loaded with 1 ml of sample, and fractions were collected. Our previous herbicide work had shown pH 7.4 loading to allow retention by the Ab phase; elution in one 1 ml fraction had been possible using low pH but had needed 50% v/v ethanol in the elution solvent. With this protocol columns had proven stable for >50 water samples although the concentration of organic modifier had to be minimized to avoid denaturation of the bound Ab. Typical HPLC traces are shown in Fig. 1 for SPE eluates from pH 7.4 samples containing, in a 'PBS' medium (phosphate-buffered saline), morphine or clenbuterol. Amounts and SPE conditions are indicated later (text and the heading to Table 1).

Developing the procedure adopted.- As for the herbicides, morphine and clenbuterol were each retained and none was eluted by up to 10 washes with pH 7.4 PBS. Table 1 shows recoveries.

Table 1. Analyte recovery (%) at different buffer (PBS) pH's - 4 successive 1 ml fractions (#1-#4), and total. No analyte eluted in the loading or in 5 1-ml wash fractions preceding #1.

		Without modifier					With modifier					
Analyte	pH	#1	#2	#3	#4	Total	Modifier	#1	#2	#3	#4	Total
Morph-ine, 25 ng	7	15	19	17	15	56	Ethanol, 40% v/v	10	65	25	0	100
	6	15	15	7	7	44		13	74	13	0	100
	5	11	16	24	18	69		33	67	0	0	100
	4	8	7	8	8	31		0	100	0	0	100
Clen-buterol, 200 ng	7	0	0	0	0	0	Methanol, 50% v/v	0	43	32	10	85
	6	0	0	0	0	0		2	50	35	13	100
	5	0	0	11	10	21		5	35	38	10	88
	4	0	50	15	3	68		0	13	50	8	71
	3	0	0	16	46	62		0	50	15	3	68
	2	0	13	68	11	92		0	108	0	0	108

The drugs are evidently being retained and, from the results in the left part of Table 1, recovered quite well at lowered pH but not in a single fraction. Total elution in a volume >1 ml is disfavoured because sample preparation time would lengthen.

Organic modifiers (methanol, ethanol and acetonitrile) in combination with PBS varying in pH have been evaluated, taking account also of interference from solvent fronts as we were injecting the IAf column extracts direct onto the HPLC column. For morphine the best results were obtained using ethanol as modifier, and for clenbuterol methanol was better (right part of Table 1). The desired 100% recovery in one 1 ml fraction was obtainable with morphine at pH 4 (with no benefit from a lower pH) and clenbuterol at pH 2.

CONCLUSIONS

The optimized immuno-extraction protocol involved elution of morphine with 40% ethanol in pH 4 PBS and of clenbuterol with 50% methanol in pH 2 PBS. Each eluted quantitatively in one 1 ml fraction. These conditions closely resemble those we have presented for the herbicides chlortoluron and isoproturon [9, 10]. Our work suggests that a generalized protocol for sample isolation on rigid particles is feasible, and it will be further explored in bioanalysis contexts. Immuno columns might suffice as the only sample preparation procedure needed, or may prove useful as one step in a complex procedure. The major limitation is the availability of antisera; but many pharmaceutical companies raise antisera to their compounds when these show promise at an early stage of development.

The analyte levels that could be measured eventually will depend principally on the detection step rather than the immuno column, which accordingly could be much smaller and so use less Ab than in the preliminary work now described.

References

1. van Ginkel, L.A., Stephany, R.W., van Rossum, H.J. & Zoontjes, P.W. (1992) *TRAC 11,* 294-298.
2. Bagati, R., Oriundi, M.P., Russo, V., Danese, M., Berti, F. & Fannelli, R. (1991) *J. Chromatog. B 564,* 493-502.
3. Farjam, A., Brugman, A.E., Lingeman, H. & Brinkman, U.A.Th. (1991) *Analyst 116,* 891-896.
4. Komatsu, S., Murata, S., Aoyana, H., Jenki, T., Ozawa, T., Tateishi, M. & Vrbanac, J. (1991) *J. Chromatog. B 568,* 460-466.
5. de Frutos, M. (1995) *TRAC 14,* 133-140.
6. Pichon, V., Chen, L. & Hennion, M-C. (1995) *Anal. Chem. 67,* 133-140.
7. Thomas, D.H., Beck-Westermeyer, M. & Hage, S. (1994) *Anal. Chem. 66,* 3823-3829.
8. Rule, S.G. & Henion, J-D. (1992) *J. Chromatog. A 582,* 103-112.
9. Shahtaheri, S.J., Katmeh, M.F., Kwasowski, P. & Stevenson, D, (1995) *J. Chromatog. A 697,* 131-136.
10. Shahtaheri, S.J., Kwasowski, P. & Stevenson, D. (1996) *J. Chromatog.,* in press.
11. Pawula, M., Shaw, P.N. & Barrett, D.A. (1994) *J. Chromatog. B 653,* 106-111.
12. Botterblom, M.H.A., Feensta, M.G.P. & Erdstieck-Ernste, E.B.H.W. (1993) *J. Chromatog. B 613,* 121-126.

Note by Ed.- IAf-SPE arts. in earlier vols. [ed. E. Reid *et al.;* Roy. Soc. Chem.] include a general account (embracing antiserum generation) by L.F. Statham *et al.* (Vol. 22, pp. 297-304), and coupling to HPLC (A. Farjam *et al.;* cf. ref. 3. above). In all vols. the index entry 'Affinity' may be helpful.

#D-7n

A Note on

SAMPLE COLLECTION: ISSUES, PROBLEMS AND SOLUTIONS

†H.M. Hill, K. Tennant, T.A.G. Noctor & S. Pratt

Corning Hazleton (Europe), Dept. of Biopharmaceutical Analysis, Harrogate, N. Yorks. HG3 1PY, U.K.

Plasma/serum drug concentrations are generally regarded as the most biologically relevant parameter in evaluating drug absorption and exposure. Yet there are occasions when whole blood may provide a more useful indicator or where collection conditions preclude separation of plasma from blood. Sample collection and processing procedures may be evaluated/validated only at the end, largely due to the need for a validated assay before evaluation can be done. Problem areas in sample collection, processing and storage have been well defined*; yet their manifestation can often cause surprises. They include:-

- stability: light, temperature and oxidation;
- contamination: glassware (detergents etc.), plasticware (plasticizers, modifiers, antioxidants), handling (plastic gloves, hands, clothing).

Problems are not confined to blood, but it is the present focus.

Instances where blood is the preferred analytical matrix include preferential partitioning of the drug into red blood cells (RBC's) or where these may be the pharmacological target. Moreover, in instances where the drug is rapidly metabolized by blood or plasma enzymes, this process must be blocked by rapid inhibition of the enzymes. Some inhibitors such as fluoride ions and a wide range of cholinesterase blockers can reduce the hydrolysis of esters; but only complete denaturation of the enzyme can assuredly eliminate activity. Catechols which may be methylated by an RBC transferase (COMT) can be preserved by diverse agents, e.g. pyrogallol (a competitive inhibitor).

Addition of acetonitrile to blood (1:1; but 2:1 preferable - art. #A-6) is effective in denaturing enzymes, as is sodium metabisulphite (antioxidant; 5% w/v). Thus dopexamine, a catechol used in surgery as a vasodilator, is stabilized by the addition of sodium metabisulphite and EDTA (metal chelator). Following addition of metabisulphite the blood is haemolysed. Dopexamine must therefore be quantitated in whole blood [1].

---

†addressee for any correspondence

**Note by Ed.*- Problems and remedies appear in numerous arts. in this book series (pertinent Index entries include: Blood; Lability; Enzymes; Erythrocytes; Storage; Vessels); the present art. serves as a brief survey of the area.

By chelating metal ions which are known to foster this oxidation process, addition of EDTA helps preserve catechols. Addition of sufficient EDTA can minimize gel formation, which frequently occurs after freezing blood samples. This problem can be minimized by centrifugal removal of the RBC ghosts prior to freezing.

Binding of drugs to RBC's can become manifest in various ways. Thus, for FK506 (a macrolide lactone immunosuppressant) [2], which is bound to RBC's and which needs a long and complex sample preparation procedure, the initial protein precipitation step can be critical; as the recovery of FK506 is temperature-dependent, failure to allow the sample to reach room temperature (equilibrium) can result in variable recovery.

Similarly, a potential anti-migraine agent, compound 'S', binds to RBC, giving a concentration double that in plasma. While equilibration *in vitro* takes <30 min, the *in vivo* ratio varies significantly over the first 90 min post-administration. Thus quantitation of the drug in whole blood obviates any need to correct for RBC/plasma equilibration. This phenomenon was observed to have a practical analytical consequence during the spiking process. It was difficult to obtain a reproducible bulk spike at very low concentrations, viz. <1 ng/ml, using a standard solution of compound 'S' in water (containing small quantities of DMSO) which introduced <3% water into the blood.

This difficulty was resolved by spiking each calibration level individually. This observation is not uncommon at low spike concentrations for drugs which partition into/onto RBC's. The mechanism is uncertain; but it is likely that following addition of a small amount of drug it rapidly adheres to a small number of RBC's - which are then difficult to disperse homogeneously through the bulk sample.

Although blood is not an ideal matrix to handle or process, its use can obviate *ex vivo* stability problems during sample collection, processing and storage.

References

1. Baker, P.R., Gardner, J.J., Lockley, W.J.S. & Wilkinson, D. (1995) *J. Chromatog. B 667,* 283-290.
2. Perotti, B.Y.T., Prueksaritanont, T. & Benet, L.Z. (1994) *Pharm. Res. 11,* 844-847.

Comments on #D-1, Anderson: HPLC WITHOUT PROTEIN PRE-REMOVAL
#D2/3/4n, Poole/Westwood/Noctor: SPE including DISC USE
#D-6n, Stevenson: IMMUNOAFFINITY EXTRACTION
#D-7n, Hill: BLOOD COLLECTION

Roger S-H. Hsu asked Anderson if analytical column life-time decreases as pre-column particle size increases. **Reply**: the converse occurs [Table 1 in his art.], probably because smaller particles, having greater surface area, retain more protein which eventually elutes onto the analytical column and impairs its performance. **Reply to J.F. Stobaugh.**- We don't know whether with multiple injections serum/plasma proteins adversely affect autosampler tubing, needles, syringes etc.

Points concerning **SPE: Poole and others.**- Some packings such as C-8 and C-18 are, as **R.J. Simmonds** remarked, more reliable than (e.g.) CN, DEA or CBA, such that when developing valid methods one should line up alternatives, e.g. CN/C-2, CBA/diol or, if only one type chosen, at least two sources: as many cartridges as possible should be tried out. Cartridges of disc type (cost ~£1) - which resemble ordinary cartridges but have a disc (at the foot) instead of a packing - in the experience of **I.D. Wilson** have inadequate capacity for some assays. **Remark by Stobaugh to Westwood**: for multi-sample assays with SPE it may be better to operate sequentially rather than batch-wise unless the analyte is unstable either in its matrix or, if applicable, after derivatization; physico-chemical properties and reactivity obviously govern the choice of approach.

Replies by Stevenson.- To R. Venn: we do give our antisera one pre-treatment - ion-exchange purification - before coupling to the support. **To Anderson**: omission of organic modifier from the pH 2 elution buffer gives a wider peak such that one has to collect several fractions. **To H. de Bree**: problematic availability discourages trial of receptors instead of Ab's.

For a 'companion' to Stevenson's Forum contribution, see below, following the Editor's citations.

Blood collection.- **Reply by Hill to M. Bayliss**: mostly we use EDTA rather than heparin for whole blood although sometimes we encounter particular problems. **D. Dell remarked** that EDTA is becoming more widely used, probably because it can inhibit most blood enzymes. **S. Westwood**: too little lithium heparin in whole blood, when frozen and thawed, can lead to gelling. **R. Mitchell**: SDS prevents analyte degradation by esterases and makes blood homogeneous and easy to handle; it can overcome gel formation from frozen blood.

Citations contributed by Senior Editor For journals, Chr denotes Chromatography

Circumventing pre-RP-HPLC deproteinization (cf. art. #D-1)

Alkyl-diol silica restricted access media (raM; preparation described) were used in a pre-column/back-flush system for bio-samples: Boos, K-S. et al. (1995) Fresenius' J. Anal. Chem. **352**, 684-690; likewise with plasma using, as model compound, 8-methoxypsoralen (methoxysalen): Vielhauer, S. et al. (1995) J. Chr **B 666**, 315-322. In similar work, with ropivacaine and bupivacaine in plasma, the pre-column was of semipermeable surface (SPS) type and the main HPLC column was Hypercarb Ph.- Yu, Z. et al. (1994) J. Chr **B 654**, 221-230. The deproteinization efficacy of an unorthodox pre-column packing, hydroxy-apatite, was demonstrated for plasma containing theophylline and diazepam.- Iwase, H. et al. (1994) J. Chr **B 655**, 73-81.

Protein removal on-line with the chromatographic system can be done in other ways. A flat-bed dialyzer has been used with an ASTED (Gilson) system for plasma containing anti-epileptic drugs.- Johansen, K. et al. (1995) J. Chr **B 669**, 281-288: this follows up K.E. Rasmussen's detailed art. in Vol. 23 of this series. In a switching-valve technique, protein precipitated in an acetonitrile stream was removed by paired filters prior to RP-HPLC (SDS flushing stage).- Asafu-Adjaye, E.B. et al. (1994) J. Chr **B 652**, 35-42. Sect. #ABC has many refs.

A membrane stratagem for liquid-liquid extraction

Employing amperoxide and des-N derivatives as model analytes (strongly protein-bound), an automated ASTED-based 'supported liquid membrane' system, linked to HPLC-UV, has been devised for extraction (then back-extraction into dilute acid) with dihexyl ether, optionally containing trioctylphosphine oxide; LOD ~0.05 µg/ml plasma: Lindegard, B. et al. (1994) Anal. Chem. **66**, 4490-4497 (& J. Chr **665 A**, 259-268; Jonsson, J.A. et al.)

Sample preparation by (immuno)affinity (IAf) methods*

On-line immuno systems in HPLC contexts including detection: de Frutos, M. & Regnier, F.E. (1993) Anal. Chem. **65**, 17A-25A.

For GS-MS analysis of β-sympathomimetics in liver or urine, IAf extraction was followed by another SPE step: Hooijerink, H. et al. (1994) J. Chr **B 660**, 303-313. In screening urine for, e.g., propranolol by HPLC-MS, Protein G extraction and an RP trapping column were used: Rule, G.S. & Henion, J.D. (1993) J. Chr **B 582**, 103-112. For a benzothiazole and oxygenated metabolites in plasma, IAf-SPE preceded RP-HPLC-UV.- Miyairi, S. et al. (1994) J. Pharm. Biomed. Anal. **11**, 389-395.

*See next p. for conventional dexamethasone assays (cf. #D-5) and the Calcitonin entry in #ABC, Group 7, for an IAf example. Ways to obtain and immobilize Ab's were described in Vol. 22, this series [eds. & publisher as now], by **L.F.** Statham *et al.*

For HPLC of plasma or urinary sulphonamides, an immobilized carbonic anhydrase method (Chromatographia **23**, 727) selected those lacking a substituent on the sulphonamide group: Ohta, T. et al. (1994) Biomed. Chr. **8**, 184-188.

Relating to Seymour's use (#D-5) of IAf for dexamethasone: in a 'conventional' method (embracing betamethasone too) for plasma, SPE (C-18) was followed by fluorescent derivatization and HPLC (C-18).- Wu, S-M. et al. (1995) Anal. Chim. Acta **307**, 103-107. In a C-18 (UV) method for neonates, less onerous than RIA, liquid-liquid extraction was used.- Schild, P.N. et al. (1994) J. Chr **B 658**, 189-192.

Various sample-preparation approaches including discs

NMR was used to study the surface chemistry (cf. art. #D-2) of seven C-18 SPE phases that differed widely in behaviour.- Albert, K. et al. (1994) J. Chr **A 665**, 253-258.

Automated SPE (ASPEC system) may be interfaced with GC using a trapping column, as shown for tradozone in plasma (put into glass, not plastic, ASPEC vials): Louter, A.J.H. et al. (1995) Chromatographia **40**, 500-506.

Supercritical fluid extraction (SFE; CO_2) tried with model compounds (a flavone and ketorolac) in plasma was effective but gave less clean extracts than liquid-liquid extraction: Liu, H. & Wehmeyer, K.R. (1994) J. Chr **B 657**, 206-213.

For RP-HPLC of basic drugs (e.g. benzodiazepines) the first step was protein precipitation by generating $NiWO_4$ and extraction into a low-density solvent mixture.- Bernal, J.L. et al. (1994) Chromatographia **38**, 617-623. For iophenoxic acid (strongly protein-bound), C-18 analysis was preceded by serum deproteinization with tungstate and methanol.- Jones, A. (1994) J Chr **B 654**, 293-296. Consult #ABC for kindred approaches.

Clenbuterol screening in biofluids using IAf-SPE (cf. #D-6n): SUMMARY of a Forum contribution by **P. Kwasowski** (ClifMar Associates, Sch. of Biol. Sciences, Surrey Univ., Guildford)

Abusive use of clenbuterol in the meat industry and in sport calls for a rapid and specific sample clean-up capability prior to screening or confirmatory analysis. This has been achieved, with analyte concentration, by automated IAf-SPE, integrated with a semi-automated IA; daily throughput can be hundreds of samples. Careful production/selection of the Ab's used for the IAf columns permits tailoring of performance characteristics. Thus, depending on the choice of Ab's, highly or non-specific, either a particular analyte or a 'structural family' can be extracted; the latter applies in our method for β-agonist growth promoters in urine, serum or bile. We use Ab's raised with clenbuterol, immobilized on a controlled-pore glass affinity matrix , in 3 ml column format, enabling use of a vacuum box. Finally we do a screening assay, e.g. by an ELISA specifically developed for this purpose; sub-ppb levels are measurable.

FORUM DISCUSSION on the foregoing presentation (P. Kwaskowski)

Replies by Kwasowski.- To R.J. Simmonds: Our cartridges for IAf extraction are re-usable ±20 times depending on their assay performance. **To D. Anderson.-** Depending on the assay, the Ab amount is typically 1-10 μg — sufficient to bind ~100 ng of an analyte having mol. wt. 200-300; the particle diameter for the clenbuterol work was 100 μm. **To H. Birnböck.-** The Ab for the β-agonist ELISA was different from that used in the IAf purification. **To M. Seymour.-** GC-MS is used to confirm a positive result from an IAf/ELISA assay. **To A. Gulaid.-** Stereospecificity of enantiomer distribution awaits study.

MS-MS(TSP) for clenbuterol and other β-agonists in spiked urine, using SPE but no HPLC, was studied by van Rhijn, J.A. et al. (1995) J. Chr **A 712**, 67-73. A. Adam & co-authors investigated clenbuterol assay by IAf extraction followed by EIA or confirmatory GC-MS [J. Immunoassay: (1990) **11**, 329; (1991) **12**, 207; also Analyst (1994) **119**, 2663], and obtained better sensitivity with post-dose calf hair than with urine or tissue [J. Agric. Food Chem. (1995) **43**, 2716-2718].

Section #E

APPROACHES FOR ANALYTICAL SEPARATION AND DETECTION

The suffix '**n**' as in #E-4n distinguishes 'NOTES' from main articles. COMMENTS (starting on p. 300) comprise Forum discussion remarks and Editor's 'annotations'.

#E-1

IMPACT OF ATMOSPHERIC PRESSURE IONIZATION ON HPLC APPROACHES

D. Thomas†, S.D. Clarke, H.M. Hill, T.A.G. Noctor & ▫J.N. Robson

Corning Hazleton, Otley Road, Harrogate, N. Yorks. HG3 1PY, U.K.

MS(API) and especially MS-MS(API) for HPLC detection, with electrospray and APcI options, has distinct advantages. It suits for polar/ionic compounds not amenable to other methods. Being highly specific, HPLC-MS-MS(API) needs minimal sample work-up and gives fast run times (<2 min). MS allows an isotopic form of the actual analyte to be used as i.s. In some MS(API) systems the ion source is removable for maintenance without disturbing the vacuum. Disadvantages include lack of structural information on analytes (less so with MS-MS), the need to avoid alkali metal buffers, and ion suppression due to interferences which call for investigation of altered extraction and HPLC conditions.*

Since the marketing in 1981 of an API* mass spectrometer‡ the MS analysis of therapeutic drugs has been revolutionized [1-3], due to ease of HPLC coupling, the ruggedness of the API-MS interface, with ESP and APcI choices, and the inherent sensitivity of MS. It can serve as a general detector (in full scan mode), a selective detector (aided by RIC's for full scan; otherwise SIM), or as a specific detector when full-scan MS-MS is used to elucidate structures or, with MRM, for quantitation. Merits and demerits of different ways of bringing HPLC eluates into an MS system have been discussed [4]. Thus, analyte volatility and/or thermal stability problems may beset the moving-belt interface, which allows both electron-impact and chemical ionization, thereby giving fragmentation patterns for structure elucidation or pseudo-molecular ions for mol. wt. information.

Atmospheric pressure ionization (API, including APcI)

The commonly available interfaces are APcI and ESP - 'pure' or, as now described, pneumatically assisted (ISP; Fig. 1A).

▫now with Perkin-Elmer, Post Office Lane, Beaconsfield HP9 1QA
**Abbreviations.-* For MS (MS-MS = tandem): AP(c)I, atmospheric pressure (chemical) ionization; CID, collisionally induced decomposition; ESP, electrospray (ISP, ionspray); RIC, reconstructed ion chromatogram. MRM/SRM, multiple/selective reaction monitoring; SIM, single ion monitoring. LLOQ, lower limit of quantitation (LOD, ...detection); i.s., internal standard. LC signifies HPLC.

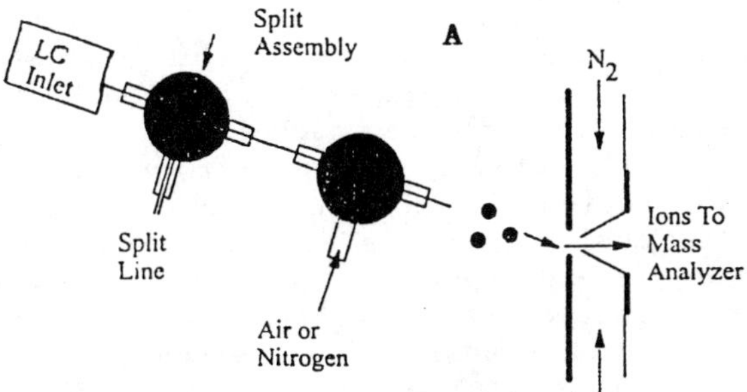

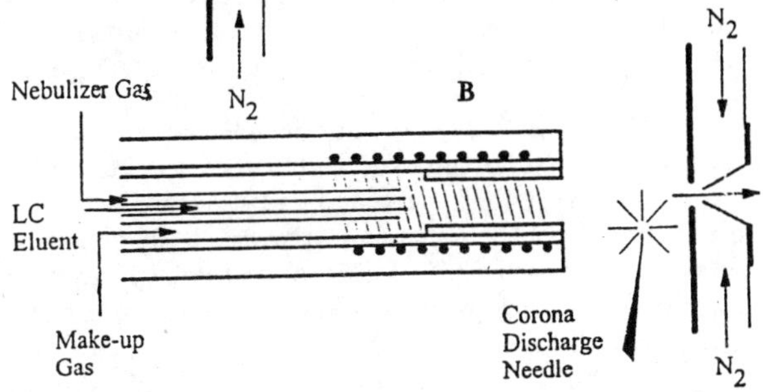

Fig. 1. Schematic representations: **A**, a pneumatically assisted ESP source (ISP); **B**, an APcI source.

A fused silica line connected to the LC column passes through two zero-dead-volume T-connectors (the first of which can be used to split the eluate flow to the ion source) and terminates flush with or just inside the ESP needle. A high voltage (e.g. 5 kV for positive-ion work) at the needle tip in conjunction with a nebulizing gas (air or N_2) causes the eluate to be sprayed from the needle tip and thereby form a mist of highly charged droplets. As these move towards a small orifice in the ion plate separating the API chamber from the mass spectrometer solvent evaporates from the droplet surface and at some point ions evaporate into the gas phase; droplet evaporation is aided by a counter-flow of N_2 which also keeps the vacuum chamber clean. Solvent molecules clustered to the analyte ions are removed by gentle collisions with background gas molecules in intermediate-pressure regions between the orifice and the mass spectrometer vacuum. ESP offers high sensitivity and compatibility with both CE and HPLC, and uses very little sample.

Another interface lately available is the Turbo ISP source which focuses at the ESP needle a stream of hot gas (typically 550°) which, by increasing solvent evaporation from the droplet surface, causes increased ion evaporation; thereby more ions are available to be sampled into the spectrometer. 'Pure' ESP is best at low flow rates (<10 µl/min), while ISP can operate at rates up to 200 µl/min [5]. Hence if the mobile phase flow rate is 1 ml/min the eluate will have to be split for entry into the ion source. The Turbo ISP source allows sampling of up to 1 ml/min without splitting, and shows higher sensitivity at all flow rates compared with ESP - as exemplified (Fig. 2) by a comparison, using LC-MS with SRM, between ISP and Turbo ISP for an anti-diabetic agent; Turbo ISP is 50-fold better.

APcI is a technique of considerable use in bioanalysis [6, 7]. The interface (Fig. 1B) comprises a narrow-bore stainless steel tube traversed by the eluate which is then nebulized into a fine mist by a high velocity jet of air or N_2. The droplets pass through a heated quartz tube (~120°) aided by a make-up gas, where the mobile phase is evaporated and sample vaporized. The make-up gas and vaporized sample molecules pass into the spectrometer's reaction region where ionization occurs by charge transfer and proton transfer from reagent ions in a plasma initiated by a corona discharge needle at (e.g.) 5 kV.

This interface accommodates a flow of up to 1 ml/min of aqueous mobile phase without splitting, and allows the spray vaporization of relatively involatile and thermally labile compounds with minimal thermal decomposition. However, care must be exercised with potentially labile metabolites (e.g. **N**-oxides) or conjugates because of possible breakdown inside the interface. Also, low mol. wt. compounds (<200 amu) can give rise to both the molecular ion and the protonated molecule [8].

Advantages of API

API is a notably efficient mode of ionisation even though the fraction of API-source ions that can be transported through the orifice is limited to ~10^{-4}. As all the ion current is distributed around a single ion (generally $[M+H]^+$), LC-MS(API) and, nowadays, LC-MS-MS(API) provide LOD's and LLOQ's* that are usually appreciably better than with conventional systems such as LC-UV or GC-MS. Fig. 3 shows, for a steroid, an LC-MS(APcI) chromatogram (using MRM) and a good calibration plot.

MS can furnish varying degrees of specificity depending on detection mode (RIC, SIM, etc.); the most specific mode, MS-MS, enables separation of the compound of interest from co-eluting species. Fig. 4A shows a total ion chromatogram (TIC) from m/z 100 to 600 for an LC-MS(ISP) analysis of human plasma spiked with a 5-HT antagonist and subjected to protein precipitation. Evidently there is a considerable amount of eluting material that masks the presence of the drug. In comparison, the same sample run in the SRM mode shows a single, clearly defined peak with a very high signal-to-noise ratio (Fig. 4B).

As amplified by R.J. Simmonds & co-authors in a companion article, with MS-MS detection very simple sample preparation usually suffices, with the further benefits that the method is more likely to be rugged and easier to trouble-shoot because there are fewer stages than for classical high-sensitivity methods, and that automation is feasible. These factors lead to improved assay precision at the LLOQ; thus at the 100 pg/ml level 5-10% is commonly obtained. Protein precipitation is the

*LOD, limit of detection; LLOQ, lower limit of quantification.

preferable approach because of its ease and speed (~3 h for 60 samples). Typically it entails acetonitrile addition (at least 1 vol.), mixing and centrifugation, evaporation (N_2) and reconstitution if needed. If it fails, e.g. due to matrix effects or too much particulate material reaching the HPLC column, SPE or one-step liquid-liquid extraction should be more selective.

Development time with API-MS can be very short. A day may suffice to optimize MS parameters for quantitation, given a full-scan mass spectrum to relate to mol. wt. and to disclose any adduct ions such as $[M+Na]^+$ or $[M+K]^+$. The drug's MS-MS spectrum is acquired, and collision energy and orifice voltage optimized to give an intense parent-fragment transition for the assay. With MS, MS-MS and (see below) HPLC optimized for the drug, the i.s. is run and the LOD established. It may take a day or weeks to develop a robust extraction method of minimal complexity. Importantly, this standard approach may not solve traditional problems to which the analyst must still be alert, e.g. related to metabolites, stability or protein binding.

Operational aspects.*- Options for the MS mode are dealt with above. Whereas with (e.g.) UV detection the HPLC must separate the i.s. from the analyte, MS can measure each with the other present. Because methods may be simplified, generic approaches can be proposed for particular classes of compounds such as anti-inflammatory steroids; for these a standard SPE protocol may be used, then a C-8 column with acetonitrile/pH 4 10 mM ammonium acetate (1:1 by vol.) and, routinely giving LLOQ 10 pg/ml, negative-ion MS-MS(APcI) of the $[M+acetate]^-$ ion.

Coupling to MS(API) minimizes the HPLC step. The column, with an appropriate 3 µm silica-based packing and pre-column, is short (say 50 × 4.6 mm). With isocratic elution at 1 ml/min the MS inflow is splitless for APcI and split to ~40 µl/min for ISP or 200 µl/min for Turbo ISP; column-conditioning between analyses is not required, and the run time can be only ~2 min as in Fig. 5 which shows baseline separation of a drug and two metabolites by HPLC with MRM(ISP). However, it must be ensured that the analytes do not elute too close to the solvent front amid poorly retained endogenous material; elution away from the solvent front can minimize possible matrix effects, usually API-source saturation by large amounts of endogenous material. With the various advantages, throughput can be ~300 samples/day - beneficial to decision-making during the drug's development.

The i.s. being an isotopically labelled version of the analyte, it serves in different species with no worry about potentially different matrix effects, minimizing evaluation as needed if an analogue is used. However, care must be taken to produce an isotopically pure i.s., preferably not overlapping with the analyte in the isotopic distribution of $[M+H]^+$ ions.

*Here and earlier the authors' text has been condensed.-*Ed.*

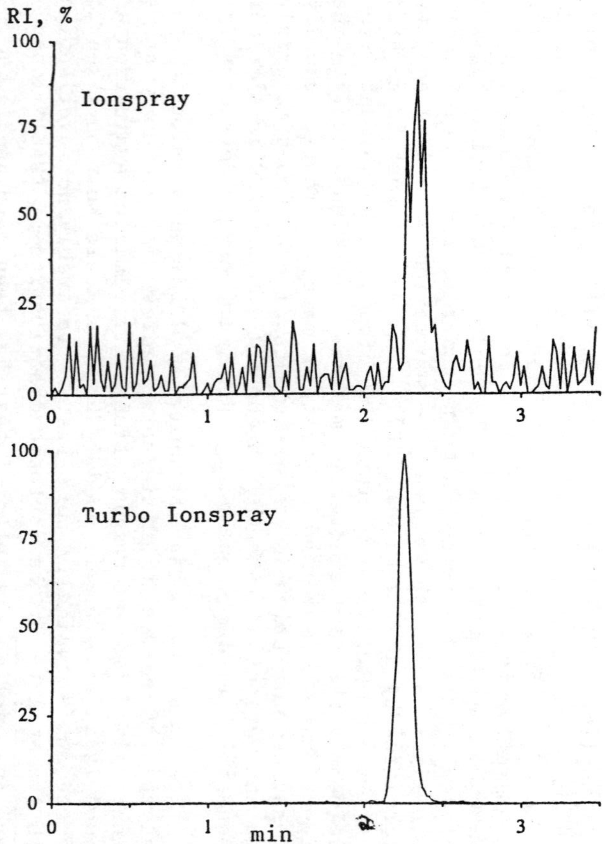

Left:
Fig. 2. HPLC-SRM of an anti-diabetic agent: Ionspray *vs.* Turbo Ionspray (equal amounts at 40 µl/min).

Right:
Fig. 3. HPLC-MRM(APcI) for a steroid (10 pg/ml) and, *below*, a calibration curve showing linearity over 10-2000 pg/ml; $r^2 = 0.9994$.

For mass spectra, RI = relative intensity. Perkin-Elmer market ISP as 'IonSpray'® (CAP. letter).

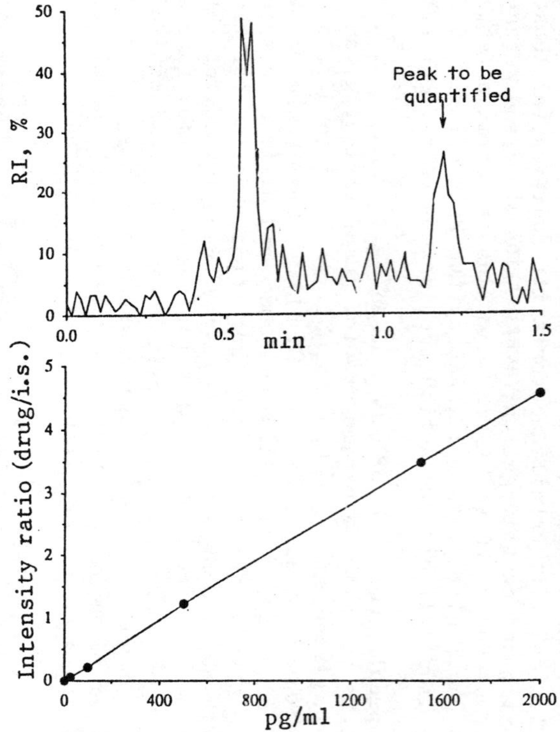

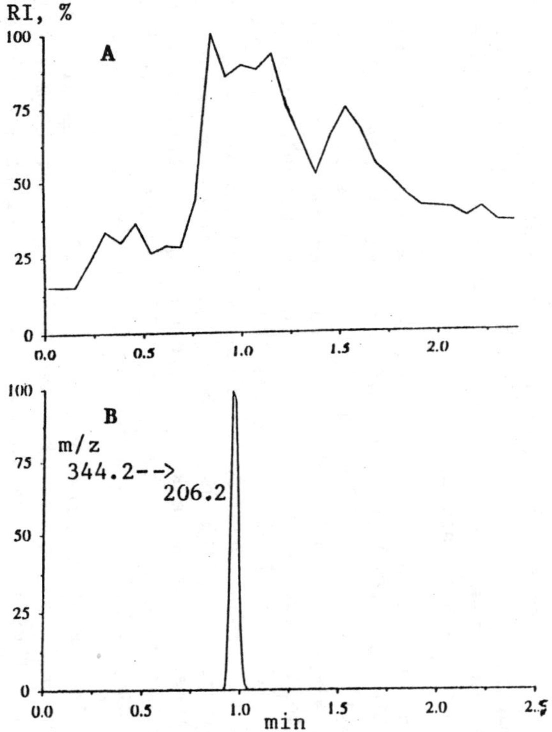

Left:
Fig. 4.
HPLC of a 5-HT antagonist; ISP. **A**: total ion current (TIC), m/z 100 to 600; **B**: SRM.

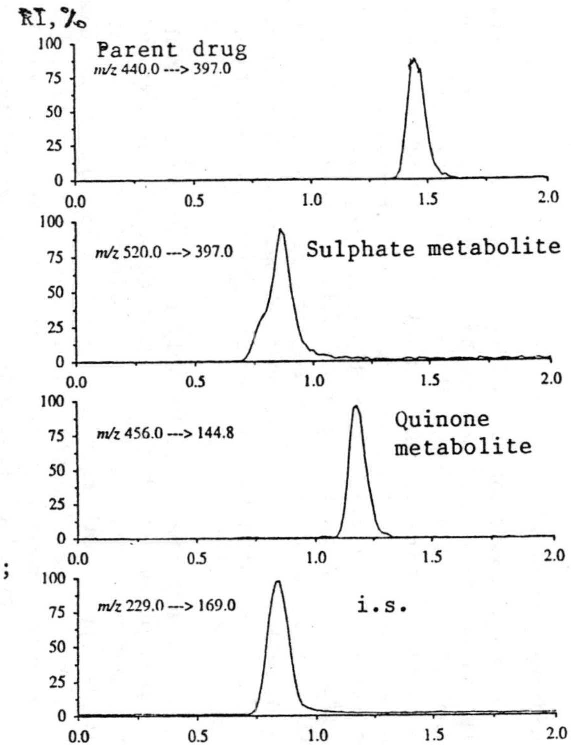

Right:
Fig. 5.
HPLC-MRM (ISP) of an anti-diabetic agent and 2 metabolites. Column: BDS C-8 (50 × 4.6 mm; 3 µm); mobile phase acetonitrile/pH 4 10 mM amm. acetate (60:40), split: 40 µl/min to the mass spectrometer. Same drug as in Fig. 2.

Disadvantages of API

The main constraint with LC-MS(API) is the cost of the system. This can be up to 10 times higher than for LC with conventional detection. Because the API technique yields only one kind of ion (usually $[M+H]^+$), experiments with MS-MS or CID in the ion source have to be done if structural information or increased specificity is needed. MS-MS capability can add significantly to the capital cost, whereas with in-source CID the analyst must be sure that the observed ions are due only to the analyte and do not originate from solvent-mediated background ions in the APcI mode or arise from electrochemical oxidation of Fe, Cr or Ni in the ESP needle.

Matrix effects in LC-MS(API) may, as mentioned above, be a serious problem; if they are undetected, results may be biased [cf. companion art. by R.J. Simmonds *et al.*- *Ed.*]. It is vital that all potential matrix effects be assessed and dealt with during method development and subsequent validation [9]; if one is detected, reduction or elimination can be sought by several approaches including i.s. selection, enhanced sample clean-up, modification of HPLC conditions, or reducing injection volume.

Fig. 6 shows calibration curves obtained by LC-MS(ESI) for a protease inhibitor spiked (50-5000 ng/ml) into rat or human plasma and prepared by protein precipitation. The rat curves were linear in some cases but in others were similar to that shown in Fig. 6. Changing to liquid-liquid extraction removed the matrix effect and rat curves became consistently linear. Matrix effects can be routinely found and evaluated by spiking the analyte at three concentrations into the same matrix from at least six different individuals. Following processing and results comparison, any non-linearities in response and/or variability increases >20% should be regarded as a matrix effect and addressed accordingly [9].

Buffers (for sample preparation and mobile phase) that are compatible with LC-MS(API) comprise a limited range. Thus, with a volatile acid a problem was found during extraction and evaporation steps. $NaHCO_3$ added before evaporation decreased the acid's volatility but was found to be incompatible with the ion source of the particular mass spectrometer used. Moreover, the injection solvent has to be totally compatible with the mobile phase used. With small-column geometries there is an increased tendency for a peak to split during its traversal of a column if the injection solvent is different from the mobile phase; a change in solvent make-up can abolish peak splitting.

Adsorption problems [discussed in Vol. 23 by H.M. Hill *et al.*, on p.398 -*Ed.*] may occur with the low LLOQ that LC-MS(API) can furnish. Thus the drug may adsorb to endogenous material or tubing, or within the injector such that carry-over must be watched for. Autosampler-associated adsortion may be obviated

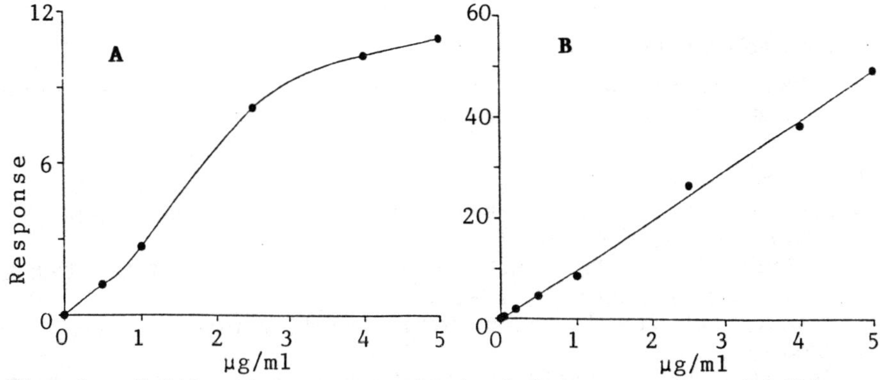

Fig. 6. Calibration curves obtained for a protease inhibitor by LC-MS(ESI) in (**A**) rat and (**B**) human plasma. The rat curve reveals matrix effects, overcome by extraction-stage changes.

by optimizing the needle wash solvent (e.g. reduce the polarity to reduce hydrophobically driven adsorption), or else a change from one manufacturer's autosampler to another may be effective.

IMPACT ON HPLC METHODS

Whilst, as already considered, MS(API) benefits sensitivity and drug-development time, it is crucial to address additional validation and pre-validation issues in order to ensure that a robust and reliable method emerges. Two major issues largely resulting from the minimal pre-MS approaches used are matrix effects and metabolite degradation to the parent compound.

LC-MS(API) has enabled the detection and quantitation of analytes that were not previously possible using conventional methodologies. The fast throughput allows data to be evaluated in clinical and toxicokinetic studies prior to the next dosing. It has impacted on staff aspects, in that specialist staff are needed to conduct the method development and trouble-shooting, supported by technical staff with well developed manipulative skills capable of processing hundreds of samples in a day. So as to accomplish this goal, sample preparation is becoming more simplified and automated. The large amounts of data generated must be processed and checked in real time in order to minimize the accumulation of unchecked results and to provide timely feedback to the scientists. Fortunately, LC-MS-MS(API) chromatograms are generally simpler and easier to evaluate than those obtained using classical chromatography with single-parameter detectors, e.g. LC-UV.

CONCLUDING COMMENTS

Methods based on LC-MS can be developed much faster than classical methods and per-sample time reduced at least 10-fold, placing pressure on the analyst to rapidly process and evaluate the data. Sample preparation may be the rate-limiting stage, and urgently needs further simplification and automation.

Acknowledgements

The authors thank Dai Games, Bob Boyd, Mike Morris, Steve Lowes and Nigel Brown for their advice.

References

1. Henion, J.D., Thompson, B.A. & Dawson, P.H. (1982) *Anal. Chem. 54,* 451-456.
2. Gilbert*, J.D., Hand, E.L., Olah, T.V. & Covey, T.R. (1992) *Biol. Mass Spectrom 21,* 63-68.
3. Moseley III, M.A. & Unger, S.E. (1992) *J. Microcol. Sep. 4,* 393-398.
4. Garcia, J.F. & Barceló, D. (1993) *J. High Resolu. Chromatog. 16,* 633-641.
5. Hopjgartner, G., Bean, K. & Henion, J. (1993) *J. Chromatog. A 647,* 51-61.
6. Fraser, I.J., Clare, R.A. & Pleasance, S. (1994) in *Biofluid and Tissue Analysis for Drugs including Hypolipidaemics* [Vol. 23, this series] (Reid, E., Hill, H.M. & Wilson, I.D., eds.), Royal Society of Chemistry, Cambridge, pp. 113-120. [See also *Proc. 42 ASMS Conf. Mass Spectrom.* (1994) 323 - Pleasance & Fraser.]
7. Taylor, L.C.E., Johnson, R.L., St. John-Williams, L., Johnson, T. & Chang, S.Y. (1994) *Rapid Comm. Mass Spectrom. 8,* 265-273.
8. Thomas, D., Sim, P.G. & Benoit, F.M., *as for* 7., 105-111.
9. Clarke, S.D., Hemsley, M.J., Hill, H.M., Noctor, T.A.G. & Thomas, D. (1996) *Pharm. Sci.,* in press.

---

*His arts. in this vol. and in Vol. 23 (p. 157) are pertinent - *Ed.*

#E-2

ADVANCES IN PLANAR CHROMATOGRAPHY

P. Martin, J. Taberner, I.D. Wilson† & ▫W. Morden

Dept. of Safety of Medicines, Zeneca Pharmaceuticals, Mereside, Alderley Park, Macclesfield SK10 4TG, U.K.

and ▫ICI Chemicals and Polymers, P.O. Box 8, Runcorn WA7 4QE

TLC, especially the high-performance (HP) mode, remains a potentially valuable technique for analysis of drugs and metabolites in biofluids. The capabilities of HPTLC with scanning densitometry for bioanalysis of plasma and urine are discussed. TLC-MS and TLC-MS-MS for analyte identification, directly from the sorbent, is described with emphasis on MS-MS advantages.*

TLC is a robust and flexible separation technique, but in many laboratories remains essentially a qualitative rather than a quantitative tool. TLC is still widely regarded as slow, low resolution, unsuited to quantitative analysis and insensitive. The acknowledged ruggedness and simplicity of planar chromatography means that it continues to be popular for screening and metabolite profiling, e.g. in radiolabel studies. However, the persistence of the view of TLC as an old-fashioned and 'low tech' type of chromatography has limited more general application. Such a view gives scant regard to the advances which have been made in instrumental HPTLC [1, 2]. With investment in state-of-the-art instrumentation, TLC can produce precise, accurate and sensitive bioanalytical results. However, TLC should be seen as a complementary technique to HPLC, GC etc. and not as a competitor.

TLC capabilities relevant to bioanalysis

The circumstances where HPTLC rather than other chromatographic techniques could be used beneficially result from the inherent differences between planar and column chromatography. These should be considered when deciding the situations in which TLC can be used to advantage for bioanalysis.

By its nature TLC serves best for batch analysis, putting multiple samples on the same plate whereas column-based methods perforce entail sequential operation. Thus, assuming a typical

---

†addressee for any correspondence

Abbreviations*.- AMD, automated multiple development; DAD, diode array detection; OPLC, over-pressure layer chromatography. 'TLC' usually implies **HPTLC. MS modes (MS-MS = tandem): EI, electron impact; FAB, fast atom bombardment.

HPLC run time of 10 min and a TLC development time of 30 min, only for speed of analysis is TLC disadvantageous when single samples are analyzed. However, a plate with a batch of 10-20 samples still needs only the same run time as a single sample whereas, inherently, HPLC time increases in proportion to the number. Clearly, then, the more samples are loaded for TLC the less the analysis time per sample; the running being parallel, TLC is very apt for multi-sample routine analysis.

Since, moreover, operations such as sample loading, chromatography and chromatogram evaluation in TLC are done 'off-line' there is notable flexibility in handling high and/or variable sample throughput - this being limited only by the number of plates that can be spotted and by tank availability, since the evaluation is done separately. The fact that the developed TLC plate 'stores' the separation also brings benefits. Thus, the analyst can rank the plates in order of priority for evaluation or else defer this until convenient (without regard to sequence of the runs). Visual inspection of the plate prior to scanning densitometry is easy, and any with poor chromatography can be rejected forthwith. A modern scanning densitometer can then provide an evaluation of the developed, acceptable and prioritized plate in a few minutes.

As scanning densitometry is non-destructive, quantification can be followed by spectrophotometric analysis to confirm/identify the analytes (see below). The plate being storable, chromatography and evaluation can be carried out in different locations; thus one densitometer (or radioactivity detector) can serve several laboratories.

PRACTICAL ASPECTS OF INSTRUMENTAL TLC

TLC plates and sample application

The HPTLC plates, 10 × 10 or 10 × 20 cm, have coating layers (0.1-0.2 mm) of sorbent comprising 5-15 μm particles. The more efficient chromatography provided by these plates compared to conventional TLC plates reduces development times and solvent consumption and, by providing more compact spots/bands, gives better sensitivity (ng quantities by UV absorption and low-pg amounts by fluorescence). Good sample application is crucial for accurate quantification, and care must be taken not to damage the layer during this step. Spot size should be kept to a minimum. Automated methods are the only practical way to do this reproducibly, as achieved in respect of positioning and dispensing the sample by systems from several manufacturers*.

A wide range of stationary phases, similar to those for HPLC, are produced for HPTLC - e.g. bare silica, C-18, C-8, C-2, cyanopropyl, aminopropyl, diphenyl and diol - besides specialty chiral ligand exchange material. However, in practice, and in

*notably Desaga, CAMAG and Shimadzu

contrast to HPLC, most TLC separations are NP using silica gel, probably reflecting both tradition and non-appreciation of the advantages of bonded phases, e.g. diol, in some circumstances.

Development techniques

The simple ascending mode with glass tanks that is still customary remains the general method of choice. For a development distance of, say, 6 cm on an HPTLC plate it affords ~5000 theoretical plates [3]. This number is rather small compared with HPLC but often suffices. Horizontal development done simultaneously from both sides towards the middle of the TLC plate can double the number of samples run per plate. Better resolution, if needed, is obtainable by AMD or OPLC.

AMD entails solvent migration for a short, pre-determined distance; then development is terminated, the plate dried and development resumed; after the solvent has migrated a little further, development is again halted. This process is repeated till the desired degree of separation is attained. The solvent composition may be kept constant or varied with each step (to give a solvent gradient). With each development the bands for the sample components are refocused by the advancing solvent, providing sharper bands and so improving both separation and sensitivity. The cost of AMD (besides apparatus purchase) lies in analysis time: AMD is slow because the solvent migrates only under the influence of capillary action.

OPLC, in contrast, entails forcing the solvent through the layer, allowing development within minutes. As the solvent can be pumped off the end of the plate to waste, solvents of relatively low eluotropic strength can be used in order to enhance separations. Compounds can also be eluted with use of a fraction collector and further examined, e.g. by NMR or MS.

QUANTIFICATION

Modern scanning densitometers can give good precision and accuracy for coloured, UV-absorbing or fluorescent substances; plates (or gels) can be scanned in the 190-800 nm range by reflectance or transmission, or (as with plates incorporating a fluor) by fluorescence or fluorescence quenching. Reflectance is the commonest mode with silica-based TLC plates, and transparent transmission with media such as electrophoresis gels. For good results in fluorescence detection, plates without a fluorescent indicator must be used, to minimize the background.

An obvious practical consideration in TLC when choosing between reflectance and transmission for quantification is the nature of the sorbent layer's support. Aluminium-backed plates are obviously opaque at all wavelengths, while for transmission glass-backed plates can be used only above 320 nm due to absorption of UV light. Silica itself is UV-absorbent at low wavelengths. Reflectance and transmission are especially sensitive to small changes in sorbent quality or thickness, spot shape and size, eluent and development conditions; so it is essential that suitable calibration standards be used, and every effort made to get identical spot sizes. For greatest accuracy some workers favour bands rather than spots. In case plates differ, standards and unknowns should be on the same plate. HPTLC can usually furnish values with S.D.'s similar to those of other chromatographic techniques, i.e. 1-2% in favourable situations.

Densitometer features now include, for example, a means of scanning tracks on a developed TLC plate at a range of wavelengths so that the best can be selected for evaluating chromatograms. Absorption spectra for individual components can be acquired too and, if the instrument has the requisite software package, compared with 'library' spectra to help identification - the HPLC analogy being DAD. Similarly to DAD, spectra can be obtained for different segments of the spot or band (up-slope/apex/down-slope) to get an estimate of peak purity (or, rather, impurity) if the TLC zone is heterogeneous. Some instruments can scan the whole plate in <5 min to get a 2-D image.

Some densitometers can compensate, at least partially, for distortion problems (where the spots do not lie in a straight line), through optimization of the track scanning. Moreover, inevitable background noise attributable to the plate can, with suitable software, be subtracted from the final result.

A drug-measurement example that may compare favourably with HPLC-UV concerns antipyrine in human plasma after a 1 g dose (Taberner, Martin & Wilson, unpublished pilot work). The sample preparation (dichloromethane extraction from 1 ml of plasma at alkaline pH) and internal standard (i.s.) were as for HPLC. HPTLC was performed on a silica gel plate with development to 70 mm (using chloroform/acetonitrile/methanol, 75:25:5 by vol.) and then scanning at 254 nm and comparing with i.s. Antipyrine was readily detectable down to 0.3 µg/ml, ample for 'real' plasma samples which never had <7.5 µg/ml. A pharmacokinetic study gave µg/ml values - HPTLC|HPLC - as follows:

2 h: 18.4|19.7; 4 h: 18.0|18.5; 6 h: 16.3|16.9; 8 h: 15.3|15.7; 10 h: 13.1|14.3; 12 h, 12.5|13.8; 18 h: 8.69|9.15; 24 h: 7.5|8.1

hence $t_{\frac{1}{2}}$ (h) 15.6|16.0, and AUC (µg.h/ml) 455.0|491.0

These results for a volunteer show good accord between HPTLC and HPLC. Further work might well have reinforced this accord, but could not be undertaken because of changing priorities.

IDENTIFICATION

TLC has long been in qualitative use, e.g. for metabolite profiling and drug screening. Supplementing the R_f data, some colour reactions of varying specificity may help [see p. 93 in our Vol. 12 (1983; Plenum), & p. 199 in Vol. 6 (1981) of *Prog. Drug Res.- Ed.*]. Thus, for glucuronides blue spots are obtained with naphthoresorcinol (p. 164 in Vol. 12).

Where identification is the aim, scanning densitometry provides no advantage over HPLC using UV or DAD; MS data are usually needed to confirm identity. Ionization by FAB can be done directly on the plate, and over the last decade TLC-MS has become a well-established technique (review: [4]), whose useful features compared with some other chromatography-MS systems include absence of solvents and off-line operation.

Obtaining EI spectra from silica plates can be problematic because of the strong adsorption of polar analytes such that TLC-MS(EI) can usually be performed only on volatile analytes - but can be done with non-polar phases such as polyamide. There are no problems with silica when FAB, or the related technique liquid secondary ion (LSI) MS, is used, and there are reports of the use of TLC-MS-MS for metabolites [5-7]. However, other phases such as diol are also usable for TLC-MS, as shown with HPTLC-MS(FAB) for a urinary extract containing paracetamol (Fig. 1A) - metabolites of which have also been studied by TLC-MS using silica [5]. To obtain the Fig. 1A spectrum the area of the plate corresponding to the analyte was excised, mulled with glycerol and the mixture inserted into the spectrometer's ion source on the probe. Because of the high background from other ions this spectrum might hardly give really convincing evidence on identity, especially as FAB in general provides only molecular ion data. However, when MS-MS was performed on the same sample the spectrum obtained (Fig. 1B) perfectly matched that of the standard. Fig. 2 shows similar results for ibuprofen following HPTLC (diol) of a urine extract. A study of TLC-MS and TLC-MS-MS with antipyrine metabolites led to essentially similar conclusions [7].

All these results were obtained by removing the stationary phase and inserting the sample (mixed with the FAB matrix) into the spectrometer on a probe. However, some companies including Jeol and VG market custom-made TLC-MS probes whereby segments of a TLC plate can be moved through the ion source, giving results of the kind familiar to HPLC- and GC-MS practitioners.

In our experience, where analytes are present on the plate in low-µg amounts TLC-MS-MS is preferable to simple TLC-MS for analyte identification. The increasing availability of MS-MS instruments is fostering TLC-MS-MS as a viable option for the bioanalyst and, as the examples illustrate, provides a ready means of identifying metabolites with parent drug present. As an additional benefit, the instrument need not be modified.

CONCLUDING COMMENTS

HPTLC with scanning densitometry is a technically advanced quantitative technique which can deliver sensitive, precise and accurate results for drugs and metabolites in biofluids. HPTLC 'hyphenated' with MS or MS-MS can help settle the identity and structure of separated components and so aid metabolism studies.

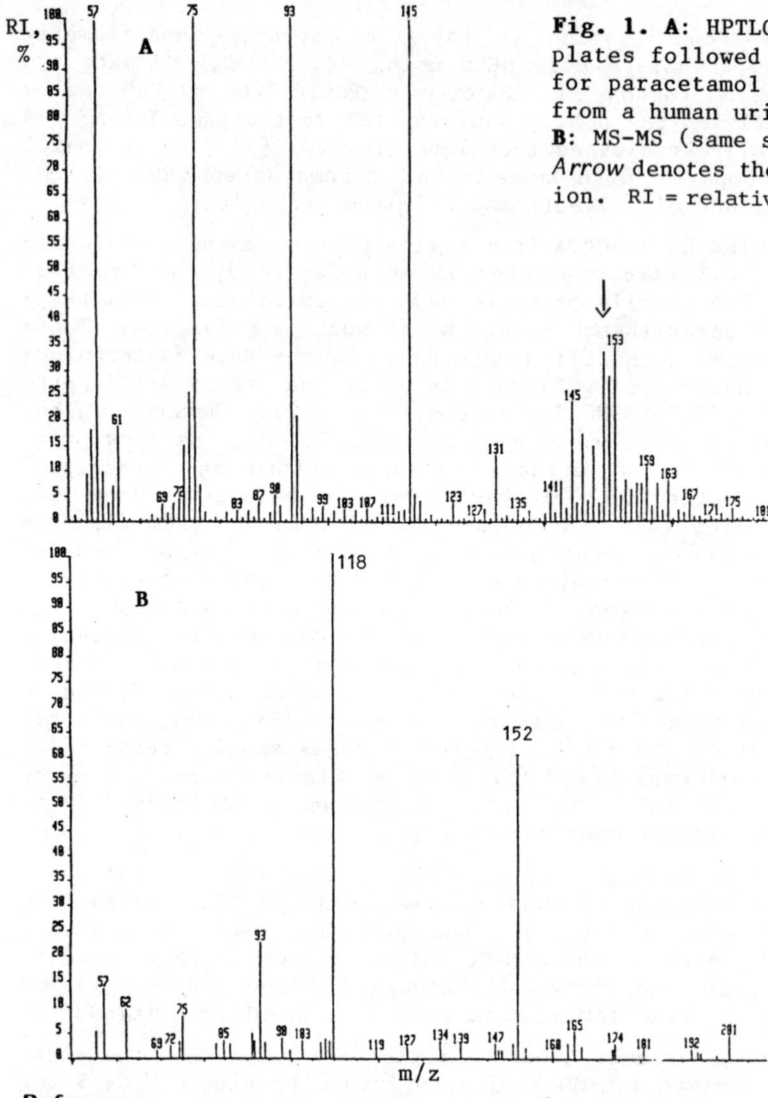

Fig. 1. A: HPTLC on diol plates followed by FAB-MS for paracetamol extracted from a human urine sample. **B**: MS-MS (same sample). *Arrow* denotes the molecular ion. RI = relative intensity.

References

1. Bertsch, W., Hara, S., Kaiser, R.E. & Zlatkis, A., eds. (1987) *Instrumental HPTLC,* Huthig, Heidelberg,
2. Geiss, F. (1987) in *Fundamentals of Thin Layer Chromatography (Planar Chromatography),* Huthig, Heidelberg.
3. Poole, C.F. & Poole, S.K. (1991) *Chromatography Today,* Elsevier, Amsterdam,
4. Somsen. G.W., Morden, W. & Wilson, I.D. (1995) *J. Chromatog. A 703,* 613-665.
5. Spurway, T., Phillips, P.J., Wilson, I.D. & Warrender, A. (1990) in *Analysis for Drugs and Metabolites....* [Vol. 20, this series] (Reid, E. & Wilson, I.D. , eds.), Roy. Soc. Chem, Cambridge, pp. 347-35

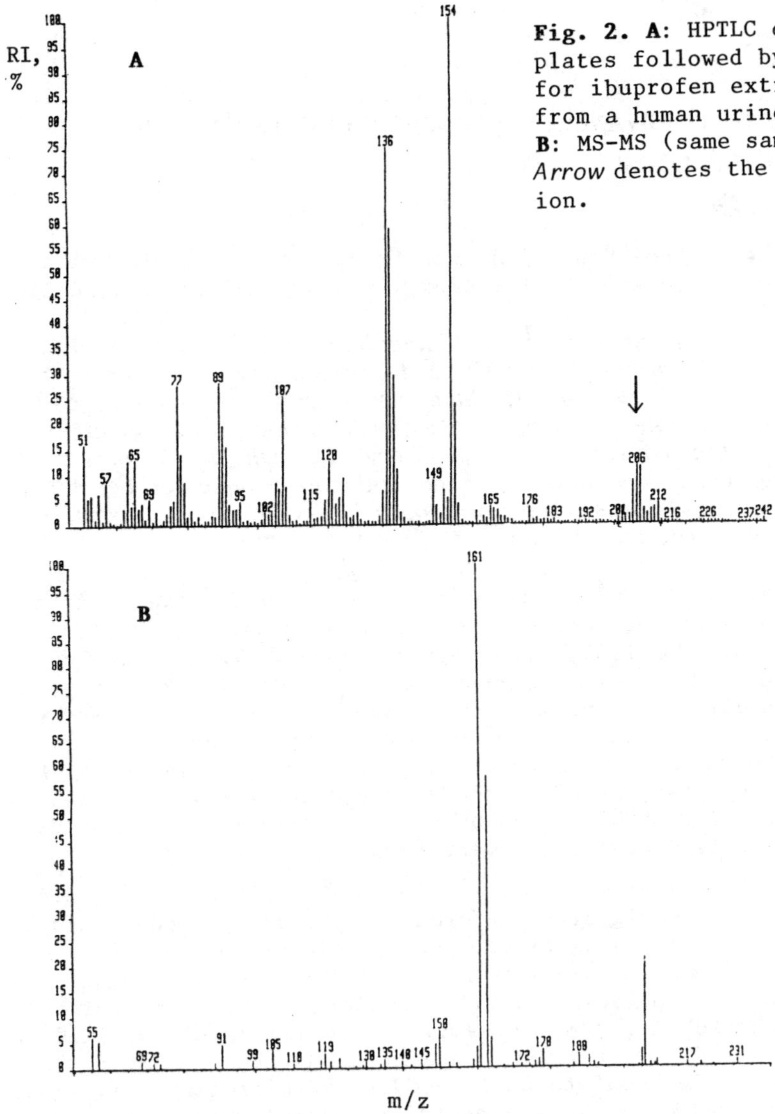

Fig. 2. A: HPTLC on diol plates followed by FAB-MS for ibuprofen extracted from a human urine sample. **B**: MS-MS (same sample). *Arrow* denotes the molecular ion.

6. Wilson, I.D. & Morden, W. (1991) *J. Planar Chromatog.* *4,* 226-229.
7. Martin, P., Morden, W., Wall, P. & Wilson, I.D. (1992) *J. Planar Chromatog.* *5,* 255-258.

#E-3

CAPILLARY-GC ADVANCES: LARGE-VOLUME INJECTION AND ON-LINE LC-GC

K. Grob

Official Food Control Authority of the Canton of Zürich (Kantonales Labor), P.O. Box, CH-8030, Zürich, Switzerland

Nowadays trace analysis in biomedical contexts is commonly done with HPLC rather than GC; but present-day GC deserves a revived role in view of its high separation efficiency, good detectors and simple coupling to MS (with access to EI spectra). Injection-volume constraints in capillary GC can be circumvented in ways which are now discussed and can bring other benefit: HPLC can be used for sample pre-separation, with on-line transfer of appropriate fractions to a GC system.*

Advances in GC by laboratories working in the field of biomedical applications have not been breathtaking, for historical reasons. In the 1960's and 1970's, drug analyses involved a large amount of GC work. However, the typically rather polar analytes required derivatization, sometimes multi-step, and yet all too often produced 'potatoes' rather than sharp peaks (also because the packed columns of that era were less inert). For most such applications HPLC indeed proved superior. Apparently GC struggles left behind such an amount of frustration that drug analysts resolved never to revert to GC.

Sufficient time has now elapsed to justifiably reconsider whether GC in today's high-performance (capillary) mode could be of some help in the drug field. Large volumes may now be injected (50-500 µl), directly or after, e.g., extraction from an aqueous into an organic phase in an autosampler (in-vial extraction). Another approach for complex samples, so far little applied in the drug field, is sample preparation by HPLC with on-line transfer of a 'window' to GC (LC-GC), exploiting the different separation principles as well as the excellent detection in GC.

LARGE VOLUME INJECTION

GC detectors are usually more sensitive than LC detectors; yet HPLC has often provided lower detection limits because only small sample volumes could be injected in capillary GC by the

---

Abbreviations: EI, electron impact (ionization, in MS); PTV, Programmed Temperature Vapourizing. LC is a shortened term for HPLC; GC here signifies the 'high performance' capillary mode.

conventional splitless technique (~2 µl compared with 50-200 µl in LC). Now, however, it has become possible to inject 500 µl or more, primarily through on-column and PTV techniques; the record stands at 20,000 µl [1]. Advantages in injecting >1-2 µl of sample include: (i) one can largely avoid loss of volatile components, as encountered during conventional solvent evaporation prior to injection; (ii) at least 50 µl must be injected so as to introduce reliably the whole sample material available; (iii) injection of large volumes is more convenient than sample re-concentration prior to injection.

On-column injection

Quantitatively, on-column injection is the most accurate and reliable method of sample injection in GC [2, 3]. Large volumes can be injected, provided that an uncoated pre-column is used that retains the sample liquid as a layer on its wall. For instance, 15 m × 0.32 mm i.d. or 10 m × 0.53 mm i.d. pre-columns retain 80-100 µl of sample liquid, provided that the latter wets the wall (a problem primarily for water and water-containing mixtures). Volatile solutes are re-concentrated by the solvent effects, whereas the 'retention gap' effect focuses the initial bands of the high-boiling components at the inlet of the separation column (Fig. 1) [4].

Still larger volumes can be injected if a substantial proportion of the solvent is evaporated during injection (partial concurrent evaporation [5]). This technique, originating from coupled LC-GC, usually involves an early solvent-vapour exit, accelerating solvent evaporation and protecting the detector [6]; it was used for the transfer of LC fractions with a volume of up to 800 µl [7]. Such sample introduction requires sample flow rate adjustment to slightly above the solvent evaporation rate, as well as correct closure of the vapour exit. Recently a dedicated GC instrument became available which, governed by pre-determined evaporation rates, sets automatically the injection speed of an autosampler and the vapour-exit closure [8].

Application of on-column injection is, however, limited to solutions containing fairly low concentrations of material that stays unevaporated, since resulting residues tend to accumulate in the uncoated pre-column and retain the solutes of interest. Peaks start tailing and are reduced in area. Water-containing samples (or RP eluents) also cause problems because they do not wet well-deactivated pre-columns [9] and impair the deactivation, rendering the pre-column adsorptive [10]. Injection *via* a vapourizing chamber is more robust for 'dirty' samples, as the higher temperature overcomes the retentiveness of the non-evaporating material. However, such techniques do discriminate against the higher boilers and easily cause degradation of labile components, owing to the higher temperature and the active material accumulated.

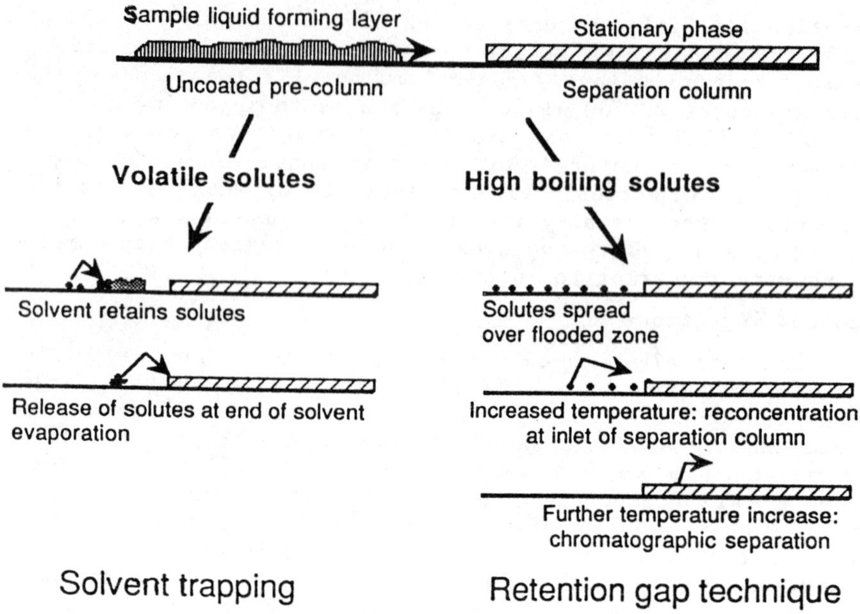

Fig. 1. The two main mechanisms of solute re-concentration using on-column injection and the retention gap technique. *Based on a diagram in ref. 4* (publ. Hüthig).

Splitless injection

Splitless injection with overflow of solvent vapours is technically the simplest approach [11-13]; the liner of the conventional split/splitless injector contains a packing, e.g. Tenax. Rapid introduction of the sample liquid locally cools the packing to the solvent boiling point, causing the solvent to evaporate first. The solvent vapours expand backwards out of the injector through a widely open septum purge outlet, so resulting in self-regulating discharge. Several hundreds of microlitres can be injected (2000 µl has been attained), but volatile solutes are lost partly at least.

Programmed Temperature Vapourizing (PTV) injection

In 1978, Vogt *et al.* [14] introduced PTV in the solvent split mode. The sample is introduced into a cool injector liner and the solvent evaporated in a carrier gas stream leaving through the split outlet. Then the split valve is closed, the injector heated and the solute material transferred into the column. The maximum sample volume was 250 µl. Volatile components may, of course, be lost through the split exit; but this is not a severe problem for typical drug analytes. Other problems concern the often inadequate inertness of the materials available for packing the vapourizing chamber [15-17].

Conventional PTV injectors contain a thin-walled liner of i.d. 0.8-1.2 mm; its packing can retain ~20 µl of sample liquid [18]. Injecting >20 µl thus requires that the sample evaporate during introduction, i.e. that the injection speed be adjusted to the evaporation rate. The reason for choosing so small a liner was a belief that very rapid heating would be important to achieve release of the solutes as sharp initial bands; but wider bore liners can be used, provided that initial bands are re-concentrated by cold trapping, i.e. sample is transferred at a column temperature clearly below the elution temperature of the solutes of interest. Using 2-3 mm i.d.liners, 100 µl at least could be injected at once, circumventing the somewhat tedious adjustment of the injection speed [19]. It was also observed that losses of volatile components could be reduced in this way, presumably because retention in the liquid sample reduced the volatility of the solutes (solvent effects). The only parameter left to be adjusted is the time for closing the split outlet; the solvent should be fully removed, but the gas should not carry away volatile solutes. With the advent of the 'at once' injection, PTV solvent split injection thus became the technically simple and attractive alternative to large-volume on-column injection for 'dirty' samples.

ON-LINE COUPLED LC-GC

Trace analysis usually requires sample preparation in order to remove the bulk material of the sample as well as interfering material. The lower the detection limit required, the more efficient must be the clean-up. Pre-separation in the liquid phase ties up well with GC analysis, because it occurs by a fundamentally different process. The SPE techniques commonly used involve new cartridges for each sample, eliminating problems regarding cleaning and re-conditioning. However, the separation efficiency attained in practice is poor, primarily due to the very broad windows of the fractions collected, often comprising the whole eluate obtained with a solvent.

On-line LC-GC substantially improves on this. High resolution LC for sample preparation increases separation efficiency, and use of an on-line UV-detector helps optimization of the step (e.g. selecting eluent and packing material) and accurate selection of the fraction of interest. On-line transfer of the whole LC fraction to GC makes best use of what is loaded into the LC, simplifies quantitation and avoids introducing contaminants. Finally, automating the process minimizes manual work in sample preparation and renders the procedure more reliable.

Applications in the drug-analysis field

There are many routine LC-GC methods in the fields of food and mineral-oil analysis [20], but few in the biomedical field. In 1986 a small pharmaceutical company started analyzing an active compound in plasma by on-line LC-GC-ECD [21]. Neither LC alone nor GC with conventional sample preparation provided sufficient sensitivity. Analogous methods for the analysis of other compounds followed [22]. In our laboratory's residue analyses, stilbestrols and chloramphenicol in animal urine are routinely determined by LC-GC-MS after a first clean-up on a disposable cartridge (methods based on that described in [23]). The main benefit was ~10-fold reduction in the detection limit.

For various components [24, 25] other workers have identified, by MS, RP-LC peaks subjected to GC, the LC eluate being extracted on-line into an organic phase before transfer to GC. For some applications the component of interest was derivatized on-line in the GC pre-column. Sample pre-treatment by immuno-affinity, with coupled GC, gave notably high selectivity [26].

Example: sterol analysis in olive oil

The potential of LC-GC is illustrated by a recent example from food analysis. The method enabled a subtle adulteration of olive oil to be detected. Olive oils are often adulterated by a breed of sunflower oil whose fatty acid composition fits well that of olive oil. The sterol compositions, however, are markedly different. To render it supposedly undetectable, such sunflower oil is specially prepared by destroying its sterols through extremely strong bleaching (desterolization). Admixture of such sunflower oil can in fact be detected only through the degradation product of the most distinguishing sterol, i.e. the Δ7-stigmastenol [27, 28].

Strong heating in the presence of bleaching earth converts Δ7-stigmastenol into an isomer with the double bond in Δ8(14) position. This isomer remained undetected up to the present because in standard methods it co-elutes with β-sitosterol and has nearly the same fragmentation pattern in MS. Its determination presupposes that LC largely removes the Δ5-sterols, which are present at ~1000 times higher concentration. As shown in Fig. 2, LC separation is not baseline; but removal of at least 95% of the Δ5-sterols is just about sufficient to achieve GC separation on a 50% phenylsilicone stationary phase (Fig. 3). As LC retention times in routine application are not constant enough (±5 sec at 10-min retention times), the fraction window is determined not by absolute retention times but by so-called 'peak detection'. The upslope of the peak of the Δ5-sterols is recognized by criteria as used for the detection of a peak by an integrator and serves, adding a ~43 sec delay, to establish the start of the fraction to be taken for transfer to GC.

There are at least three methods for introducing whole LC fractions involving 100-1000 μl of eluate into capillary GC [29]. Here the fraction was pumped through a small vapourizing

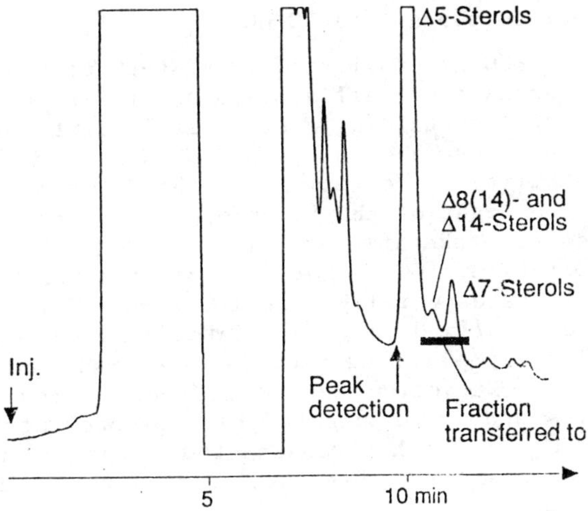

Fig. 2. HPLC pattern for a trans-esterified olive oil presumably slightly adulterated with de-sterolized sunflower oil. Silica gel column, 250 × 2 mm; 2 mm; 1% 2-propanol in pentane (300 µl/min); detection at 200 nm.
Figs. 2 & 3 are from ref. [28], courtesy of Staz. Sper. degli Oli e Grassi, 1-20133 Milano, Italy.

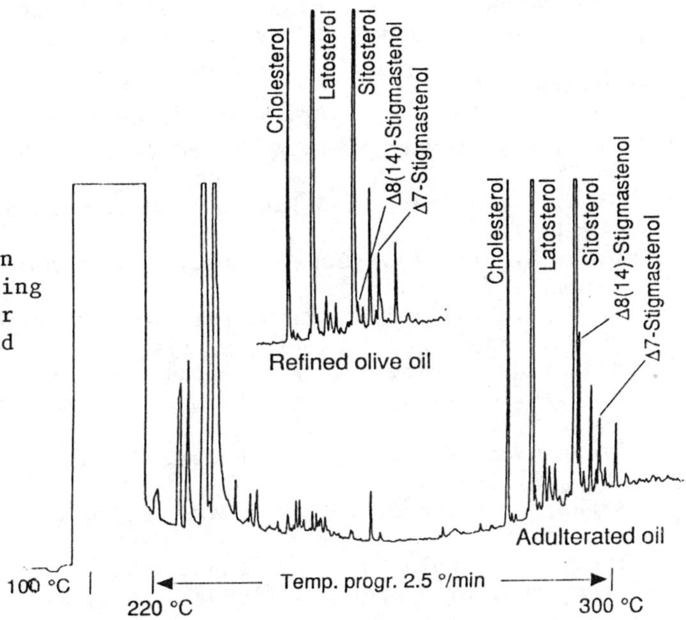

Fig. 3. LC(UV)-GC(FID) chromatogram of the adulterated olive oil shown in Fig. 2, and *(inset)* the relevant portion of a corresponding chromatogram for a normal refined olive oil.

chamber at 300°, comprising 0.32 mm i.d. fused silica capillary tubing with a piece of fused silica capillary inserted, whose ends were flame-sealed and whose polyimide was removed. Vapours passed through a 2 m × 0.32 mm retaining pre-column and left through an early vapour exit. Fig. 3 shows results for the adulterated olive oil and a pure specimen. The method required transesterification of the oil so as to cleave sterol esters. Extracts obtained were analyzed automatically (Dualchrom 3000; Fisons/CE Insts.). Hundreds of analyses were thus performed.

Reversed phase (RP) LC-GC

LC-GC generally involves NP-LC with eluents such as pentane or hexane and small admixtures of more polar solvents. Transfer of water-containing eluents is in fact more complicated. Often it has been said that, for LC-GC adoption, breakthrough hinges on ability to perform LC in the RP mode. This seems plausible in the light of the LC commonly performed for drug analytes but does not take into account that GC effects the main analytical separation. GC is not feasible for the highly polar components that are especially suited to RP-LC. For GC many components must be derivatized, e.g. silylated, and so reside in apolar organic solvents; water in fact degrades many such derivatives. Moreover, extraction into a solvent of low polarity helps remove salts and other highly polar by-products that hinder analysis by GC. For other reasons too, none of our many LC-GC applications could have been performed with RP- rather than NP-LC.

In summary, GC is a highly versatile technique and, taking account of advances now surveyed, warrants fuller exploitation.

References

[Chr. as in *J. Chr.* denotes *Chromatography]*

1. Grob, K., Schmarr, H-G. & Mosandl, A. (1989) *High Resolu. Chr.*12,* 375-382.
2. Grob, K., Karrer, G. & Riekkola, M-L. (1985) *J. Chr. 333,* 129-155.
3. Grob, K. (1991) *On-Column Injection in Capillary GC,* Hüthig, Heidelberg.
4. Grob, K. (1995) *Einspritztechniken in der Kapillargaschromatographie,* Hüthig, Heidelberg, p. 415.
5. Munari, F., Trisciani, A., Mapelli, G., Trestianu, S., Grob, K. & Colin, J.M. (1985) *High Resolu. Chr. & Chromatog. Comm. 8,* 601-606.
6. Biedermann, M., Grob, K. & Meier, W. (1989) *High Resolu. Chr.* 12,* 591-598.
7. Artho, A. & Grob, K. (1990) *Mitt. Gebiete Lebensm. Hyg. 81,* 544-558.
8. Munari, F., Colombo, P.A., Magni, P., Zilioli, G., Trestianu, S. & Grob, K. (1995) *J. Microcol. Sep. 7,* 403-409.
9. Grob, K. & Li, Z. (1989) *J. Chr. 473,* 391-400.
10. Grob, K., Neukom, H.P. & Li, Z. (1989) *J. Chr. 473,* 401-409.
11. Grob, K., Brem, S. & Fröhlich, D. (1992) *High Resolu. Chr.* 15,* 659-664.
12. Grob, K. & Brem, S. (1992) *High Resolu. Chr.* 15,* 715-722.
13. Grob, K. & Fröhlich, D. (1992) *High Resolu. Chr.* 15,* 812-814.
14. Vogt, W., Jacob, K. & Obwexer, H.W. (1979) *J. Chr. 174,* 437-442.

**J. High...* and *HRC* are variants of the title (ref. 5 has a former title).

15. Loyola, E., Herraiz, M., Reglero, G. & Martin-Alvarez, P. (1987) *J. Chr. 398,* 53-64.
16. Herraiz, M., Reglero, G., Loyola, E. & Herraiz, T. (1987) *High Resolu. Chr. & Chromatog. Comm. 10,*598-605.
17. Mol, H.G.J., Hendriks, P.J.M., Janssen, H-G., Cramers, C.A. & Brinkman, U.A.Th. (1995) *High Resolu. Chr. 18,* 124-128.
18. Grob, K. & Li, Z. (1988) *High Resolu. Chr. & Chromatog. Comm. 11,* 626-632.
19. Mol, H.G.J., Janssen, H-G., Cramers, C.A. & Brinkman, U.A.Th. (1995) *High Resolu. Chr. 18,* 19-27.
20. Grob, K. (1994) in *Proc. 16th Int. Symp. Capillary Chromatography,* Riva del Garda (Sandra, P. & Devos, G., eds.), Hüthig, Heidelberg, pp. 1-9.
21. Gianesello, V., Bolzani, L., Brenn, E. & Gazzaniga, A. (1988) *High Resolu. Chr. & Chromatog. Comm. 11,* 99-106.
22. Gianesello, V., Brenn, E., Figini, G. & Gazzaniga, A. (1989) *J. Chr. 473,* 343-350.
23. Grob, K., Neukom, H.P., & Etter, R. (1986) *J. Chr. 357,* 416-422.
24. Ogorka, J., Schwinger, G., Bruat, G. & Seidel, V. (1992) *J. Chr. 626,* 87-96.
25. Wessels, P., Ogorka, J., Schwinger, G. & Ulmer, M. (1993) *High Resolu. Chr. 16,* 708-712.
26. Farjam, A., Vreuls, J.J., Cuppen, W.J.G.M., Brinkman, U.A.Th. & de Jong, G.J. (1991) *Anal. Chem. 63,* 2481-2487.
27. Biedermann, M., Grob, K., Mariani, C. & Schmid, J.P. (1995) *Z. Lebemsm. Unters. Forsch,* in press.
28. Biedermann, M., Grob, K. & Mariani, C. (1995) *Riv. Ital. Sost. Grasse 72,* 339-344.
29. Grob, K. (1991) *On-Line Coupled LC-GC,* Hüthig, Heidelberg.

#E-4n

A Note on

A COMPARISON OF HOT AND TEMPERATURE-PROGRAMMED GC INJECTION IN RELATION TO PSEUDOEPHEDRINE

D. Browne, K. Tennant & M. Shepherd

Corning Hazleton, Otley Road, Harrogate HG3 1PY, U.K.

Requirement	A GC assay for pseudoephedrine in plasma samples from a clinical study, linear from 10 to 500 ng/ml.
End-step	GC-NPD* on a DB-5 column, using (1) a hot injection technique, or (2) temperature-programmed injection.
Sample handling	Add i.s. (diphenylamine); LLE at alkaline pH and then dry down and reconstitute in ethanol; or, in later work with injection mode (2), automated (ASPEC) SPE.
Comments	Chromatographic interferences using a published GC method [1] were reduced by altering conditions for LLE and (1), and especially by changing from (1) to (2) with automated SPE in place of LLE.

LLE and injection optimization.- In the reported method [1] the plasma was made alkaline (NaOH) and extracted with diethyl ether. As this gave a very noisy baseline, an attempt was made to improve selectivity by using ethyl acetate, chloroform, dichloromethane or 1-chlorobutane as the extractant. Chloroform gave the cleanest extract but its quality was not ideal. As it was thought that the NaOH might contain some interferants, an extraction was tried using 0.3 N NaOH that had been previously extracted with chloroform. This improved the cleanliness of the extract, but there appeared to be some instability when the compound was left for any time in alkaline conditions and/or in the presence of chlorinated solvents.

Whilst modifying the method to improve its robustness it was found that lowering the injector temperature gave cleaner chromatography. It was postulated that the chloroform and/or the NaOH was causing the compound to break down when in contact with the hot injector: if set at 225° an extracted calibrant gave a very poor peak shape (Fig. 1), but with lowering to 160° and sample re-injection there was improvement. With further lowering of injector temperature to 130° the i.s. peak height diminished; hence 160° was adopted as a compromise temperature.

---

* *Abbreviations*.- ASPEC, Automated Sample Preparation with Extraction Columns; i.s., internal standard; LLE, liquid-liquid extraction; NPD, nitrogen-phosphorus detection.

Fig. 1. Extracted standard (10 ng/ml spike) showing peak distortion at a high injector temperature (225°).

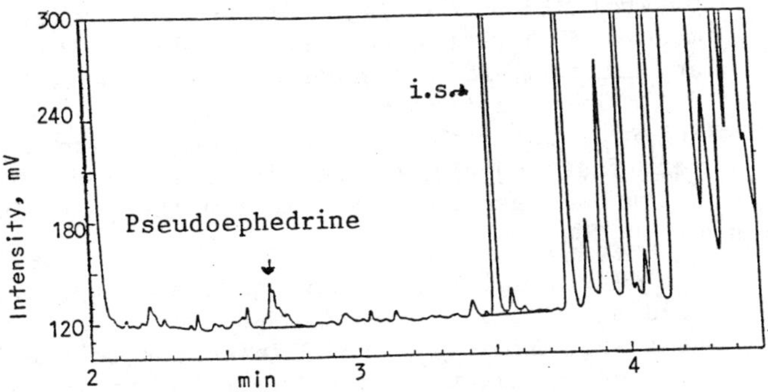

As the injector seemed to have a major influence on the GC, the liner type and injection speed were also investigated. After comparing the chromatography and response obtained using various designs of liner (packed and unpacked), the best peak shape was found with a silanized 2 mm i.d. single goose-neck liner packed with a small amount of silanized glass wool. Also, the injection speed should be as fast as feasible with the autosampler used; the HP 7673 would be ideal for very fast speed, but our instrument was equipped with a CTC-A200S.

Extraction.- The initial sample preparation problems were resolved by minimizing batch size (~10 samples) and the time from the addition of NaOH and chloroform to the reconstitution in ethanol, as the samples appeared unstable in the interim. This was accomplished using the following extraction scheme.-

To 1 ml plasma add diphenylamine (200 µl, 20 µg/ml) as i.s.; vortex-mix and add 200 µl 0.3 M NaOH and 5 ml chloroform. Mix, centrifuge (1100 **g**, 4°, 5 min), aspirate the aqueous layer and take it to dryness at 20° (N_2 stream). Reconstitute in 50 µl ethanol and inject onto the GC column.

Later the extraction was changed to move away from LLE with the need for harshly alkaline conditions and chlorinated solvents. Automated SPE was adopted, with the Gilson ASPEC system using SPEC MP3 extraction disc cartridges from Ansys®. An extra benefit from the ASPEC system was that the samples were extracted individually; so each was in contact with each reagent for exactly the same length of time, thereby obviating any possible variation due to instability during extraction.-

To 1 ml plasma add i.s. and mix, as above; add 1 ml acetonitrile, vortex-mix and centrifuge (1300 **g**, 4°, 20 min). To supernatant, reduced in volume (20°, N_2 stream; 20 min), add 1 ml pH 6 0.1 M potassium phosphate buffer and vortex. Prime cartridges with methanol then phosphate buffer, and load the sample. Wash with 1 M acetic acid, elute with dichloromethane/isopropanol/ammonium hydroxide (40:10:1 by vol.), and take to dryness (20°); reconsitute in 25 µl ethanol and inject onto GC.

Optimized injection and GC.- The injector was changed to a Gerstel CIS 3 Cool Injection System operated in solvent venting mode. The venting of the solvent at low temperature minimized any effects it may have at elevated temperatures and allowed full vapourization of the i.s., as occurs above 160°. As this injector was fitted to a Hewlett Packard instrument (HP 5890) a very fast injection speed could be used (HP 7673 auto-injector) which had been found to improve the peak shape (Fig. 2).

The final instrument conditions were as follows.- Injector held at 70° for 0.5 min, then ramped at 12°/sec to 300°; split vent shut after 0.5 min. Injection volume 5 µl. Column: DB-5, 30 m × 0.25 mm i.d., 0.25 µm film thickness; column temperature programme: initially 70°, held for 1 min; ramped at 40°/min to 300°. Detector: NPD at 300°.

In summary, the improved version of the LLE method was validated and found to give, with calibration and QC standards, acceptable results for accuracy and specificity over the full analytical range (C.V.'s $<15\%$; recoveries $>90\%$). The method was successfully used to analyze samples from a clinical study. Automated SPE and a programmable temperature injector improved it.

Reference

1. Bye, C., Hill, H.M., Hughes, D.T.D. & Peck, A.W. (1975) *Eur. J. Clin. Pharmacol. 8,* 47-53.

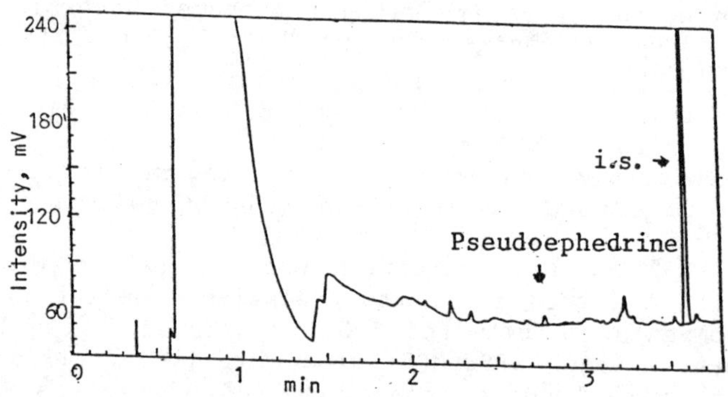

Fig. 2. Extracted standard (10 ng/ml spike) using automated SPE and, for GC, a Programmable Temperature Injector.

From this chromatogram (which was concerned with the LLOQ) the 2.6-3.0 min portion has been amplified:-

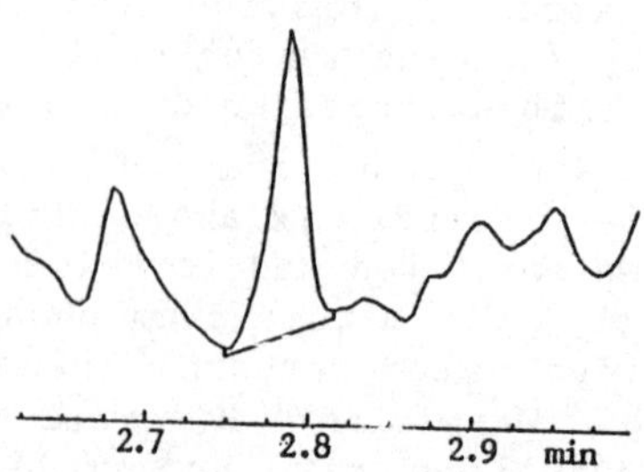

#E-5

BIOANALYTICAL APPLICATIONS OF ENANTIOSELECTIVE HPLC

Terence A.G. Noctor

Corning Hazleton Europe, Dept. of Biopharmaceutical Analysis, Otley Road, Harrogate HG3 1PY, U.K.

All chiral drugs should be analyzed enantioselectively in biological samples - at least at some point in their development. For racemic mixtures it is usually a matter of pharmacokinetic assessment of the individual enantiomers, whereas for an administered enantiomer non-racemization in vivo *may have to be verified. The armoury available to the bioanalyst to achieve the necessary enantiomeric separation is now extensive and quite mature†; but there remain some problems unique to enantiomer determination in biofluids. Current solutions to these are presented.*

Since the mid 1980's, the determination of enantiomers by HPLC, once an academic exercise, has become virtually a routine analytical task. This has been facilitated primarily by the commercialization of diverse CSP's* for HPLC, and the advent of various CDA's for diastereoisomer formation.† The analyst must first choose between the two main approaches, direct (CSP) and indirect (CDA), unless trying specific detection (see later).

Indirect enantioseparation in bioanalysis

Advantages.- Until the early 1990's, indirect methods predominated in the bioanalytical literature, and they still have an important role. This approach, with a suitable enantiomerically pure reagent, has some inherent advantages that should be kept in mind during method development [1]. These include wide applicability - many CDA's are available to react with primary and secondary amines (acid chlorides, isocyanates) and carboxylic acids (chiral amines, *via* a reactive intermediate) - and the potential to introduce a fluorophore or strong chromophore to enhance sensitivity.

Disadvantages may, taking a broad view, outweigh advantages. They include: partial kinetic resolution; problematic enantiomer purity with some CDA's; the time-consuming and

---

**Abbreviations.-* AGP, α_1-acid glycoprotein; CDA, chiral derivatizing agent; CSP, chiral stationary phase; OVM, ovomucoid.
†Reviews by I.W. Wainer in earlier vols. (Plenum; ed. E. Reid *et al.*): #16 (1986) and, for cardiovascular drugs, #18 (1988).-*Ed.*

tedious nature of the derivatization process; and the fact that diastereoisomers may differ in detector response. While one or more of these problems may indeed occur in a given instance, usually they are easily overcome. Yet the potential difficulties are such that indirect analysis is not now the preferable approach, except maybe for a few compound types where reliable generic methods exist (e.g. 2-arylpropionic NSAID's; cf. art. by Tan *et al.*). Opting for the direct route is now the norm.

Direct enantioseparation in bioanalysis

Usually the chiral selector that gives transient diastereoisomeric complexes for separation [2] is integral with the HPLC stationary phase. Far less commonly the selector is put into the mobile phase - rarely done in routine bioanalysis. Having opted for a direct separation the analyst has the daunting task of selecting the best CSP from >100 marketed for HPLC [3]; but fortunately the choice, for most drug analytes, need only be amongst a few 'core' CSP's. Two types are of greatest value, based on immobilized proteins or cellulose/amylose derivatives.

The scope of protein-based CSP's is unique, embracing bases and non-protolytic compounds with diverse structures. AGP- and OVM-based CSP's, the most widely used, are fairly similar in properties and application range, although differences indeed exist. Ideally an analyst would have access to both types; but given their high cost and the ~80% overlap in compounds that they can resolve, either one would be a useful investment. The same disadvantages apply to both: low loading capacity, hypersensitivity to mobile-phase composition and, despite high enantioselectivity, poor distinction amongst structurally similar compounds (e.g. metabolites), hard or impossible to resolve.

Of the many polysaccharide-type CSP's, the most useful have proved to be the tris(3,5-dimethylphenyl carbamate) derivatives of cellulose and amylose ('OD' and 'AD' respectively; made by Daicel) and the tris(4-methylbenzoate) derivative of cellulose, 'OJ' (Fig. 1). Fig. 2 shows the separation of the enantiomers of an anionic NSAID in human plasma with an OD CSP, after SPE.

Disadvantages of polysaccharide CSP's are their price and their limited flexibility in respect of mobile-phase composition, this being confined mainly to a few NP solvents. [Vol. 23 has an account by H.T. Smith of tribulations and remedies - *Ed.*]

The CSP's described above can resolve the enantiomers of a surprisingly large range of compounds. Other CSP's, notably the cyclodextrins [4], add different types of selectivity and are a useful addition to a CSP library. Newer phases such as those based on macrocyclic antibiotics [5] offer exciting possibilities but usefulness in routine work is not yet proven.

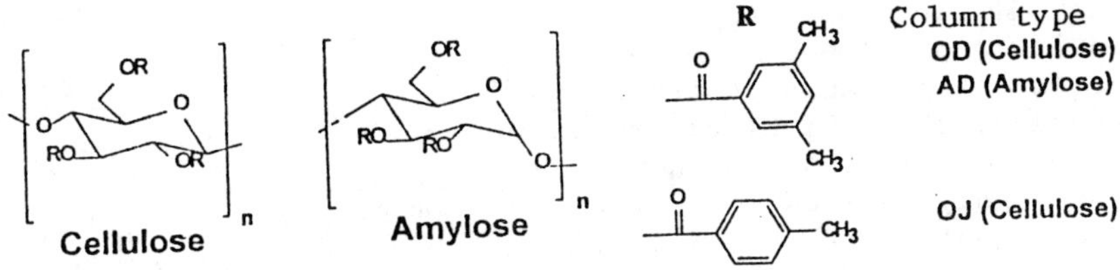

Fig. 1. Polysaccharide-based CSP's used in bioanalysis.

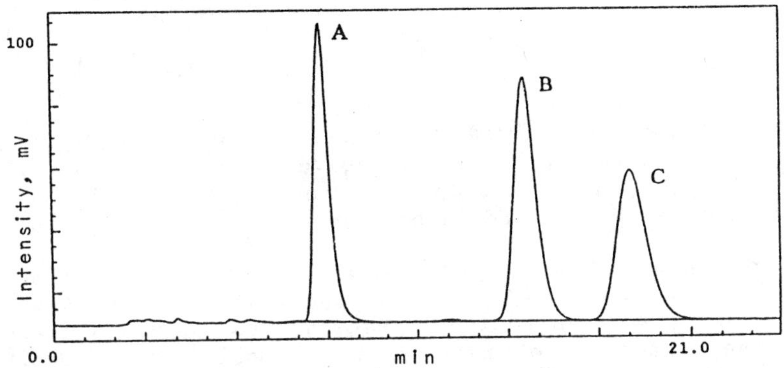

Fig. 2. HPLC, after SPE, of human plasma with a racemic NSAID spike, 4 µg/ml. Chiralcel OD column (25× 4.6 mm); hexane/ethanol /TFA (97:3:0.005 by vol.), 0.8 ml/min; fluorimetry (ex 280 nm, em 338 nm). Peaks: enantiomers (A, *R*; C, *S*) and i.s. (B).

Problems with direct separation.- Developing satisfactory resolution of the enantiomers of interest may need much effort; then the method has to be applied to extracted samples. Often there is unfortunate interference with one or both enantiomers, by a matrix component or a metabolite, and it may be impossible to get a sufficiently clean extract without losing the compound of interest; manipulating the HPLC conditions to obviate the interference may cause loss of the essential chiral separation.

One common means of overcoming this type of problem is to use two or more HPLC systems: achiral, to remove interferences, then chiral. Broadly speaking, there are three ways to do this.

Alternative modes for twin-HPLC, achiral and chiral

Approach 1: achiral and chiral runs are done sequentially using physically separate systems (Fig. 3). This approach, the commonest, has the merit that each system can be well optimized without consideration of mobile-phase compatibility. Potential also exists for several components (e.g. metabolites) to be collected individually; the peak of interest is collected after achiral HPLC and is either dried down (NP eluents) or extracted

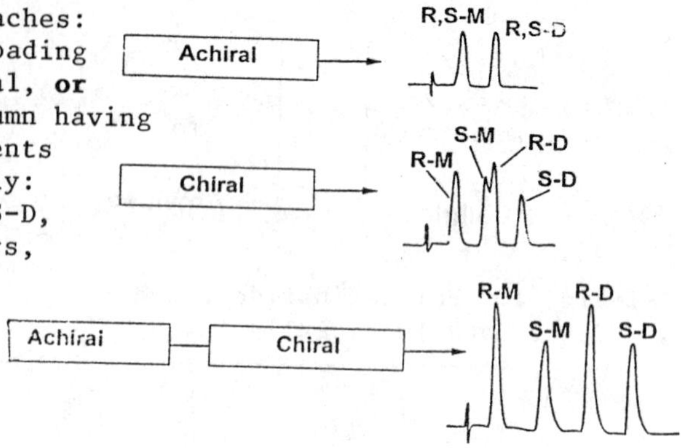

Fig. 3. Twin-HPLC approaches: **either** achiral and, re-loading minus interferants, chiral, **or** coupled, the achiral column having resolved unwanted components from analytes - notionally: drug enantiomers, R-D & S-D, and metabolite enantiomers, R-M & S-M.

(RP eluents; SPE or liquid-liquid extraction). The dry product is redissolved in a CSP-suitable solvent, usually mobile phase. Approach **1** requires considerable manual intervention and so is cumbersome, costly and maybe unsuitable for large studies.

Approach 2 entails direct coupling, in series, of achiral and chiral columns (Fig. 3). It applies only when an achiral stationary phase can be found that can resolve the interferant from the components of interest under mobile-phase conditions apt for enantioresolution on the CSP. This is a relatively rare occurrence. However, when applicable the approach is useful in separation of a metabolite from the parent compound while maintaining the integrity of a chiral separation.

Approach 3.- A switching valve may be used to effectively link the achiral and chiral systems. Only the eluate region containing the compound of interest is transferred onto the chiral column. A notable advantage of this approach is the ability to fully automate the analysis; the primary disadvantage is poor reliability in routine applications. [Vol. 22 has a notably pertinent art.: p. 191, C. Lindberg *et al.*- *Ed.*]

Specific detection

An alternative to introducing extra chromatographic selectivity to avoid interference problems in chiral bioanalysis is to employ a specific detection technique. The early promise of enantioselective detection techniques based on chiroptical properties such as circular dichroism or optical rotation has not been fulfilled. Generally these techniques are much too insensitive for bioanalytical needs. However, recent advances in HPLC-MS have made this an accessible answer to many of the problems traditionally associated with the chiral separation of drugs in biofluids. Fig. 4 shows resolution, using an AGP-CSP, of the enantiomers of a cyclic imide in plasma after SPE. With conventional rather than MS detection, endogenous interferences vitiated the separation.

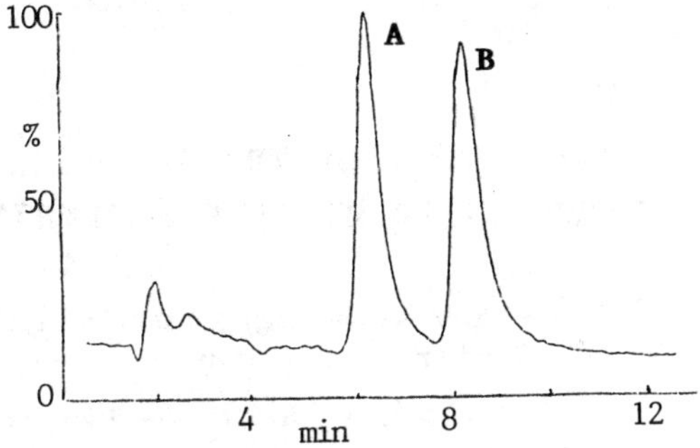

Fig. 4. Chiral HPLC-MS of SPE-treated human plasm spiked with 5 μg/ml (racemate) of a cyclic imide. Column: AGP; 10 mM ammonium acetate pH 4; MS(ESP), SRM. Peak A = *S*-enantiomer; B = *R*-enantiomer.

CONCLUDING COMMENTS

Enantioselective bioanalysis remains a difficult analytical problem that will not rapidly go away. Indirect methods that involve diastereoisomer formation remain popular as they are easy to apply, use cheap, regular HPLC packings, and can be exploited to augment detectability. However, this approach has significant disadvantages such that, particularly for routine applications, it is losing ground to direct methods. These can often be effected by using one of a small set of CSP's, drawn from commercially available protein and polysaccharide types. The direct approach is ideal if the compound of interest can be resolved without interference from endogenous materials or metabolites. Often this fails and two columns (achiral and chiral) are needed. The widespread availability of HPLC-MS instrumentation is likely to have a strong influence on the next generation of chiral bioanalysis.

References

1. Ahnoff, M. & Einarsson, S. (1989) in *Chiral Liquid Chromatography* (Lough, W.J., ed.), Blackie, Glasgow, pp. 39-30.
2. Wainer, I.W. (1987) *TRAC 6,* 125-134. [See also footnote on title p. of present art.]
3. Ahuja, S. (1991) in *Chiral Separations by Liquid Chromatography* (Ahuja, S., ed.), ACS Symp. Series Vol. 471, Washington, pp. 1-26.
4. Armstrong, D.W., Chen, S., Chang, C. & Chang, S. (1992) *J. Liq. Chromatog. 15,* 545-556.
5. Armstrong, D.W. (1994) *Anal. Chem. 66,* 1473-1484.

#E-6

CYCLIC VOLTAMMETRY TO OPTIMIZE EC DETECTION WITH HPLC, EXEMPLIFIED BY DRUG-ANALYSIS CASE STUDIES

[1]Stephen A. Wring†, [1]David Ayres, [1]Satish Dayal & [2]Geoffrey Harrison

Glaxo Wellcome Research and Development▫

[1]Park Road, Ware, Herts., U.K. [2]Greenford, Middx., U.K.

CyV is employed to determine the electroactivity of pharmaceutical compounds, and to aid the development of sensitive and robust methods based on HPLC-EC, apt for supporting pre-clinical and clinical drug development programmes. CyV observations now presented are derived from systematic studies to elucidate the EC behaviour of several pharmaceuticals, to assess the effect of solution pH, to address electrode fouling, and to optimize mobile phase composition and instrumental conditions for HPLC-EC.*

EC detection generally affords a more sensitive alternative to UV/VIS detection for HPLC. Yet HPLC-EC methods are often perceived as being markedly less robust than those employing spectrophotometric detection. Moreover, not all compounds are electroactive over the practicable range of applied potentials. Hence EC may not always be considered during method development or may be given an inappropriately superficial evaluation.

This article surveys how CyV* may be employed to determine electroactivity and optimize the solution conditions for the reliable determination of pharmaceutical compounds by HPLC-EC. In particular, examples are included to illustrate how CyV has been used systematically to:

- assess whether a compound is electroactive;
- determine the effects of solution pH on EC activity and behaviour;
- elucidate the likelihood of electrode fouling (poisoning);
- optimize mobile phase composition for EC detection;
- provide an indication of likely detector specificity.

---

†addressee for any correspondence, at Glaxo Wellcome Research Inc., 5 Moore Drive, Research Triangle Park, NC 27709, U.S.A.

▫Postcodes: Ware, SG12 0DP applies; Greenford, UB6 0HE

**Abbreviations*.- BR, Britton Robinson (buffer; see Experimental); CyV/CyVgrams, cyclic voltammetry/voltammograms [Editor's abbreviations; authors' text had 'CV', confusable with coefficient of variation]; EC, electrochemical (detection); FSD, full-scale deflection; LLOQ, lower limit of quantification; LOD limit of detection; SCE, saturated calomel electrode.

For more information on voltammetry and other electroanalytical techniques†, several monographs are recommended [1-5].

EXPERIMENTAL

Apparatus and reagents.- CyV was performed with an Eco-Chemie PGSTAT10 potentiostat interfaced to a Viglan 4DX33 PC. A 3-electrode configuration was used comprising a planar glassy carbon working electrode (cleaned by wiping the surface with a soft tissue impregnated with methanol between runs), a Pt foil counter-electrode, and an Ag/AgCl reference electrode or a reference SCE. All experiments were performed in an EG&G polypropylene micro-vessel. HPLC-EC was performed with an ESA 580 pump and Coulochem 5200 detector, in conjunction with 5020 guard and 5011 analytical cells.

Reagents were analytical grade from Merck (BDH) or Fisons; solvents were obtained from Rathburn. A stock BR buffer was prepared containing citric acid, KH_2PO_4 and *o*-boric acid [6]. Buffer solutions were taken (pH meter) to the desired pH by 0.2 M NaOH addition. Deionized water (>18MΩ) was obtained from an ELGA UHP system. Test compounds (from Chemistry Division colleagues) dissolved in an appropriate solvent (polypropylene tubes) on the analysis day were protected from light till used.

CyV procedures, all experiments being in duplicate.- The settings were typically: initial potential -0.3 V, scan rate 50 mV/sec and final potential +1.2 V. More negative potentials were avoided as dissolved O_2, present at mM concentrations, can undergo reduction at a working electrode and generate high cathodic currents. O_2 can be removed by passing purified N_2 (O_2-free) through test solutions for ~10 min prior to any measurements [2, p. 26]. However, applied potentials <-0.3 V can make HPLC-EC troublesome owing to the impracticalities in removing dissolved O_2 from test solutions. Whilst not prescriptive, it is where applied potentials can be more positive than ∿-0.3 V that EC is reckoned particularly suitable.

CyVgrams were recorded for ~0.5 mM solutions of each test compound in, as supporting electrolyte solution, BR buffer or the mobile phase employed for an existing HPLC-UV method. For some compounds it was necessary to add organic solvent (e.g. methanol) to the BR buffer to enhance solubility.

†In 1978 (Vol. 7, this series), P.T. Kissinger stated that all EC detectors which have achieved practical importance are of amperometric type, current being measured as a function of time with a constant potential applied at a fixed-electrode exposed to a moving fluid.- *Ed.* (Up-date: coulometric detectors are now often considered more sensitive alternatives.)

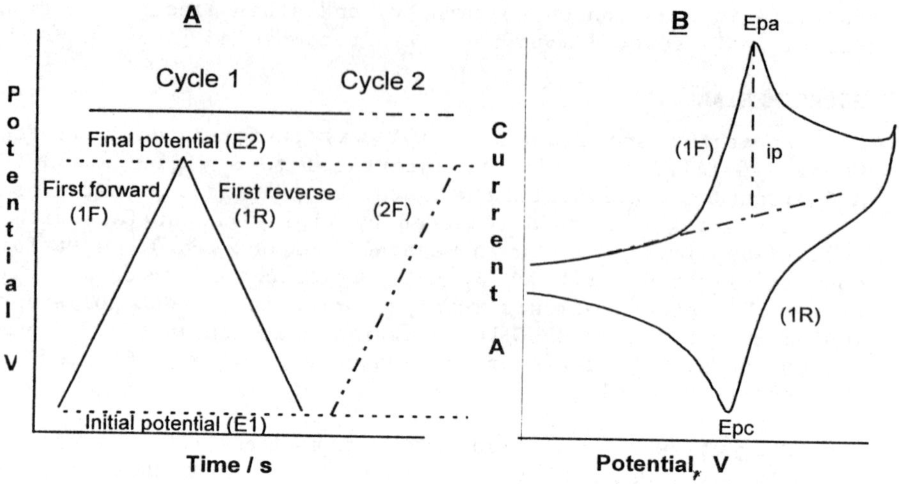

Fig. 1. CyV waveform (A) and model reversible response (B).

RESULTS AND DISCUSSION

Cyclic voltammetry (CyV)

CyV involves applying a linearly increasing d.c. potential (usually between 10 and 500 mV/sec) to a working electrode; the ramp is reversed at a pre-defined switching potential and may be thereafter cycled to form a repeating triangular excitation waveform (Fig. 1A). The resulting CyVgram can reveal mechanistic information on any redox processes occurring at the electrode surface. A typical CyVgram for a reversible redox couple is shown in Fig. 1B; the main parameters of interest are: the anodic and cathodic peak potentials (E_{pa} and E_{pc}, V) and the magnitude of the peak current responses (i_{pa} and i_{pc}, A).

The difference in applied potentials (ΔE_p) at which these peaks occur may usefully be determined, to provide an insight into the reversibility of the redox couple. For a reversible couple where both species exchange electrons rapidly, ΔE_p is ≈59/n mV (n is the no. of electrons involved in the oxidation or reduction). This separation should also be independent of scan-rate variations (ν, V/sec). The potential at $(E_{pa} + E_{pc})/2$ equals the formal electrode potential ($E°'$). Where a process is quasi-reversible, ΔE_p >59/n mV; if the reaction is totally irreversible, a peak will be observed only on the forward scan.

For quantitative determinations, or mechanistic studies, values of i_p are measured from the voltammograms. The total current measured is the sum of components formed from Faradaic processes for the analyte (and any impurities in the medium) and the generally undesirable capacity current from charging of the electrode double layer. The magnitude of the Faradaic peak current, for a reversible system measured at a planar electrode of area A (cm^2), is related to concentration C° (mol/ml) as defined by the Randles-Sevcik equation [7]:

$$i_p = (2.69 \times 10^5)n^{3/2}AD^{1/2}C^\circ v^{1/2} \quad \text{(at 25°)}$$

where D = diffusion coefficient (cm^2/sec). This equation is modified for irreversible systems:

$$i_p = (2.99 \times 10^5)n(\alpha n_a)^{1/2}AD^{1/2}C^\circ v^{1/2} \quad \text{(at 25°)}$$

where α = transfer coefficient, n_a = no. of electrons in the rate-determining step of the electrode process.

The Randles-Sevcik equation is important as it states that current (i_p) can be directly proportional to concentration providing other parameters are kept constant. Typical sensitivity achievable by CyV is ~10^{-5}-10^{-6} M [pp. 51-128 in 4; and 7]; however, the technique's performance is often detrimentally affected by an inability to separate the Faradaic currents from non-Faradaic capacity current - particularly for compounds where high applied potentials (~≥1 V) are required. Consequently, although the technique is quantitative it may have only limited value for direct methods to support pharmacokinetic studies. Nonetheless, CyV can provide considerable mechanistic and diagnostic information for studying the EC behaviour of organic compounds, and the optimization of instrumental and solution conditions for HPLC-EC.

Electroactivity

Fig. 2 presents CyVgrams for 3 compounds with different EC behaviour, including one (curve B) that lacked EC activity under the conditions used, as evidenced by the current response being virtually superimposable on the response recorded for plain buffer solution. In contrast, the electroactive compound in curve C, GR-X, underwent an oxidation giving a single anodic wave, E_{pa} ≃0.58 V *vs.* SCE, and a subsequent reduction yielding the single cathodic wave E_{pc} ≃0.41 V *vs.* SCE. The difference in the values for peak potential is ≃0.17 V, which indicates quasi-reversible behaviour. GR-X is a good candidate for HPLC-EC since the applied potential is below levels likely to induce high background currents caused by electrolysis of possible mobile-phase organic solvents, e.g. acetonitrile (<~0.9 V *vs.* solid state reference electrode in the ESA5011 analytical cell).

The compound whose CyVgram appears in curve D undergoes an irreversible oxidation (E_{pa}≃1.04 V *vs.* SCE) and, whilst being electroactive and amenable to HPLC-EC, is not such an ideal candidate because a notably higher applied potential is needed.

Effects of solution pH

Peak potential.- The EC behaviour of drugs is frequently pH-dependent, and this property has been utilized during the development of HPLC-EC methods. For oxidation of the $5HT_{1D}$ antagonist GR 127935 (Fig. 3, comparing pH 10.7 and pH 3.5),

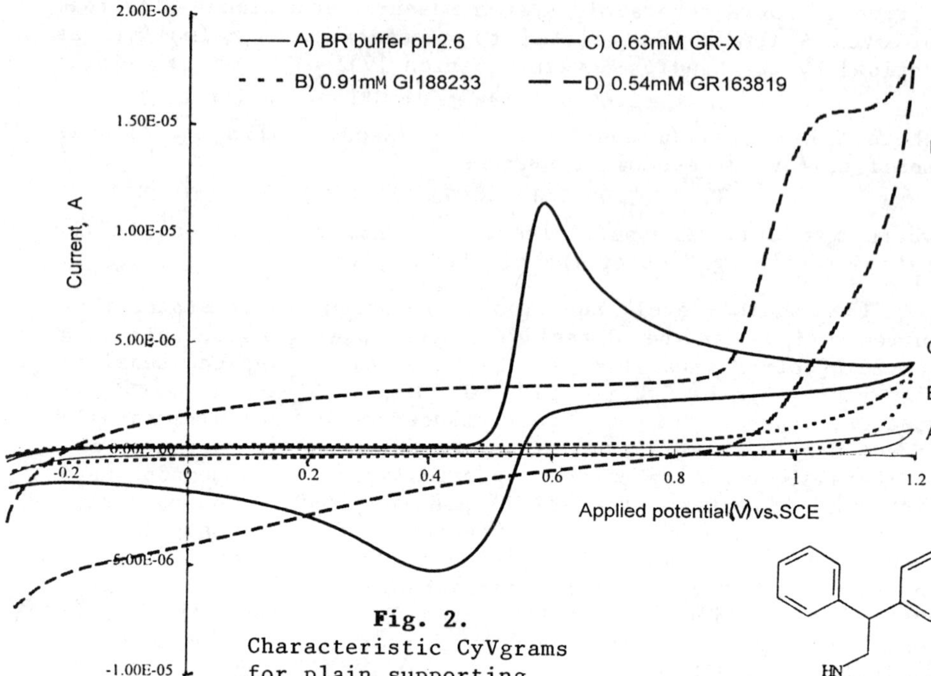

Fig. 2.
Characteristic CyVgrams for plain supporting electrolyte solution (A) and 3 compounds. GI 188233 (B) is non-electroactive [structure not relevant]. GR-X (C): a suitably electroactive candidate drug. GR 163819 (D): electroactive; needs a high applied potential.

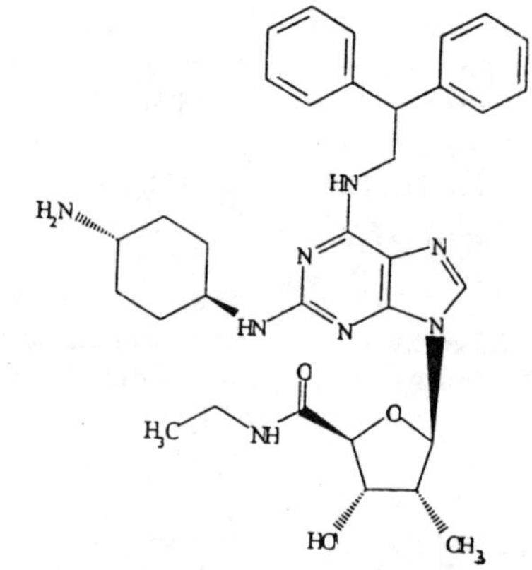

GR163819

the peak potential (E_{pa}) decreased with increasing solution pH. E_{pa} (V) varies with pH by -0.059m/n.pH (m represents the no. of protons), which corresponds to a decrease of 59 mV per unit increase in pH for processes involving an equivalent number of protons and electrons.

Fig. 4 presents graphs of E_{pa} *vs.* pH for the oxidation of several candidate drugs. The E_{pa} for GR127935 decreased from 0.99 V to 0.72 V on increasing the pH of the supporting electrolyte from 4.6 to 9.7. Within the limitations of the data, the graph displays a change in slope, and hence pH-dependency, at pH ~5. The position of this break corresponds to the pKa (or pK') of GR127935. The limited data for GR203040 also

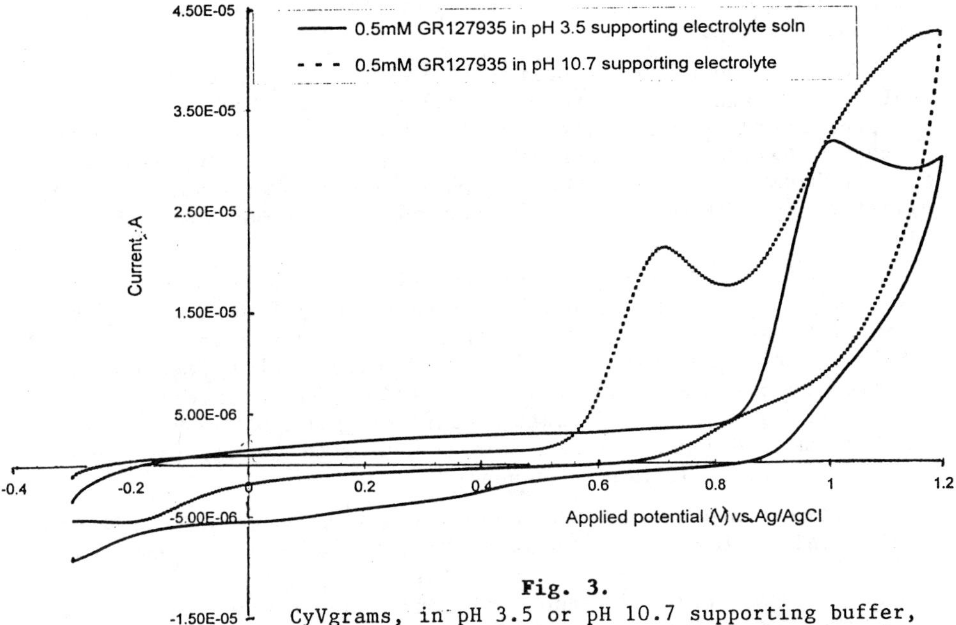

Fig. 3.
CyVgrams, in pH 3.5 or pH 10.7 supporting buffer, for 0.5 mM GR 127935 (see Fig. 7 for structure).

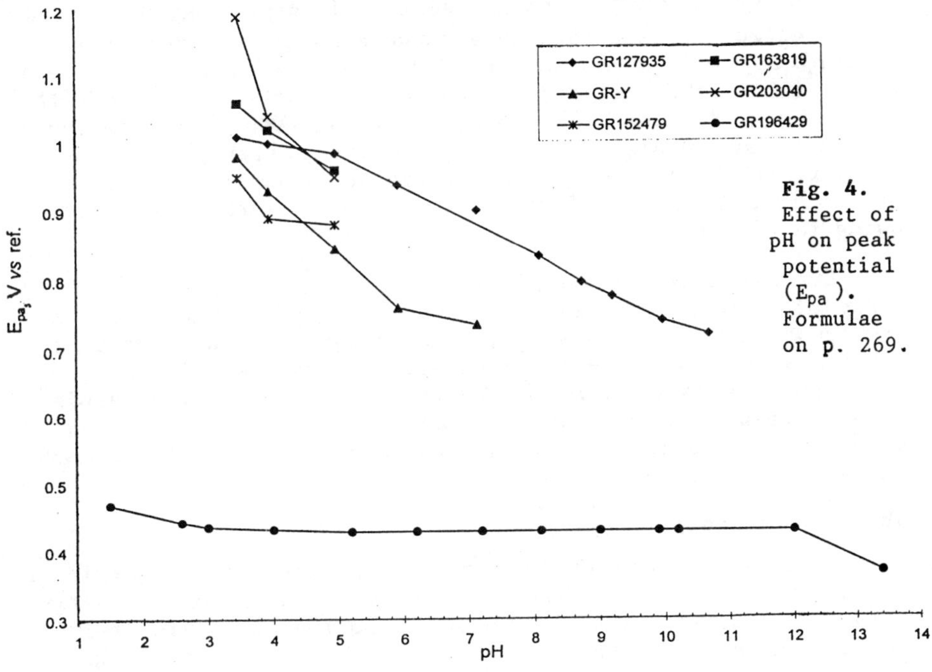

Fig. 4.
Effect of pH on peak potential (E_{pa}). Formulae on p. 269.

reveal a marked change in slope at pH 4, in agreement with the pKa of the compound, pH 4.3. For the other compounds pKa data were unavailable.

Some compounds lost their pH-dependencies after changes in ionization states. For example, in a previous study [8] one of the present authors found that all-*trans* retinol underwent a pH dependent oxidation between pH 5.2 and pH 6.74; however, the process became pH-independent at pH values >6.74. Likewise, the oxidation of GR196429 (Fig. 4) was pH-independent over the pH range 4-12.

Overall, evidently a change in mobile phase pH may in some cases be used to enhance HPLC-EC methodological sensitivity by separating the oxidation potentials of the analyte and a potentially interfering compound. Alternatively, robustness may be improved by using a mobile phase within a pH-independent region - so setting a fixed analyte oxidation potential, and hence constant method sensitivity, not susceptible to any batch-to-batch imprecision in mobile-phase pH.

Electrode processes.- Solution pH may also affect EC and/or chemical follow-up reactions. Fig. 5 presents CyVgrams for GR196429 recorded in pH 3.35 and pH 10.42 supporting electrolyte solutions. At pH 10.42, two additional anodic waves were seen for EC oxidations that occur following the primary oxidation (E_{pa}, ≃0.43 V *vs.* SCE). These additional waves may be used with coulometric detectors to enhance sensitivity. For example, setting applied potential to 0.6 V in the analytical cell would allow the primary oxidation and the first of the following oxidations; the summated current responses from these two processes would be measured as the analytical response. However, the benefit of the increased sensitivity would need to be considered in the light of any loss in method selectivity owing to the slightly higher applied potential.

Current response may also be affected by solution pH. Indeed, current decreased with increasing pH for each of the compounds in Fig. 4. However, the contrary is found for some other assays, e.g. all-*trans* retinol. Altogether, investigating the effects of solution pH on EC behaviour is valuable because results allow optimization of applied potential and current response that can lead to enhanced method selectivity and sensitivity.

Theoretical sensitivity

EC detectors for HPLC generally are either amperometric or coulometric, depending on the % conversion of electroactive compound. The former involves much less than 100% electrolysis while coulometric detection affords 100% conversion - and in principle is most conducive to sensitivity which, however, is

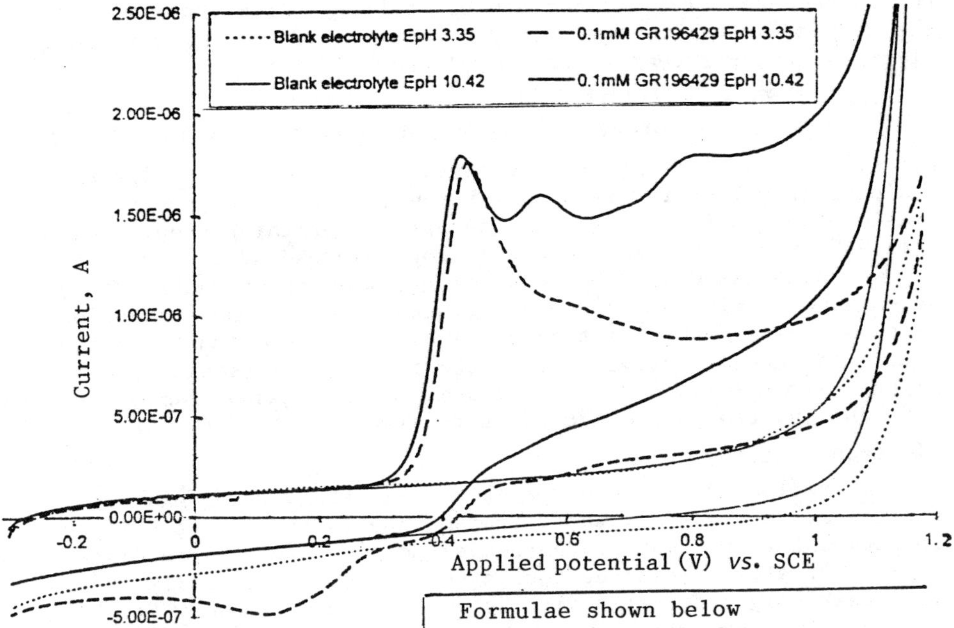

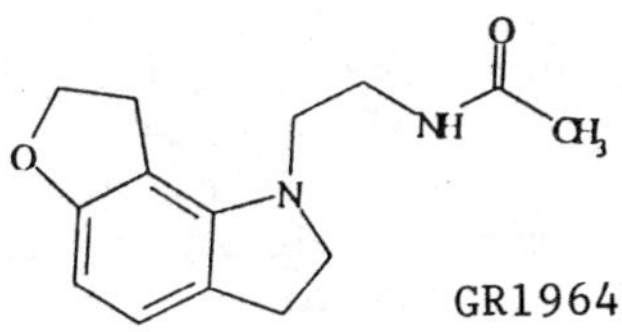

GR196429

Fig. 5. Effect of pH on the oxidation mechanism for GR196429.

See Figs. 2 & 7 for other Fig. 5 formulae.

Formulae shown below for Fig. 4 compounds

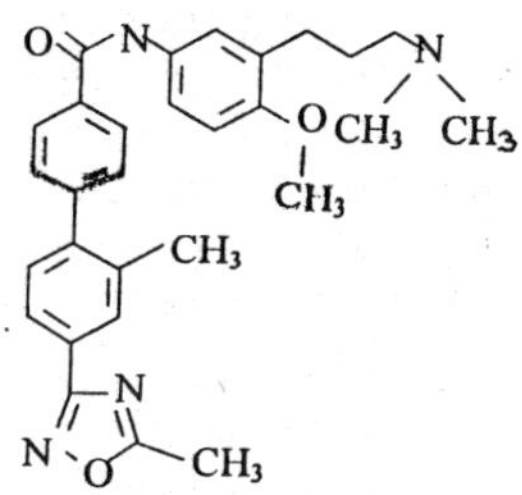

GR152479

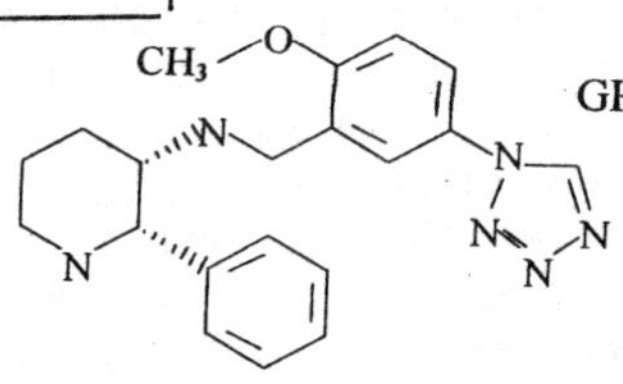

GR203040

not always attained because of limitations in the signal-to-noise ratio. Integration of peak areas following satisfactory coulometric detection yields a value for charge (Q, coulombs) that can be directly related to concentration and theoretical sensitivity by the equation $Q = nFN$ (F, Faraday constant; N, no. of moles). This equation indicates that, for a constant N value, peak area depends on the number of electrons involved in the EC process. This number can be estimated from the slope of the E_{pa} plot against pH and by calculating values for the electron transfer coefficient (αn_a) from CyVgrams:

$$\alpha n_a = 0.048/(E_{pa} - E_{pa/2})$$

where $E_{pa/2}$ is the potential corresponding to $i_{p/2}$. Values for α generally, for 1-electron processes, range from 0.5 if irreversible or from 1 if reversible. They may be >1 when more than a single electron is involved in the rate-determining step. For GR127935 in the pH 5-10 range the slope was -48 mV and the mean $\alpha n_a \pm s$ value was 0.47±0.062. Seemingly 1 electron and 1 proton were involved in the rate-determining step of the oxidation process. The HPLC-EC method achieved 21 pg on-column sensitivity, in general agreement with the theoretical sensitivity of 23 pg (from Q = nFN). The measured peak response was 0.3 nA (6% FSD) and the peak width at half-height was 15 sec.

Electrode fouling

Adsorption of oxidation products onto the working electrode surface can degrade current response, due to the adsorbed layer forming a barrier that prevents analyte diffusion to the electrode substrate. Electrode fouling with HPLC-EC generally shows as a peak-height decrease with sequential injections. Fig. 6A presents voltammograms recorded for two drugs that demonstrate the presence and absence of electrode fouling, respectively. Those obtained on the first and second scans for troglitazone are virtually superimposable, although i_p commonly decreases on the second and subsequent scans since the analyte diffusion to the electrode becomes rate-limiting and limits current response whereas the second of those for salmeterol lacks the oxidation wave, not restorable by stirring the solution between scans.

Electrode fouling by oxidation products may be prevented in several ways; we have had success with an addition/increase/change of organic solvent within the supporting electrolyte or mobile phase solutions. Alternatively, modifying the solution pH may be effective. Fig. 6B shows that in pH 7.0 solutions the response for the oxidation of GR 163819 was attenuated on the second scan, whilst at pH 2.6 fouling was prevented and the responses for the second and third scans were commensurate with a diffusion-limited process.

An alternative analytical cell design may also be used with HPLC-EC to prevent fouling, as was shown [8] with all-*trans*-retinol (842 ng/ml standard solution): 3 successive injections caused current response at a thin-layer amperometric cell to decrease by 65%, suggesting that the final oxidation product adsorbed strongly onto the cell's glassy carbon surface, causing passivation. This was avoided by use of a cell with a wall-jet configuration; the solution impinged directly onto the electrode surface, which had the effect of continually cleaning it.

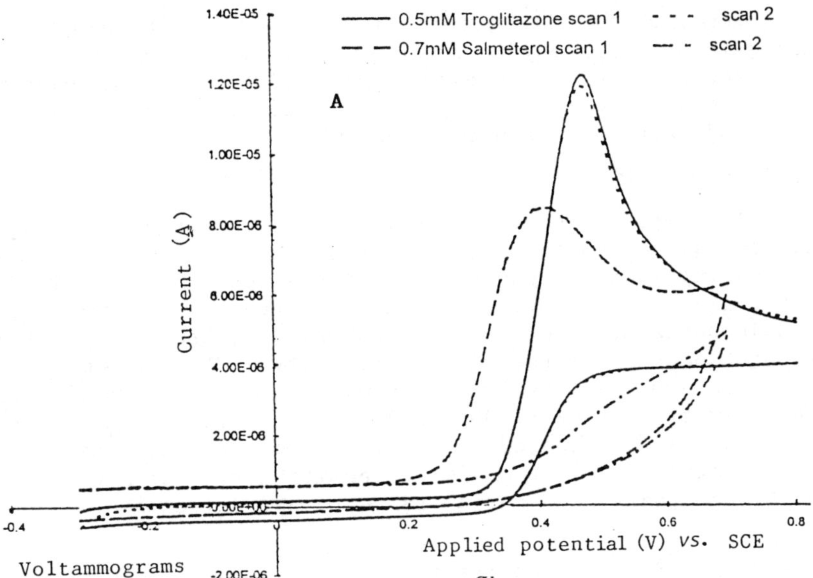

Fig. 6. Voltammograms relating to electrode fouling during analyte oxidation. **A**: Salmeterol (fouling) and [formula shown] troglitazone (no fouling). **B**: Overcoming fouling, by suitable pH choice, for 0.5 mM GR163819 [formula in Fig. 2] in BR buffer.

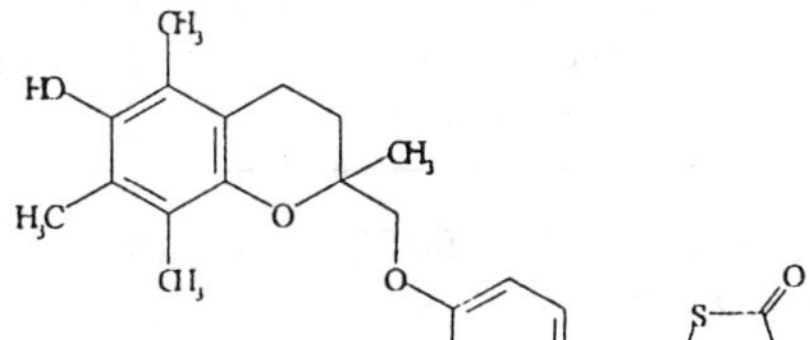

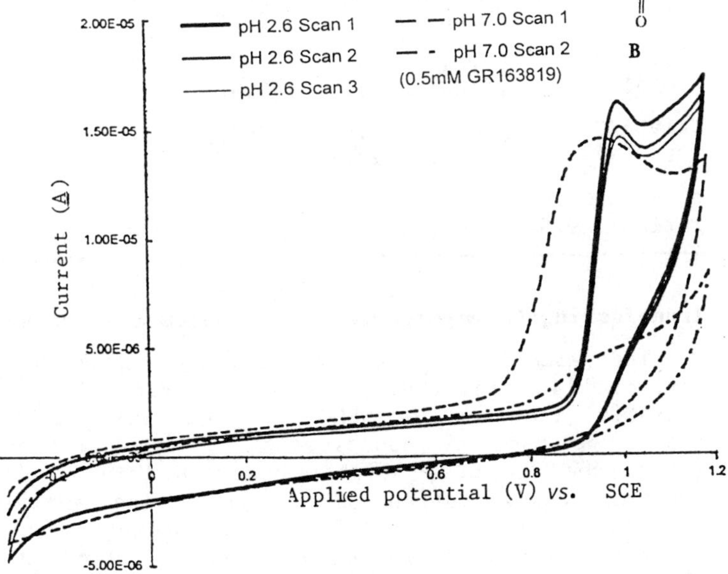

Fouling may also occur owing to non-analyte components in the sample. This may be overcome by modifying the mobile phase as described above; however, if fouling cannot be prevented a pulsed wave-form may be employed to electrochemically clean the electrode after a measurement step.

Ionic strength

Ionic-strength effects were slight for each of 4 compounds (Table 1), although the background current tended to rise with increasing ionic strength in solutions containing acetonitrile. Typically, 0.025 M buffer solutions were employed for preparing mobile-phase mixtures as these afforded low background currents without overly compromising sensitivity. Nonetheless, current response was marginally dependent on ionic strength for the determination of GR127935 (~24% current increase) and all-*trans*-retinol (~16% increase). Hence for some analytes ionic-strength optimization may help maximize sensitivity. To minimize background currents high-purity reagents and solvents are requisite.

Table 1. Summary results for the effects of ionic strength on voltammetric response. Supporting electrolyte solution: MeCN, acetonitrile; MeOH, methanol; compositions are by vol.

Compound & mM concn.	Buffer/organic component	Buffer range	Corresponding ranges, μA	
			$i_{background}$	i_{pa}
GR127935, 0.5	McIlwain pH 3 /MeCN, 1:1	0.006 -0.12 M	3.74-7.09	32.16-40.01
GR163819, 0.41	phos. pH 2.25 /MeCN, 2.2:1	0.024 -0.48 M	2.51-7.74	12.25-10.01
GR196429, 0.06	phos. pH 4.0 /MeOH, 1.22:1	0.025 -0.25 M	0.165-0.170	0.74-0.75
all-*trans*-retinol, 0.5	acetate pH 5.0 /MeOH, 1:19	0.013 -0.074 M		36.1-42.6

Transferring to amperometric and coulometric detectors

The value for E_{pa} determined by CyV should be similar, once corrected for reference electrode potential, to the applied potential required for HPLC-EC - the final selection of which should be made by constructing a hydrodynamic voltammogram (HDV) from the peak heights measured for the detection of a constant concentration of analyte using applied potentials spanning the E_{pa}. The value corresponding to the largest peak current is generally selected; however, it may be necessary to use lower values to enhance selectivity in the presence of possible interfering compounds.

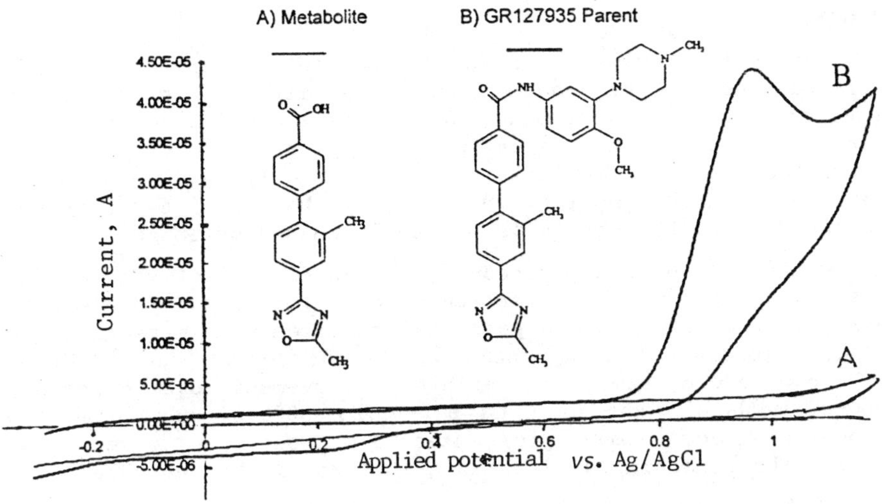

Fig. 7.
CyVgrams for 0.5 mM GR127935 and 0.49 mM GR152007, the main metabolite identified in toxicokinetic studies.

Indication of detector selectivity

Comparing metabolites and parent drug for EC behaviour can provide an indication of the possible specificity of HPLC-EC methods. Clearly, non-electroactive metabolites (Fig. 7) should not interfere. The likelihood of co-eluting compounds in the HPLC-EC method can be investigated by comparing the HDV constructed for standard solutions of the parent drug to a similar HDV for the parent drug peak measured from an *ex vivo* sample. The peak-height ratio at each potential should be constant for both test solutions; if different for the *ex vivo* sample, the presence of a co-eluting peak should be suspected. This technique is applicable for all compounds with different E_{pa} values but is unlikely to detect co-eluting peaks for compounds with similar E_{pa} values.

OUTCOME

Our systematic CyV approach has served to characterize and optimize HPLC-EC mobile phase conditions with due consideration of chromatographic criteria. CyV allowed rapid assessment of electroactivity and thence optimization of solution conditions; it has been used for mechanistic studies and to develop ways to overcome electrode fouling. Overall, CyV has enabled sensitive and robust HPLC-EC methods to be developed and optimized.

Of 16 compounds now screened, 13 had electroactivity and 3 lacked it. Of the 13, 2 were unsuitable for method developent because either the applied potential required was impracticably high or electrode fouling could not be overcome. When 8 HPLC methods were evaluated, all offered improvements over existing HPLC-UV methods, and 4 (with sub-ng/ml LLOQ's) were used to support drug development programmes; 3 methods were not applied and for the remaining analyte an HPLC-fluor method was used.

Acknowledgement.- The authors thank colleagues at Glaxo Wellcome Research & Development for their interest and support.

References

1. Wang, J. (1994) *Analytical Electrochemistry*, VCH, New York, 198 pp.
2. Hart, J.P. (1990) *Electroanalysis of Biologically Important Compounds*, Horwood, Chichester, 213 pp.
3. Smyth, W.F. (1992) *Voltammetric Determination of Molecules of Biological Significance*, Horwood Chichester, 133 pp.
4. Kissinger, P.T. & Heineman, W.R. (1984) *Laboratory Techniques in Electroanalytical Chemistry*, Dekker, New York, 75 pp.
5. Riley, T. & Wilson, A. (1987) *Polarography and other Voltammetric Techniques*, Wiley, Chichester, 282 pp.
6. McKenzie, H.A. (1969) in *Data for Biochemical Research* Dawson, R.M.C. *et al*., eds.), Clarendon Press, Oxford, ~600 pp.
7. Hart, J.P. (1984) in *Investigative Micro-Techniques in Medicine and Biology*, Vol. 1 (Chayen, J. & Bitensky, L., eds.), Dekker, New York.
8. Wring, S.A., Hart, J.P. & Knight, D.W. (1988) *Analyst 113*, 1785-1789.

#E-7

DIODE LASER-INDUCED FLUORESCENCE DETECTION IN HPLC AND CAPILLARY ELECTROPHORESIS (AFTER PRE-COLUMN DERIVATIZATION)

H. Lingeman†, A.J.G. Mank & C. Gooijer

Free University, Department of Analytical Chemistry,
De Boelelaan 1083, 1081 HV Amsterdam, The Netherlands

In liquid-phase separations, notably HPLC and CE, detection is commonly by absorption or fluorescence, advantageously laser-based if very high sensitivity is paramount. However, LIF has disadvantages including restricted applicability, paucity of discrete emission lines, and cost. Some of these drawbacks are circumventable by using diode lasers, which are very stable, inexpensive and easy to use. Since, however, lasers only emit wavelengths >600 nm, derivatization is usually a necessity.*

Basic aspects of diode lasers and modes of use for detection are surveyed, and particular attention given to HPLC (absorption and fluorescence). Emphasis is put on the use of DIO-LIF after pre-column derivatization. Experimental set-ups for HPLC and CE are given, and examples of the derivatization of thiols, amines, aldehydes and carboxylic acids. Detection limits are 25-250 amol in HPLC and 0.5 amol in CE using dicarbocyanine-based labels.

DIODE LASER-INDUCED FLUORESCENCE DETECTION

The principles of LIF - now an accepted means of selective and sensitive detection in HPLC and CE - have been explained within this series [1] and in recent reviews [2-4]. Especially in miniaturized systems, e.g. CE, laser-based techniques have some advantages: the fluorescence signal is proportional not only to the detection pathlength, as for absorbance, but also to the irradiation intensity (W/cm^2). Thereby in CE a gain of 4 orders of magnitude can be obtained comparing laser-based with conventional techniques, just by focusing the beam on a spot that can be accommodated inside a CE capillary. In HPLC this gain is normally <100-fold. In CE and HPLC respectively:
- Injection vol. typically 2 nl *vs.* 25 µl;
- Detector-cell vol. 1 nl *vs.* 10 µl;
- Absorbance pathlength 20-fold smaller in CE
- LIF irradiation area: 400-fold smaller.

†addressee for any correspondence

**Abbreviations.-* CE, capillary electrophoresis; (DIO-)LIF, (diode) laser-induced fluorescence. AlPcSn (Pc = phthalocyanine), MBT and HNE: see sections on detection, thiols and aldehydes resp.

This comparison shows the importance of compensating a smaller spot diameter by a higher light intensity. However, it should be kept in mind that the absolute laser power is certainly not the most critical parameter. For example, detection limits of 10^{-8} M in CE can be obtained using an excimer laser with an average laser power of 1 W, which furnishes 100 mW after passing a dye laser and 5 mW after frequency doubling; to prevent damage to the capillary only 10% of the light is used.

The positive features of LIF, using visible or UV lasers, are selective excitation, small detection volumes and strong signals. As the most important limitations, this technique is limited to fluorescing compounds, and the lasers traditionally used are expensive and relatively unstable.

IR and near-IR diode lasers have recently become available [5-7]. Diode lasers, emitting radiation at wavelengths >620 nm, are stable (intensity fluctuations <0.02% or, with feedback stabilization, <0.003%), cheap <$100 for a 10 mW laser emitting at 670 nm), small (<10 mm^3 for the whole package) and long-lived (>25,000 h). Moreover, the background signal due to fluorescence from optical impurities in the eluent is negligible, and the contribution of scatter to this signal is relatively small. Rayleigh and Raman scatter are proportional to λ^{-4} and thus 50-fold lower at 670 nm compared with 250 nm. This means that, applying excitation at 670 nm, the Raman scatter of eluent components (e.g. acetonitrile, methanol, water) is primarily observed at wavelengths >800 nm.

For this approach with wavelengths >600 nm the main limitation is that the applicability is even more restricted than with visible or UV lasers. However, a combination of DIO-LIF and a pre- or post-column derivatization allows the analysis of compounds lacking appropriate absorption, through introducing a label with red or near-IR absorption. At present there is the constraint that hardly any commercial labels are available.

APPLICATIONS OF DIODE LASERS

Although diode lasers are normally used for fluorescence detection after HPLC or CE separations they can also be applied for absorbance detection [8, 9].

Direct diode laser-based absorption detection

This approach provides HPLC detection limits of 3×10^{-9} M (~1.2 ng/ml) for an anti-cancer drug in urine, **mitoxantrone** (a compound that fluoresces poorly), with no need for clean-up of the sample [8]. Separation is performed with an isocratic RP system (Fig. 1), here using a ratioing system in combination with a 670 nm diode laser. A fibre-optic 10-mm Z-type detector cell, with a gradient-index lens on the light-guiding fibre, is a robust alternative to a normal commercially available absorption detector cell and provides a 20-fold gain in sensitivity.

Fig. 1. HPLC of urine, blank (**I**) and spiked with 15 nM mitoxantrone (**II**), using diode laser-based absorption detection at 670 nm. C-18 column (250 × 3.1 mm), 5 µm. Acetonitrile/50 mM phosphate pH 5.9 (1:1 by vol.). *From ref. [8], courtesy of the publisher.*

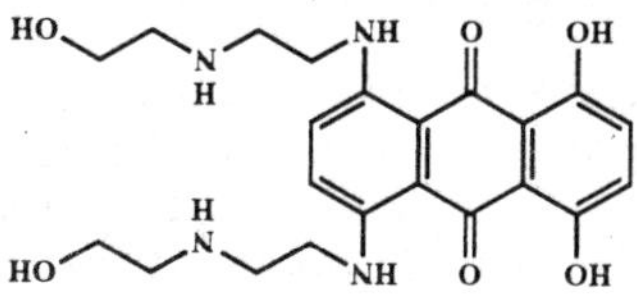

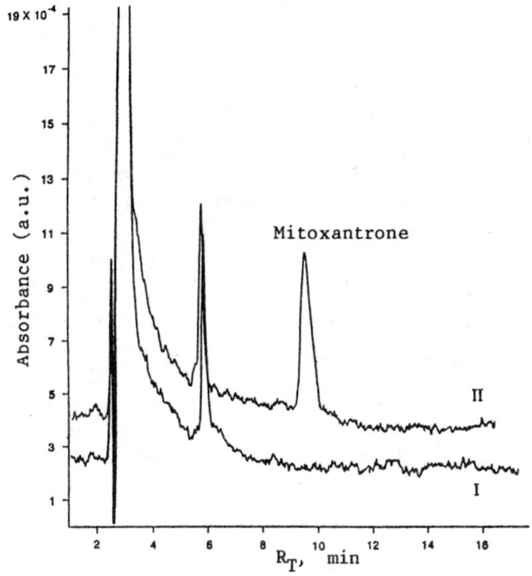

A second possibility with detection by absorption is to increase sensitivity by introducing intra-cavity enhancement [10]. Here the detector cell is positioned inside an extended laser cavity, which is created by the high-reflection coated back-facet of a medium-power diode laser and an external mirror [9]. Light travelling inside this cavity passes the detector cell several times before it is detected. This results in an effective increase in absorption pathlength (Fig. 2); moreover the laser operation as such is influenced by the absorption, because the effective gain in the cavity is reduced, and the resulting shift in lasing threshold gives a further increase in sensitivity. The results are rather disappointing in the case of mitoxantrone: the detection limit for a standard solution is 0.7 µM, which is only 4.5-fold better than with direct absorption detection. The results show that no significant multipass effect is present, which can be explained by the low coupling efficiency of the external cavity and the diode laser. It can, then, be stated that DIO-laser intra-cavity detection is not as yet really reliable but that it can perform better than double-beam absorption detection.

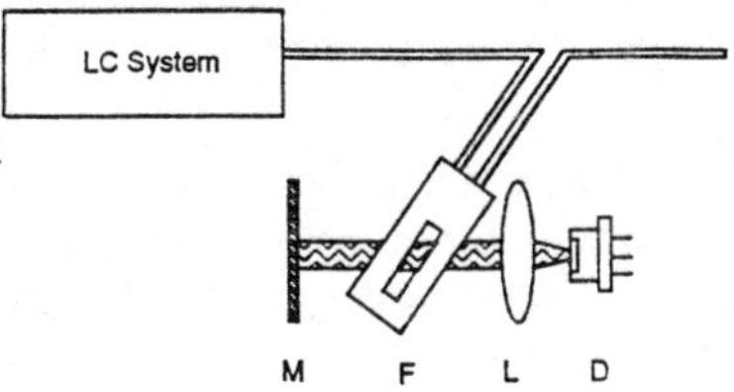

Fig. 2. On-line DIO laser-based intra-cavity absorption detection set-up (schematic). D, diode laser with integral photodiode; L, lens; F, flow-cell; M, mirror. *From ref. [9], by permission.*

Direct diode laser-based fluorescence detection

Rather few organic compounds absorb at wavelengths >600 nm. Exceptionally, sulphonated aluminium phthalocyanines (**AlPcSn**) - a family of fluorescent photosensitizers used in photodynamic therapy - absorb at >650 nm [11, 12]. Photosensitizers introduced into blood can be selectively taken up by tumours which, when a plateau level has been attained, absorb red laser irradiation whose excess energy furnishes short-lived but cytotoxic singlet oxygen [13, 14]. Optimization of such therapy, which is of growing interest, calls for selective and specific bioanalytical methods to minimize the amount of sample needed and ascertain which AlPcSn isomer is most effective.- Both the degree of sulphonation (AlPcS2 is disulphonated) and the isomer distribution affect the photosensitizer's activity.

HPLC with conventional-size columns and DIO-LIF suits well. The detection limit was 7 pM, 20-fold better than with ordinary fluorescent detection, for urine and (shown, with conditions, in Fig. 3) for PBS-extracted faeces. Evidently administration of AlPcS2 leads to several products, with time-dependence.

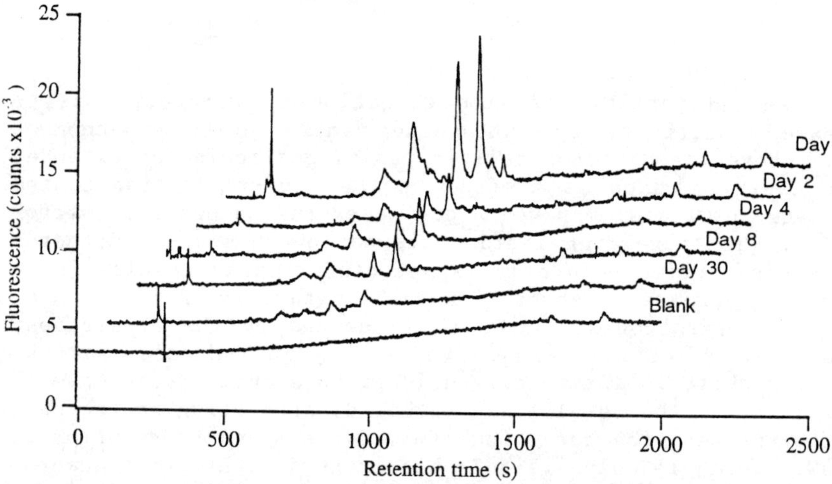

Fig. 3. HPLC of PBS-extracted faeces samples following injection of 1 µmol/kg AlPcS2. Suspensions, with 0.1 M NaOH added, kept 4 h at 50°, centrifuged, neutralized (0.2 M HCl), filtered, and diluted with methanol. RP column (250 × 3.2 mm), 5 µm, with a methanol-water gradient. For the blank trace the scale is exaggerated ×4. *From ref. [12], by permission.*

With this method the distribution of disulphonated AlPcSn isomers could be studied in an oral squamous-cell carcinoma model; seemingly the kinetics of some AlPcSn isomers differs amongst various tumour models studied [14, 15].

DIO-LIF DETECTION AFTER PRE-COLUMN DERIVATIZATION

Many analytes cannot be detected spectrophotometrically since they absorb only in the 200 nm region. The disadvantages of using wavelengths in the deep-UV region are strong interferences and limited selectivity and sensitivity. The remedies are to use indirect detection and a derivatization procedure.

Covalent derivatization with a red-absorbing fluorescent probe offers interesting scope for improving both selectivity and sensitivity. However, currently the number of organic solutes with adequate fluorescent properties is rather limited and few are commercially available. Also, the spectroscopic properties of the labels and the corresponding derivatives are similar, implying that derivatization must precede separation. Starting with a suitable cyanine molecule, various functionalities can be attached to allow selective labelling of thiols, amines, aldehydes and carboxylic acids (see below).

Detection set-ups for HPLC and CE

The set-ups for HPLC and CE are somewhat different. The light-path in a DIO-LIF detection system is shown in Fig. 4a. In most cases a 10 mW diode laser is focused at a distance of 10 cm in combination with a quartz flow-cell of external size 4.0 × 4.0 × 10 mm and bore 1.1 mm. The emitted fluorescence is collected *via* a quartz fibre placed within the flow-cell. The diode laser is placed on the same optical bench as the detector cell at ~10 cm distance. Fig. 4b shows a side-view of the cell and the HPLC connections.

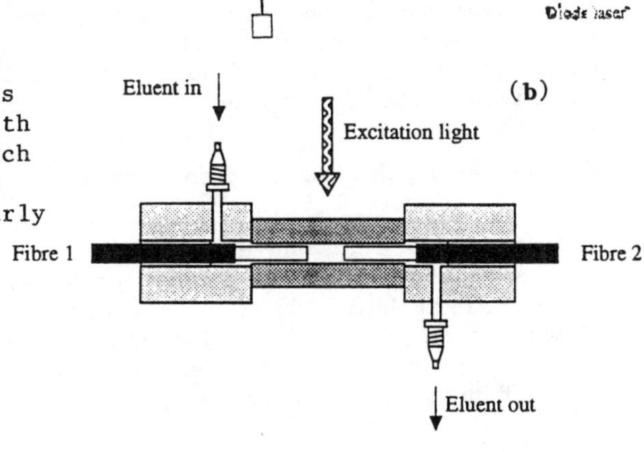

Fig. 4. (**a**) DIO-LIF light-path. A, excitation light from laser to flow-cell; B, collection of emitted light by optical fibre; C, focusing of light onto the photomultiplier tube. (**b**) Detector cell side-view with the two fibres inserted and secured with finger-tights; the $\frac{1}{16}$-inch connections to HPLC are positioned perpendicularly to the eluent stream through the flow cell. *From ref. [16], by permission.*

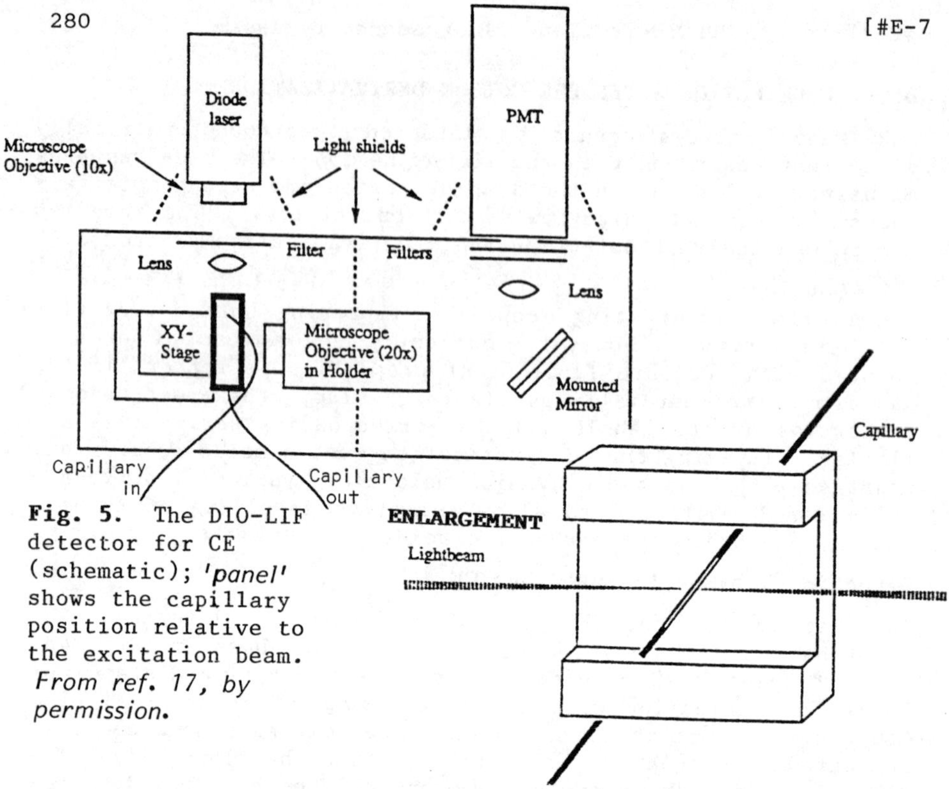

Fig. 5. The DIO-LIF detector for CE (schematic); *'panel'* shows the capillary position relative to the excitation beam. *From ref. 17, by permission.*

Because of the small i.d. of the CE capillary a slightly different set-up for the detection device is used (Fig. 5): a number of light shields are inserted to minimize stray light and an f/1 biconvex lens, with 1 cm focal length, to focus the emitted light, collected by the 20× microscope objective onto the photomultiplier. Excitation is performed with a 10 mW diode laser, equipped with a 10× microscope objective. The excitation filter comprises a 670 nm band-pass filter combined with a standard polarization filter, to remove spontaneously emitted light. For optimal excitation, the major axis of the light-beam should be oriented along the capillary. This allows the greatest amount of light to be brought into the flow-cell. Furthermore, in this geometry the incident laser light is *p*-polarized, which ensures a high transmission of laser light through the flow-cell contents and the flow-cell itself.

Red-absorbing derivatization agents

An early application of red-absorbing labels was to allow indirect detection where intrinsic detectability is poor, as discussed above [5]; DIO-LIF is probably the best choice, since diode lasers have all the characteristics of a perfect light source in the red or near-IR region, wherein interferences from matrix components are minimized. A second approach with this type of probe is to form ion-pairs, e.g. extraction of charged

Fig. 6. Cyanine structures. X, Y = O, S, $C(CH_3)_2$ or CH=CH
n = 0 – 4 and R = $(CH_2)_n$–CH_3 or $(CH_2)_n$–SO_3^-

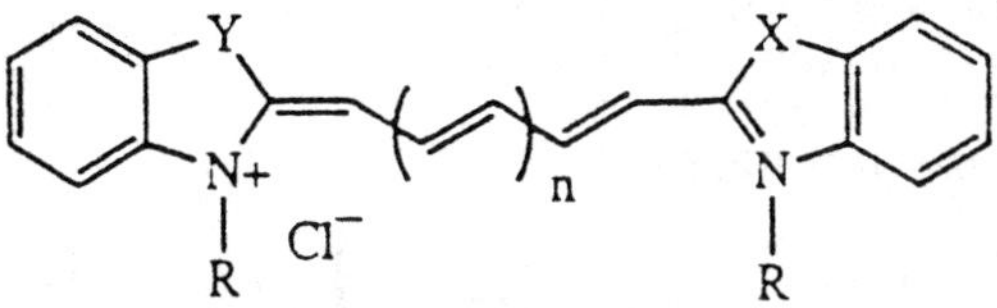

solutes into an organic phase after adding a near-IR absorbing compound as the counter-ion. Compounds usable for this purpose are rhodamine 800, oxazine 750, thionine, nile blue and methylene blue. Detection limits of ~0.1 µM are obtainable [18].

For pre-separation derivatization, the focus so far has been on synthesizing and applying compounds of (dicarbo)cyanine and squaraine type (Fig. 6). They have maxima at 620-720 nm for excitation and 640-780 nm for emission, and molar absorptivities (ε) at least 100,000 L/cm.mol and fluorescence quantum yields (Φ_f) >0.1 in water/acetonitrile or /methanol mixtures.

Large shifts in the excitation maxima can be obtained by extending the polymethine chain in a cyanine with one double bond. Dicarbocyanines have excitation maxima close to 780 nm. Smaller spectral shifts are obtainable by changing the substituents on the heterocyclic rings [X, Y = O, S or $C(CH_3)_2$].

Derivatization of thiols

Thiols can be derivatized selectively by introducing an iodoacetamide group into the basic cyanine fluorophore. Thus 2-mercaptobenzothiazole (**MBT**) can be reacted with a sulphonated cyanine containing an iodoacetamido reactive group [16]. Labels that can be applied are a dicarbocyanine (CY5.4a.IA) with 665 nm excitation maximum, an ε value of 140,000 and a Φ_f of 0.25, and a tricarbocyanine (CY7.3.IA) for which the values are 752 nm, 195,000 and 0.15. The Φ_f values are determined in an acetonitrile/water mixture (50:50 by vol.).

The derivatization procedure is quite simple. Methanol/water (70:30 by vol.) is the medium for analyte and 50 µM label solution, from which 500 µl is taken. After vial closure the mixture is heated for 1.5 h at 65°. Subsequent conditions are given in the legend to Fig. 7, which depicts HPLC separation. MBT can be determined in urine by this method. Other urinary thiols, e.g. L-cysteine, elute too early to interfere with MBT detection, for which the limit is 8 pM although for assay the limit is 1 nM since lower concentrations cannot be derivatized quantitatively.

An HPLC procedure using alumina as the stationary phase in both the pre-column and the analytical column serves for the determination of WR-1065, the active metabolite of the amino- and thiol-containing anti-cancer drug WR-2721 [19]. After

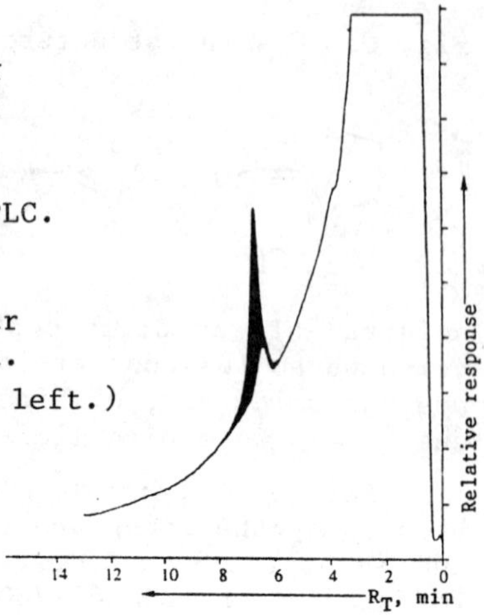

Fig. 7. HPLC of urine, blank and *(dark peak)* spiked with MBT, 0.5 μg/ml, after 500-fold dilution and derivatization with CY5.4a.IA; excess label removed by reaction with L-cysteine before HPLC.

Column: C-18 (250 × 3.1 mm); mobile phase methanol/100 mM phosphate buffer pH 6.8 (65:35 by vol.). For LIF see text. (The elution trace runs from right to left.)

From ref. [19], by permission.

pre-column derivatization of the thiol group, the analyte is determined by HPLC with DIO-LIF. Selective removal of excess reagent is achieved by means of column switching. Using this procedure 5 nM WR-1065 can be detected in aqueous medium or in deproteinized plasma diluted 10-fold.

Derivatization for CE, overcoming inadequate sensitivity, can be done with a di- or tri-carbocyanine label before the run - as shown with a mixture of thiols (reduced GSH, 2-mercapto-acetic acid, L-cysteine, D,L-homocysteine, captopril; Fig. 8) [20]. Separation was done using an 85 cm capillary of 75 μm bore and, as run buffer, methanol/100 mM borate pH 11.1 (1:1) with 27 kV applied voltage; captopril was detectable at 25 nM.

Derivatization of amines

Amine-selective succinimidyl esters can be used under both aqueous and non-aqueous conditions. Interesting features of this derivatization procedure are that the reactivity of the agent is not significantly affected by its fluorophore and that both squaraine- and cyanine-based labels can be applied.

For HPLC dicarbocyanine-based labels are apt, since they allow efficient separation of the derivatives of structurally related compounds [21]. In acetonitrile, with a tertiary amine as an activating base, n-octylamine can be derivatized (taking 45 min at 60°) at concentrations as low as 20 nM. In admixture with n-butylamine, isocratic RP-HPLC gave separation within 20 min; there was complete baseline separation between pairs.

This method has served for determining **1-adamantanamine** in urine. On 500 μl plus 500 μl 1 M NaOH, toluene extraction (1 ml; 96±7% recovery) is performed. The residue from drying down 500 μl is redissolved in 500 μl 50 μM CY5.11a.OSuc (the dicarbocyanine probe) in acetonitrile. After 45 min at 60° HPLC is performed, as illustrated in Fig. 9.

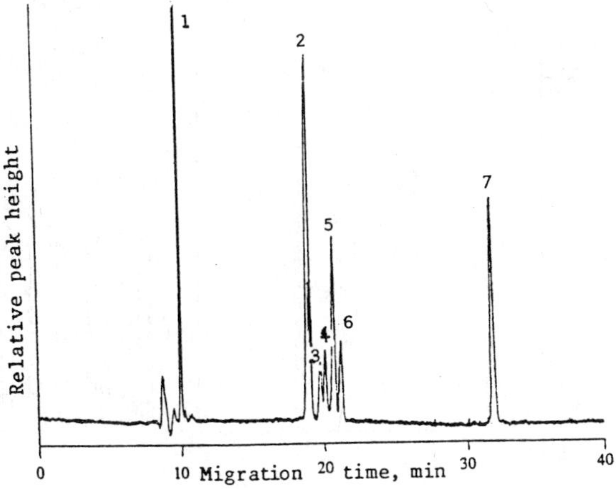

Fig. 8.
Electropherogram of thiols labelled with CY5.3a.IA. Peaks: **1**, unreacted label; **2**, captopril; **3**, D,L-homocysteine; **4**, L-cysteine; **5**, 3-mercaptopropionic acid; **6**, 2-mercaptoacetic acid; **8**, reduced GSH.
From ref. [20], by permission.

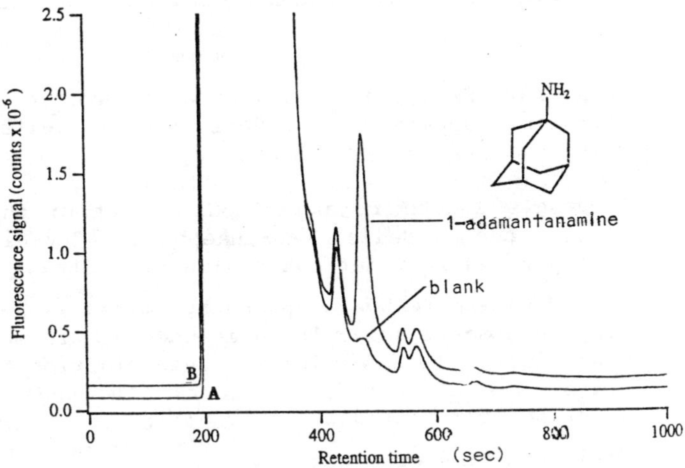

Fig. 9.
HPLC with DIO-LIF on derivatized (CY5.11a-OSuc) urine, blank (**A**) or spiked with 0.5 μM 1-adamantanamine† (**B**; fluorescence trace off-set by 0.1 axis units). Eluent: 70% (v/v) aq. acetonitrile with 1% trichloroacetic acid.

†amantadine, an anti-viral drug with a $-NH_2$ group.

For CE, derivatization with cyanine-based labels can offer a promising means of measuring amino acids and small peptides. Despite the bulky derivatization moiety, 18 amino acids can be separated at attomole levels after introducing a dicarbocyanine label containing the succinimidyl ester functionality described above (CY5.11a.OSuc). Derivatization should be performed at concentrations of 0.25 μM or higher, since in aqueous solutions hydrolysis of the label competes with the process of derivatization [17]. The micellar electrokinetic separation is performed with a 75 cm capillary (33 μm i.d.), a potential of 30 kV and a run buffer of methanol/20 mM borate pH 9.0 (28:72) containing 17.5 mM SDS (Fig. 10).

Not only amino acids can be thus separated, but also small peptides, e.g. enkephalins. Here CE surpasses HPLC insofar as, depending on the buffer pH, all tyrosine-containing enkephalin

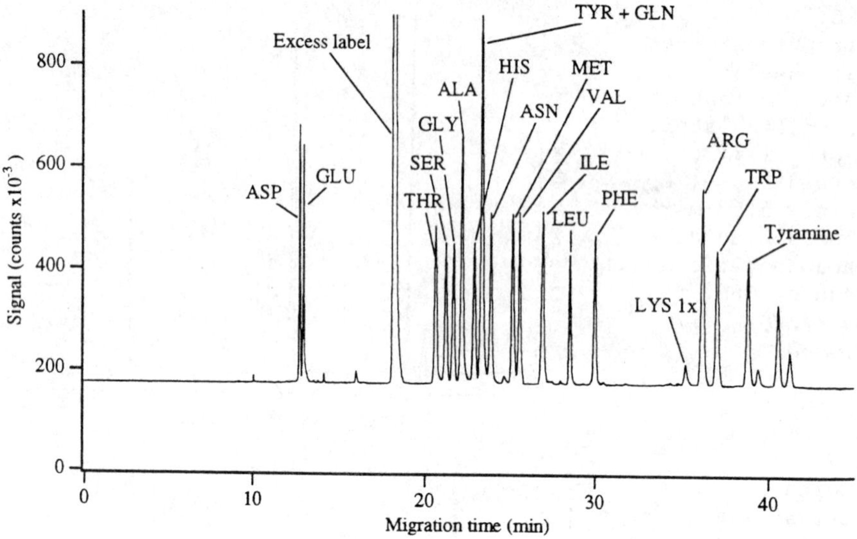

Fig. 10. Electropherogram of a mixture of 18 amino acids and tyramine, obtained after derivatization with CY5.11a.OSuc. *From ref. [17], by permission.*

fragments locate in a particular part of the electropherogram [17]. Even labelled Leu-enkephalin (Tyr-Gly-Gly-Phe-Leu) can be separated from Met-enkephalin (Tyr-Gly-Gly-Phe-Met).

Tyramine (4-hydroxyphenethylamine) is an indirectly acting sympathomimetic amine found in cheese, fermented foods and red wine [22]. The C.V. for the determination of 5 μM tyramine in urine is 4.1% (n = 5) using the same analytical procedure as described for peptides and amino acids.

Derivatization of aldehydes

The hydrazide label for aldehydes is based on CY5.3a.IA, a sulphonated dicarbocyanine with an iodoacetamido group. The labelling procedure in this case is relatively fast, the label/analyte mixture in acetonitrile/acetic acid (70:30 by vol.) being kept for 5 min at 20°.

A mixture of closely cognate n-alkylaldehydes (C5-C10) can be separated, after derivatization, within 750 sec by RP-HPLC with gradient elution [23]. The derivatization limit for n-octylamine, set by interfering impurities, is 70 nM. Citral and citronellal, two rather similar compounds, are completely separable when derivatized, by isocratic elution with 50% (v/v) aqueous acetonitrile containing 0.3% trichloroacetic acid.

Aldehydes present in biological materials as a result of microsomal lipid peroxidation are thermally labile and occur in only small amounts. 4-Hydroxy-2-nonenal (**HNE**) is a notably

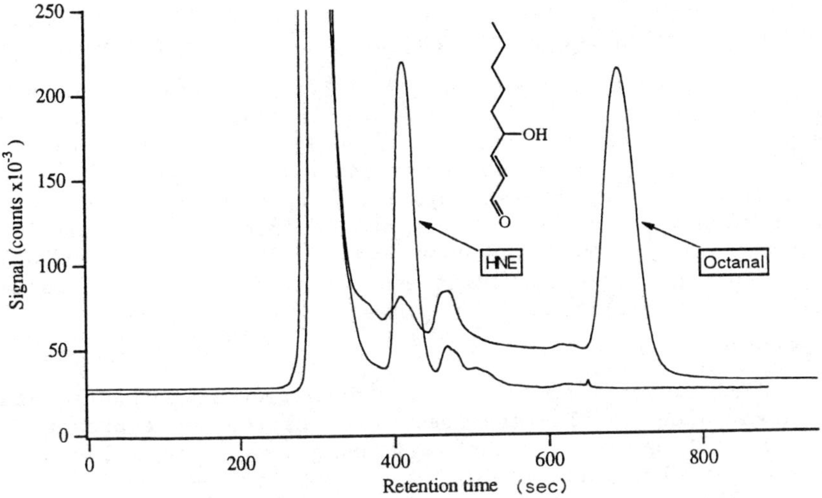

Fig. 11. HPLC of CY5.3a-hydrazide derivatives of 4-hydroxy-2-nonenal (HNE) and n-octaldehyde, prepared from a standard mixture (each 5 µM; 500 µl).

important product [24]; hence a sensitive and selective analytical method, with mild conditions, should be available. Pre-column derivatization with CY5.3a-hydrazide (10 min at room temp.) and detection by DIO-LIF offers such a method (Fig. 11).

Derivatization of carboxylic acids

Carboxylic functions are rather difficult to derivatize because of the presence of a competing carbonyl and hydroxyl group on the same carbon atom. Several approaches have been described for the pre-column labelling of carboxylic acids in combination with DIO-LIF detection [25]. Red-absorbing labels containing hydrazide, bromoacetamide or iodoacetamide functionalities can be used.

A mixture of n-alkylcarboxylic acids (C5 to C11) can be separated, after derivatization with CY5.4a-hydrazide, by RP-HPLC with a gradient [25]. The detection limit for n-octanoic acid is 0.2 µM. The large red-absorbing label does not interfere seriously with separating one acid from another. Derivatization is performed by adding 10% pyridine in acetonitrile (100 µl), a water-soluble carbodiimide (2M; 50 µl) and the hydrazide label in acetonitrile (100 µl) to the carboxylic acid in water (200 µl), then vortexing and keeping for 30 min at 45°.

The same acids can also be derivatized using the iodo- or bromoacetamide analogue of CY5.4a-hydrazide [25], with separation as above; n-octanoic acid is detectable at 30 nM. For derivatization the analyte in 100 µl of acetonitrile is dried down and 50 µl of acetonitrile containing 1 mg/ml 10-crown-6 and

2 mg/ml $KHCO_3$ are added; then 50 µl of label in acetonitrile with ultrasonication before and after; finally 30 min at 60°.

This procedure was used to determine naproxen in saliva [25]. Since use of CYS.4a-IA for labelling carboxylic acids needs water-free conditions, the analyte is first extracted by SPE [26]. Fig. 12 shows a representative DIO-LIF HPLC trace. Several of the small peaks are probably due to the presence of other carboxylic acids, being absent from water standards and blank samples. SPE recovery was 87 ±5% (n = 6) with 0.5 µM naproxen; the detection limit and linear range were the same as with standard solutions of naproxen in acetonitrile.

CONCLUSIONS

Diode lasers can be used for detection in both HPLC and CE by absorbance or fluorescence. Derivatization approaches are described along with analytical examples. Diode lasers are, advantageously, quite stable, small and inexpensive but, as the main limitation, emit only at 600 nm or higher. Generally pre-separation derivatization procedures are applied in combination with DIO-LIF. The probes normally used are based on dicarbocyanine or squaraine structures, which have good spectroscopic properties, are removable efficiently from the analytes after labelling, and allow selective labelling of several functional groups - thiols, amines, aldehydes and carboxylic acids.

Whilst labelled analytes can be detected at load levels of 1-10 pM (25-250 amol) in HPLC or 0.1 nM (0.5 amol) in CE, in practice the derivatization, which governs the detection limit, needs concentrations 100- to 1000-fold higher. This is because of several factors: label hydrolysis, limited reactivity and separation interferences (side products or excess label).

The synthesis of new labels with more reactive groups and the availability of diode lasers that emit below 600 nm will extend the applicability of DIO-LIF in HPLC and CE.

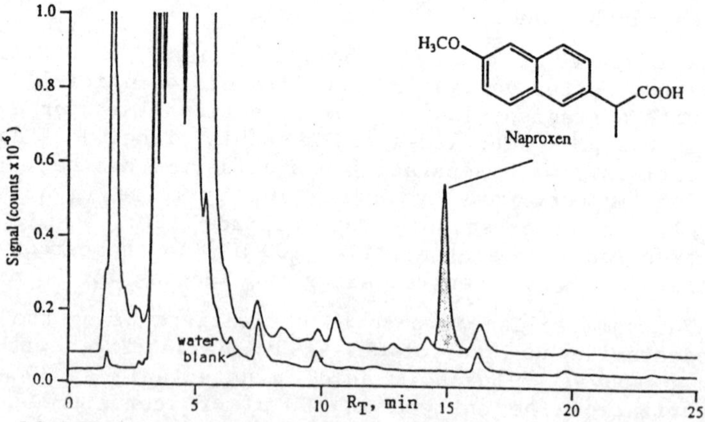

Fig. 12. LC-DIO-LIF of saliva, blank and (same trace; grey peak) spiked with 5×10^{-4} mM naproxen, extracted and derivatized (CYS5.4a-IA).

References [*Chr.* denotes *Chromatography*] Initials omitted for: Brinkman, U.A.Th.; Gooijer, C.; Lingeman, H.; Mank, A.J.G.; Velthorst, N.H.

1. Lingeman, van de Nesse, R.J., Brinkman, Gooijer & Velthorst (1990) in *Analysis for Drugs and Metabolites including Anti-Infective Agents* [Vol. 20, this series] (Reid, E. & Wilson, I.D., eds.), Royal Society of Chemistry, Cambridge, pp. 355-363.
2. Lingeman, Gooijer, Velthorst & Brinkman (1991) in *Analytical Applications of Spectroscopy II* (Davies, A.M.C. & Creaser, C.S., eds.), Royal Society of Chemistry, Cambridge, pp. 189-206.
3. van den Beld, C.M.B. & Lingeman (1990) in *Luminescence Techniques in Chemical and Biomedical Analysis* (Baeyens, W.R.G. *et al.*, eds.), Dekker, New York, pp. 237-316.
4. van de Nesse, R.J., Velthorst, Brinkman & Gooijer (1995) *J. Chr.* ***A 704,*** 1-25.
5. Mank, Lingeman & Gooijer (1992) *TRAC 11,* 210-217.
6. & 7. Mank, Velthorst, Brinkman & Gooijer (1995) *J. Chr.* ***A*** ***695:*** 6. 165-174; 7. 175-183.
8. Mank, de Nijs, H., Lingeman, Brinkman, Velthorst, & Gooijer (1996), in press.
9. Mank, Larsen, O., Lingeman, Gooijer, Brinkman, & Velthorst (1996) *Anal. Chim. Acta*, in press.
10. Shirk, J.S., Harris, T.D. & Mitchell, J.W. (1980) *Anal. Chem.* ***52,*** 1701-1704.
11. Mank, Gooijer, Lingeman, Velthorst & Brinkman (1994) *Proc. SPIE 2084,* 255.
12. as for 11. (1994) *Anal. Chim. Acta* ***290,*** 103-120.
13. Wagner, J.R., Ali, H., Langlois, R., Brasseur, N. & van Lier, J.E. (1987) *Photochem. Photobiol. 45,* 587-596.
14. Chan, W.S., Marshall, J.F., Svensen, R., Phillips, D. & Hart, I.R., as for 13., 757-761.
15. Chan, W.S., West, C.M.L., Moore, J.V. & Hart, I.R. (1991) *Br. J. Cancer 64,* 827-834.
16. Mank, Molenaar, E.J., Lingeman, Gooijer, Brinkman & Velthorst (1993) *Anal. Chem. 65,* 2197-2203.
17. Mank & Yeung, E.S. (1995) *J. Chr.* ***A*** *708,* 309-321.
18. Imasaka, T., Joshitake, A. & Ishibashi, N. (1984) *Anal. Chem. 56,* 1077-1079.
19. Mank, Molenaar, E.J., Gooijer, Lingeman, Velthorst & Brinkman (1995) *J. Pharm. Biomed. Anal. 13,* 255-263.
20. Mank, Lingeman & Gooijer (1994) *J. High Resolu. Chr.* ***17,*** 797-799.
21. Mank, van der Laan, H.T.C., Lingeman, Gooijer, Brinkman & Velthorst (1995) *Anal. Chem.* ***67,*** 1742-1748.
22. Causon, R.C. & Brown, M.J. (1984) *J. Chr.* ***310,*** 11-17.
23. Nuijens, M.J., Zomer, A.J.G., Mank, Gooijer, Velthorst & Hofstraat, J.W. (1995) *Anal. Chim. Acta* ***311,*** 47-55.
24. Bringmann, G., Gassen, M. & Schneider, S. (1994) *J. Chr.* ***A 670,*** 153-160.
25. Mank, Beekman. M.C., Velthorst, Brinkman, Lingeman & Gooijer (1995) *Anal. Chim. Acta* ***315,*** 209-216.
26. Steijger, O.M., Lingeman, Brinkman, Holthuis, J.J.M., Smilde, A.K. & Doornbos, D.A. (1993) *J. Chr.* ***A 615,*** 97-110.

#E-8

ANALYSIS OF ASPIRIN METABOLITES IN URINE AND PLASMA BY MECC* WITH DIRECT SAMPLE INTRODUCTION

Mark R. Taylor[¤], Steven A. Westwood, Karl Woodward & †David Perrett

Horseracing Forensic Laboratory Ltd.,
P.O. Box 15, Snailwell Road, Newmarket, Suffolk CB8 7DT, U.K.
and †Dept. of Medicine, St. Bartholomews Hospital Medical College, London EC1A 7BE, U.K.

MECC is a useful means of determining drugs and metabolites in biofluids, as they can be analyzed directly with little or no sample pre-treatment; but it has some limitations. Generally MECC has lower sensitivity than HPLC. Hence sample preparation procedures are necessary for dilute samples; clean-up may be done using dialysis with trace enrichment before MECC. Plasma proteins may co-elute with and mask detection of some analytes.*

Aspirin (acetylsalicylic acid) is a widely used analgesic, possessing anti-inflammatory and anti-pyretic activity. Aspirin metabolism has been compared in humans and horses using MECC. The main metabolite in the horse is SA. Salicylates are common constituents of horse urine, originating from the diet, and hence an internationally agreed maximum accepted level exists for SA in post-race urine, higher levels being treated as doping offences. A comparison of MECC and HPLC in the quantitation of urinary SA using direct sample injection is presented.

Applications of MECC in the direct analysis of drugs in biofluids are increasing rapidly [1-3]; but relatively few publications present results of full validation procedures [4, 5]. Concerns over problems with reproducibility, reliable automated operation over long time periods, unexpected interferences and capillary fouling have to be overcome before CE becomes accepted as a routine method in assay laboratories [4].

The development of modern CE instrumentation and use of effective capillary conditioning routines should allow reproducible and reliable quantitative CE assays to be performed. The major limitation of CE in the direct analysis of biofluids is

---

[¤]addressee for any correspondence
**Abbreviations*.- ASTED, Automated Sequential Trace Enrichment of Dialysates; C(Z)E, capillary (zone) electrophoresis; i.s., internal standard; MECC, micellar electrokinetic capillary chromatography; SA, salicylic acid/salicylate; & see Fig. 1. Sensitivity: LLOQ/LOD, (lower) limit of quantitation/detection.

the poor sensitivity compared with HPLC. The main advantage of MECC over HPLC and CZE is that proteinaceous matrices such as plasma, serum and saliva can be analyzed directly with no need for deproteinization by ultrafiltration, dialysis or precipitation. Plasma proteins are solubilized by surfactants, e.g. SDS, in the MECC buffer and are prevented from adsorbing to the capillary wall.

In MECC of plasma the migration-time gap between unretained material and plasma proteins is effectively a separation window [6, 7]. This could be expanded by use of high concentrations of SDS (H. Watzig & D.K. Lloyd†); but this approach is limited by current generation and consequent Joule heating. In direct plasma assays by MECC, capillary equilibration time was reduced if capillary rinsing between analyses were with SDS-containing buffers rather than commonly used acid/base washes.

Here we compare validation data from measurement of SA in equine urine by MECC and HPLC with direct sample introduction. The main metabolite of aspirin in the horse is SA [8], whereas in man it is SHA [9] (Fig. 1); SA glucuronides and GA are minor metabolites. Excretion kinetics are highly dependent on urine pH [8] and dose [10].

†Presentation at 20th Int. Symp. on Chromatography (Bournemouth, U.K.; 1994). *Added in proof:* **now** published - ref. [22].

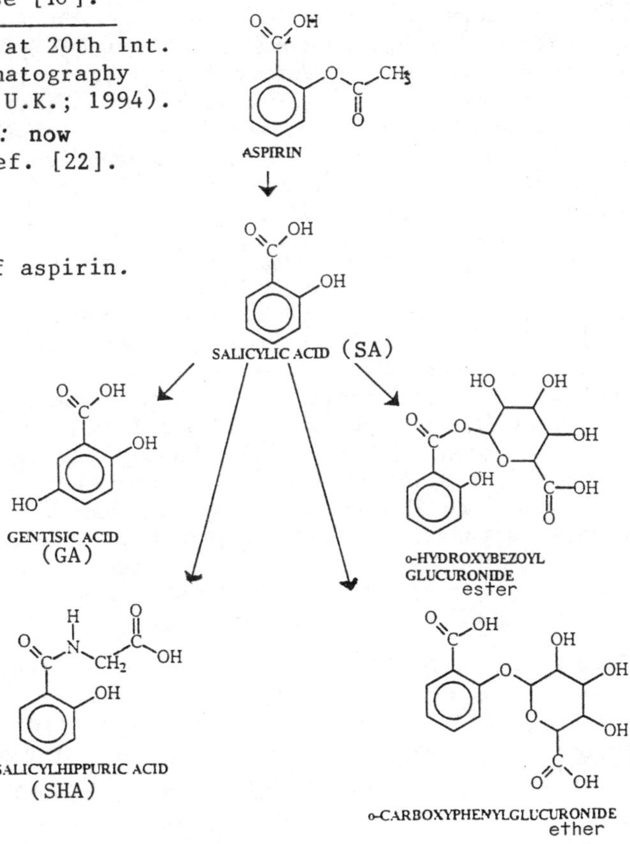

Fig. 1.
Metabolites of aspirin.

SA is a common constituent of horse urine. It is thought to arise mainly from diet, although its origin in part as a possible metabolite product from gut flora cannot be excluded. Under the rules of racing in Britain and many other countries SA is regarded as a 'prohibited substance' because of its anti-inflammatory action. Due to its common occurrence a maximum permitted threshold of 750 μg/ml SA in post-race equine urine is allowed by some Racing Authorities including the British Jockey Club. The threshold was set after GC-MS quantitation of naturally occurring levels of SA in 114 thoroughbred racehorses and post-administration levels from two others [11]. Measured levels of SA above this are reckoned as doping offences. GC-MS is not an appropriate technique for quantitation of such high levels of SA since its high sensitivity entails large dilutions and extraction into a GC-amenable solvent.

EXPERIMENTAL

Materials.- ENV+*(polystyrene-divinylbenzene; 100 mg/3 ml reservoir) and C-18 (end-capped; 200 mg/3 ml) Isolute cartridges were from International Sorbent Technology (Hengoed, U.K.), and methanol and hexane (HPLC grade) were from Rathburn Chemicals (Walkerburn, U.K.). SA (Na salt), GA, 5-chlorosalicylic acid (i.s.), *o*-hydroxyhippuric acid , SDS, Trizma base, trichloroacetic acid and orthophosphoric acid (AnalaR grade) were from Sigma (Poole, U.K.). A Purite RO50 water purifier furnished deionized water. Blank equine urine and plasma were obtained from thoroughbreds stabled at our Newmarket laboratory. Urine samples, naturally voided, were collected from a thoroughbred racehorse and a human volunteer at 0.75 and 2.5 h after oral administration of aspirin (12.5 mg/kg).

HPLC.- The apparatus comprised a Hewlett Packard HP1090 Series II instrument with a diode array detector, with 305 nm for monitoring (4 nm bandwidth) and 550 nm as reference wavelength (100 nm). Peak-controlled spectral data acquisition was done over the range 230-400 nm (2 nm step, 640 msec sampling interval). Separations (100 μl injected) were performed isocratically at 40° on a Hypersil C-18 column (5 μm; 100 × 4.6 mm; Jones Chromatography, Hengoed). The mobile phase (1.5 ml/min) was methanol/0.1 M acetic acid, 30:70 by vol., filtered through a glass sinter (0.45 μm) and helium-degassed. Relative peak areas were calculated by HPLC Chemstation integration software which was calibrated in each assay with standard solutions. Typical retention times were 3.6 and 7.3 min for SA and i.s. Total analysis time was 10 min per sample.

MECC, using an HP^{3D}CE instrument with diode array detector.- Monitoring was at 300 nm (4 nm bandwidth) with 450 nm for reference (80 nm). Peak-controlled acquisition was done over the range 230-400 nm (2 nm step, 100 msec sampling interval).

*not available during the ASTED/C-18 work; tried later, on urine

Injections were done by positive pressure application to sample vial (50 mBar, 2 sec), and separations in an extended lightpath fused silica capillary (48 cm × 50 μm i.d.; Hewlett Packard, Macclesfield, U.K.). The MECC electrolyte was 100 mM Trizma base adjusted to pH 7.8 with 1 M phosphoric acid and containing 100 mM SDS, passed through a Millex-HV syringe filter (0.2 μm; Millipore, Watford, U.K.) and replenished every 12 runs. MECC was done at 15° with 30 kV separation voltage. New capillaries were conditioned by sequential 10-min flushes at 940 mBar with 1 M then 0.1 M NaOH, equine urine diluted 1:4 with deionized water, and - also done (4 min) between runs - MECC electrolyte; the flush with urine was found to improve reproducibility by reducing migration-time drift. Relative corrected peak areas (area/migration time) were calculated using the CE Chemstation integration software, calibrated with standards in each assay.

Sample processing

HPLC (urine).- SA standards were prepared in deionized water (0.15 mg/ml), and calibration and control samples (levels as in a Table below) in 10 μl blank urine; the controls were prepared by an independent analyst using a separate standard. Methanolic i.s. was put (2.5 mg/ml) in all samples except one urine blank. All samples were water-diluted (1:100) and mixed prior to HPLC.

MECC on urine.- From aqueous SA (2.5 mg/ml), standards were prepared as for HPLC but with 50 μl urine; the i.s. (none in one urine blank) was at 1 mg/ml. The 1:4 diluted and mixed samples were centrifuged prior to MECC analysis.

Plasma.- Blank equine plasma was spiked (500 μg/ml) with methanolic (1 mg/ml) SA, GA and SHA standards. For dialysis by a Gilson ASTED XL apparatus (Anachem, Luton; 15 kDa membrane), 237 μl spiked plasma was pre-mixed with 80 μl of a conditioning reagent, 0.7 M trichloroacetic acid/0.8 M tris, and 50 μl of acetonitrile. Recipient solution, 5 mM aqueous hexanesulphonic acid, was pumped in 5 × 1 ml pulses and dialysate was collected for off-line trace enrichment by SPE.

SPE.- Samples acidified to pH 1 (HCl) were passed through Isolute C-18 cartridges pre-conditioned under vacuum with 1 ml methanol and 1 ml water. Cartridges were washed with 0.5 ml water and 1 ml hexane, dried under ~0.7 Bar vacuum for 2 min, and eluted with 1 ml methanol. The eluates were evaporated to dryness in a Genevac SF50 centrifugal evaporator; the residues, reconstituted in 50 μl water, were centrifuged before MECC.

RESULTS

Optimization of MECC separation.- After a detailed study of the main MECC parameters, separation of SA and metabolites was achieved (Fig. 2). In initial work, 0.1 M NaOH and MECC electrolyte were used for between-injection capillary rinsing.

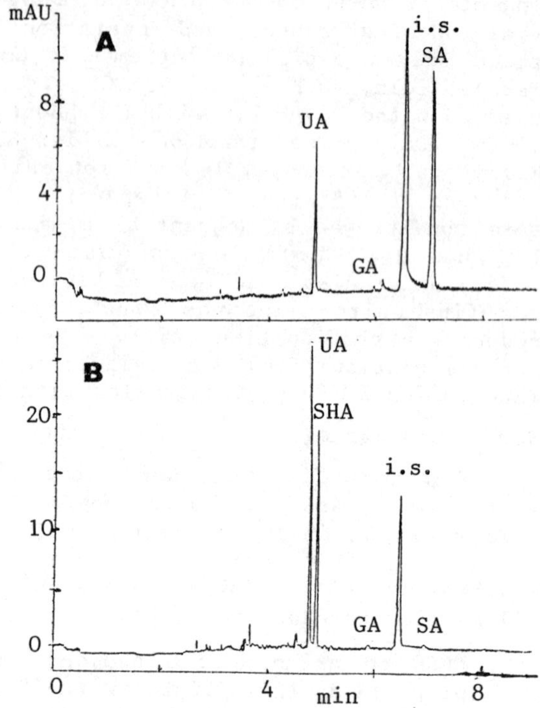

Fig. 2.
MECC analysis of (**A**) equine and (**B**) human 1:4 diluted urine taken 2.5 h after an aspirin dose. For conditions see text. **UA** = uric acid; see Fig. 1 for other abbreviations.

A significant, negative migration-time drift was observed, reduced if the NaOH rinse were omitted. As mentioned above, a diluted-urine flush of new capillaries aided reproducibility. Presumably a 'steady state' is achieved at the capillary wall, not affected by repeated injection of diluted urine samples and rinses with MECC electrolyte. The quite aggressive NaOH rinse disturbed the ion equilibrium at the capillary wall that is necessary for stable electro-osmotic flow - a phenomenon seen previously in MECC and CZE assays developed in our Newmarket laboratory, particularly with phosphate-containing buffers.

Comparison of HPLC and MECC urinary assays with direct injection

Linearity.- Proportionality between HPLC and MECC responses to analyte concentration was demonstrated. Peak-area ratios of SA to i.s. were plotted against concentration and subjected to linear regression analysis. Corrected peak areas (area ratio/ migration time) were used in the MECC calibration. Linearity was ascertained over the concentration range 375-2250 μg/ml using 5 concentration levels (besides zero; n = 36). For HPLC the result was y = 1.34x - 0.008 (r = 0.9986). For MECC it was y = 1.36x - 1.97 (r = 0.9997).

Table 1. Quantitation of SA in equine urine by HPLC and MECC.

Nominal (set) level, μg/ml:		375	750	1125	1500	1875	2250
A: C.V. (RSD), % (n = 6)	HPLC	4.5	1.3	2.2	1.0		0.8
	MECC	5.4	3.4	1.7	1.6		4.4
B: % bias (n = 12)	HPLC		-1.5			+0.4	
	MECC		-0.5			-1.1	
C: F-test*, comparing variances		1.32	6.84	1.69	2.42		2.94
D: t-test*, comparing mean results (n = 6)	S, pooled	18.8	19.2	21.8	19.7		25.3
	t	1.5	0.2	0.6	-0.6		2.8

*2-tailed; at 95% probability level F = 7.146 (n = 6, 6), t = 2.23

Table 1A shows inter-assay **precision** from 6 calibration lines, prepared by 4 (HPLC) or 5 (MECC) different operators. As evidence of **selectivity**, no significant interference peaks were observed from urine taken from 10 different horses and analyzed using both HPLC and MECC.

Accuracy is evidenced by control-sample results for bias (%; Table 1B). The **LLOQ** (sensitivity) was evidently at least as good as 375 μg/ml, which is half the threshold level.

Method comparisons (Table 1, C & D) showed good accord between HPLC and MECC in respect of variances and (using pooled S.D.'s) mean results from precision experiments.

Stability of SA and i.s. standard solutions stored 7 days at 4° was satisfactory. There was evidence of slow decomposition of SA, up to 21% loss, in spiked calibration samples after 8 days at room temperature. Samples from the same calibration curve were affected similarly, although rates of degradation varied significantly between calibrations.

MECC analyses after aspirin administration

Urine samples, both equine and human, were assayed for SA, which was manifest as the main metabolite in the equine samples with trace levels of GA (Fig. 2). The SA values found (μg/ml; n = 2) were 176 at 0.75 h - below the validated LLOQ - and 404 at 2.5 h. In the human, SHA was evidently the main metabolite, with traces (<LLOQ) of GA and SA (shown in Fig. 2 for 2.5 h).

Equine and human 2.5 h samples (2 ml) were subjected to SPE using Isolute ENV+ cartridges†, and the eluates analyzed by MECC. The major metabolites were overloaded; but GA was detected in the equine sample, and SA and GA in the human sample (Fig. 3). Recovery in SPE was estimated as 78% for SHA, 71% for GA and 64% for SA. Peaks were identified by reference to migration times and UV spectra for standards.

†tried late in the study as likely to have a high capacity for polar analytes such as SA, which may be difficult to retain on C-18.

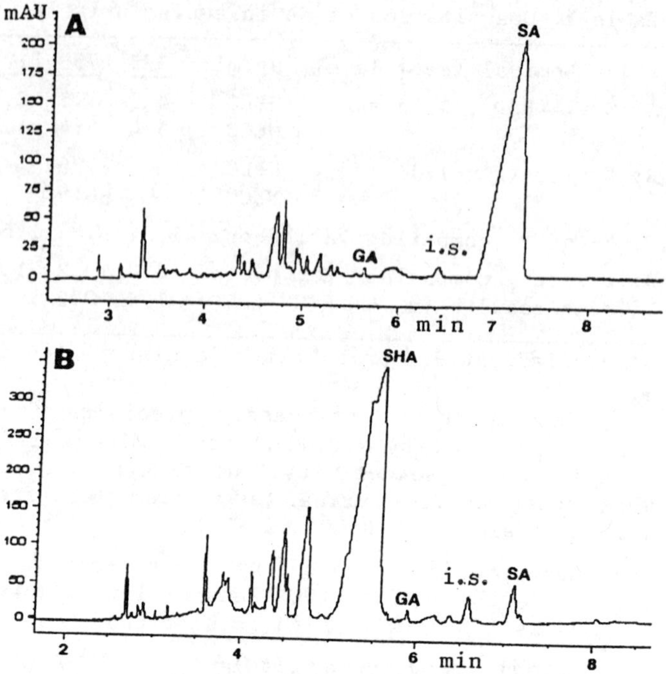

Fig. 3. MECC analysis of (**A**) equine, and (**B**) human 1:4 diluted urine, collected 2.5 h after an aspirin dose and trace-enriched by SPE. For conditions see text.

MECC analysis of spiked equine plasma

Analyses were performed on undiluted spiked plasma as in the quantitative SA procedure but with a monitoring wavelength of 205 nm. Plasma proteins ran well clear of SHA and GA, but some co-eluted with SA. When ASTED dialysates were analyzed by MECC after off-line SPE trace enrichment using C-18 cartridges, the plasma-protein interference was clearly reduced (Fig. 4).

DISCUSSION

Analyses on urine, and their validation

The measurement of inter-assay precision during validation of the quantitative procedure is especially important where a threshold level is applicable. The validation exercise was set up to generate the maximum possible variation of results which could occur in normal routine use of both quantitative procedures, HPLC and MECC. A 'worst case' scenario is created which provides a basis upon which to decide if a result is over or under the threshold (within a specified limit of confidence). taking account of the maximum expected limit of variation [12].

As can be seen from Table 1, HPLC and MECC hardly differ in the inter-assay precision of the results. Statistical comparison of the validation data indicates that the HPLC and MECC data do not differ significantly at the 95% probability level. The t-test revealed a difference at the highest calibration level

Fig. 4. MECC analysis of spiked equine plasma, demonstrating effect of removing plasma proteins: (**A**) direct analysis; (**B**) dialysis and trace enrichment initially.

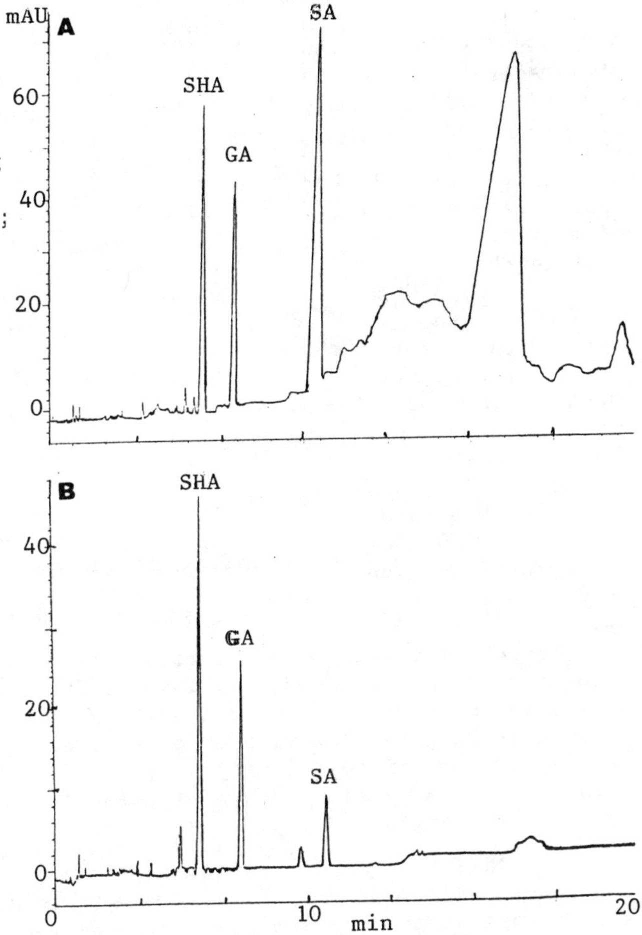

(2250 μg/ml); but the C.V. was <1.5% for results from both methods at this level. The linear correlation coefficient was slightly better for MECC than HPLC, but was >0.998 for both. The linear range may well be extendable for both procedures; but this hardly matters when dealing with a threshold level.

It should be noted that calibration samples, controls and unknowns were all prepared in the same matrix (urine diluted with deionized water) to allow for possible matrix effects. To prevent bias in results it is recommended that samples giving a result greater than the highest calibration standard be diluted with the same matrix to bring them into the validated linear range.

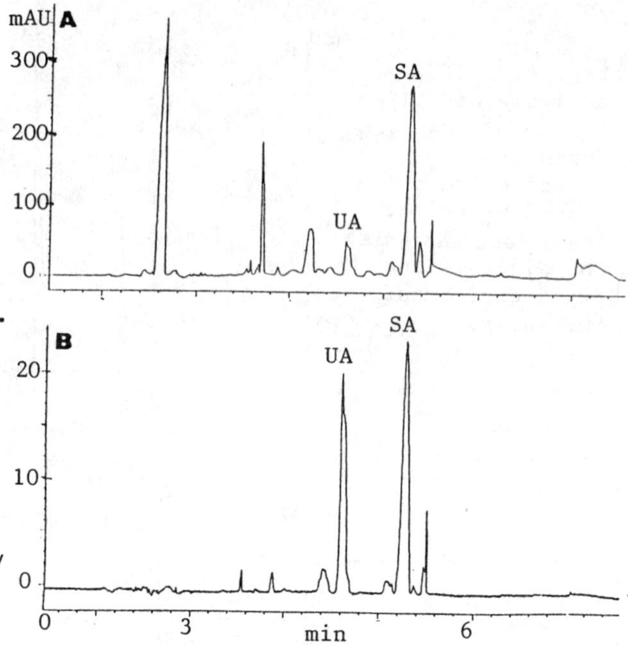

Fig. 5. MECC analysis of a spiked equine urine sample (500 µg/ml SA), demonstrating that selectivity can benefit from appropriate choice of monitoring wavelength: (**A**) 205, (**B**) 300 nm. Conditions (see text) were as for Fig. 2 except that the capillary was 50 cm × 50 µm i.d. fused silica and the electrolyte was 100 mM pH 8 borate/ 100 mM SDS; 50 mBar/ 4 sec for injection by pressure.

The chromatograms are relatively free of interference from endogenous urinary compounds. This is due to improved selectivity offered by monitoring at a high UV wavelength (300-305 nm) - the peak height of SA with a lower wavelength being increased at the expense of selectivity (Fig. 5). It is beneficial to dilute urine samples as much as possible to reduce viscosity and ionic strength and improve stacking.

The MECC assay of post-administration samples provided a sub-threshold result for SA in the equine urine samples, not surprisingly since the administered dose was approximately half that usually used for SA therapy in horses [11]. Trace levels of the minor metabolites GA (horse and human) and SA (human) were detected more easily with SPE trace-enrichment than with direct MECC analysis of urine. The glucuronide conjugates of SA were not studied as reference materials were unavailable.

Analyses on plasma, and sample pre-concentration

Analysis of spiked plasma by MECC as in Fig. 4A demonstrates the migration-time window and the problem of interference with detection of some compounds, e.g. plasma-protein overlap with SA. The window could not be widened by increasing SDS concentration due to concerns over Joule heating.

Dialysis by the ASTED served to remove interfering plasma proteins. The use of microdialysis as a sample preparation technique for CE has been reported [13-16], but we believe we are the first to report application of the ASTED in sample preparation for CE. Off-line SPE trace-enrichment was used to provide adequate sensitivity for detecting the drugs and metabolites in ASTED dialysates. The interference by plasma protein was clearly reduced, demonstrating the potential of dialysis with trace-enrichment as an MECC sample preparation technique for strongly retained hydrophobic analytes.

On-line trace-enrichment would improve sample throughput and reduce the cost of analysis, provided that regeneration of the trace-enrichment system were feasible. Isotachophoretic pre-concentration prior to CZE separation can be done [17, 18]; however, this technique could not easily be applied to an MECC separation system. Packed inlet pre-concentration capillaries have been described [17, 19]. Analytes have been enriched on a solid phase packing integral to the separation capillary prior to CZE separation. Due to slow solute-desorption kinetics the preliminary results demonstrate poor resolution and efficiency compared to conventional CZE. The application of such devices to trace-enrichment of strong acids such as SA (pK_a = 3.0) prior to MECC would be difficult to achieve in a single capillary since sample loading would need to take place at a pH of 1.0 to prevent analyte ionization and promote retention. The difficulty lies in replacement of the low-pH analyte-depleted matrix in the capillary with a suitable MECC run buffer and elution of the analyte whilst maintaining a sharp sample zone.

Morita & Sawada [20] developed, for pre-concentration, a system with three capillaries (pre-concentration, drain and separation; 50 µm i.d.), connected *via* a Y-union. Biological samples were electrokinetically loaded onto the packed pre-concentration capillary with the drain capillary as the sample waste outlet; elution was likewise performed electrokinetically with the voltage applied across the separation capillary. Serum samples spiked with propranolol were concentrated on a protein-coated ODS phase prior to CZE. The LOD was thus reduced by two orders of magnitude and injections of up to 1 µl were performed. This system overcomes the problem of the analytical capillary being filled with waste sample matrix prior to elution. No decrease in resolution was observed; but peak efficiencies, in terms of theoretical plate numbers, were not quoted.

Debets *et al.* [21] developed a two-capillary switching valve assembly with an internal micro pre-concentration column; band broadening was reduced by using a rapid zone-cutting desorption system. Pre-concentration of sample volumes up to 100 µl was achieved on-line prior to CZE analysis. Calibration plots were linear over two orders of magnitude. The system was constructed using 200 µm i.d. capillaries which would seriously limit the choice of buffer, separation voltage and hence the efficiency achievable in the CE separation.

If the system could be scaled down to use capillaries of 50-75 µm i.d. then this method may prove highly useful for on-line pre-concentration of trace analytes prior to a wide range of CE (including MECC) and micro-HPLC applications.

Concluding comments.- MECC, now validated alongside HPLC, allows reliable assay of drugs and metabolites in biofluids if the levels suffice for current CE detectors. Automated dialysis and trace-enrichment (potentially on-line) allow direct analysis of plasma, for which a micro-system would save pre-MECC time and effort and improve the LOD and range of biofluid applications.

References *[Chr.* as in *J. Chr.* denotes *Chromatography]*

1. Perrett, D. & Ross, G. (1992) in Vol. 22, 269-277. | This series, ed. Reid & Wilson or (V.23) as now; publisher as now (RSC).
2. Lloyd, D.K. (1994) Vol. 23, 41-50. | This series, ed. Reid & Wilson or (V.23) as now; publisher as now (RSC).
3. Deyl, Z., Tagliaro, F. & Miksik, I. (1994) *J. Chr. B 656*, 3-27.
4. Thormann, W., Lienhard, S. & Wernly, P. (1993) *J. Chr. 636,* 137-148.
5. Zhang, C-X. & Thormann, W. (1994) *J. Cap. Elec. 1,* 208-218.
6. Nakagawa, T., Oda, Y., Shibukawa, A. & Tonaka, H. (1988) *Chem. Pharm. Bull. 36,* 1622-1625.
7. Lloyd, D.K. (1992) *Anal. Proc. 29,* 169-170.
8. Murdick, P.W., Ray, R.S. & Noonan, J.S. (1968) *J. Am. Vet. Res. 29,* 581-585.
9. Bowman, W.C. & Rand, M.J. (1980) *Textbook of Pharmacology* (2nd edn.), Blackwell Sci. Publ., Oxford, pp. 16.19-16.24.
10. Boreham, D.R. & Martin, B.K. (1969) *Br. J. Pharmacol. 37,* 294-300.
11. Moss, M.S., Blay, P., Houghton, E., Horner, M.W., Teale, P., Smith, R.L. & Sloan, T. (1985) *Proc. Int. Conf. of Racing Analysts and Veterinarians, 6,* 97-99.
12. Taylor, M.R. & Westwood, S.A (1995) *J. Chr. A 697,* 389-396.
13. Hogan, B.L., Lunte, C.E., Stobaugh, J.F. & Lunte, S.M. (1994) *Anal. Chem. 66,* 596-602.
14. Hernandez, L., Escalona, J., Verdeguer, P. & Guzman, N.A. (1993) *J. Liq. Chr. 16,* 2149-2160.
15. O'Shea, T.M., Weber, P.L., Bamel, B.P., Lunte, C.E., Lunte, S.M. & Smyth, M.R. (1992) *J. Chr. A 608,* 189-195.
16. Linhares, M.C. & Kissinger, P.T. (1993) *J. Chr. B 615,* 327-333.
17. Swartz, M.E. & Merion, M. (1993) *J. Chr. 632,* 209-213.
18. Hjerten, S., Liao, J-L. & Zhang, R. (1994) *J. Chr. A 676,* 409-420.
19. Tomlinson, A.J., Braddock, W.D., Benson, L.M., Oda, R.P. & Naylor, S. (1995) *J. Chr. B 669,* 67-73.
20. Morita, I. & Sawada, J-I. (1993) *J. Chr. 641,* 375-381.
21. Debets, A.J.J., Mazereeuw, M., Voogt, W.H., van Iperen, D.J., Lingeman, H., Hupe, K.P. & Brinkman, U.A.Th. (1992) *J. Chr. 608,* 151-158.
22. Watzig, H. & Lloyd, D.K. (1995) *Electrophoresis 16,* 57-63.

#E-9n

A Note on

FREEZE-THAW SWITCHING IN ELECTROPHORETIC CAPILLARIES†

C.D. Bevan & I.M. Mutton
Structural Chemistry Dept., Glaxo Wellcome plc
Medicines Research Centre, Stevenage, Herts. SG1 2NY, U.K.

Flow-switching effected by valve can benefit separation efficiency and selectivity in HPLC or GC. Such benefit has been denied to CE* and electrochromatography since no suitable switching valve is marketed, nor multi-detector instrumentation. 'Nano-switching technology' recently reported by us [1] entails freezing, with a liquid CO_2 jet (Fig. 1), the section of the capillary where flow is to be stopped, and thawing to restore the flow. This stratagem (FTS*) does not need any moving parts, obviates disturbances, is especially effective with small-scale systems (near-instantaneous freezing, with arrest of electro-osmosis and ion transport despite maintained high voltage), and serves also for packed or coated columns. Other FTS advantages include low dead volume and tolerance of corrosive buffers.

Fig. 2 shows a two-limb (A, B) CE system used to verify separation effectiveness and recovery, tested with two anionic dyes [1]. When the first dye has passed the detector (D) and is judged to have reached the junction T, freezing limb A at point F diverts the flow into B. When the dye is clear of T, A is thawed and collects the other dye, B being frozen. With electrophoresis stopped, the contents of each limb are flushed by syringe (applied to the anode end, but at the cathode for briefly run standards) into an Eppendorf tube for analysis [1].

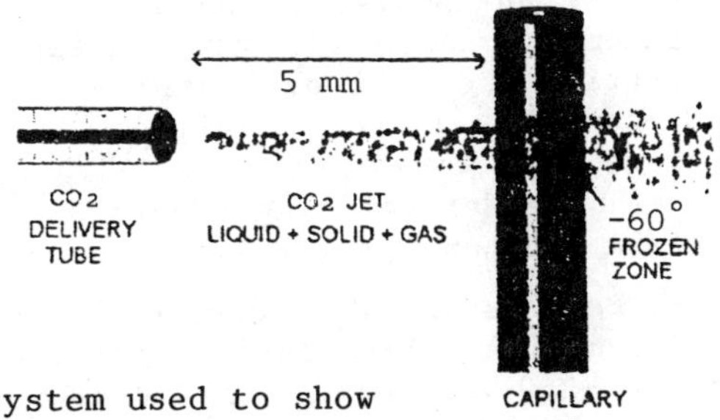

Fig. 1. Freezing CE capillary contents, for FTS, by a liquid jet of CO_2 introduced from a liquid CO_2 cylinder *via* PEEK tubing and stop valves.

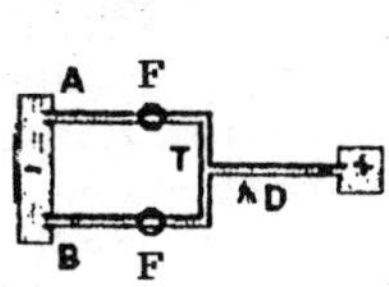

Fig. 2. The two-limb CE system used to show separation of two dyes, migrating from right to left (see text). T, verified to have <5 nl dead volume (as claimed), comprised a 'µ-Tee' from LC Packings, Amsterdam.

References

1. Bevan, C.D. & Mutton, I.M. (1995). (a): *Anal. Chem. 67,* 1470-1473. (b): *J. Chromatog. A 697,* 541-548.

†Title and text editorially shortened (cf. the citations).
*CE, capillary electrophoresis; FTS, freeze-thaw switching.

Comments on #E-1, D. Thomas: HPLC AS INFLUENCED BY APcI
#E-3, K. Grob: CAPILLARY-GC ADVANCES
#E-5, T.A.G. Noctor: ENANTIOSELECTIVE HPLC
#E-6, S.A. Wring: HPLC-EC WITH CYCLIC VOLTAMMETRY
#E-7, H. Lingeman: LASER FLUORESCENCE DETECTION
#E-9n, C.D. Bevan: CE WITH FREEZE-THAW SWITCHING

Thomas, replying to C.A. James: for LC-MS or LC-MS-MS ~70% of our extractions are by SPE; ~30% by protein precipitation.

Grob, to K. Jolley: quartz wool and glass wool are equivalent in behaviour.

Noctor, answering D.J. Anderson: the n value would be in the low thousands for a 15 cm CSP column, lowest if protein.

Wring, replying to S. Tan about use of pulsed techniques.- I have employed pulsed amperometric detection for determining saccharides at a gold electrode. The pulse serves to generate a fresh electrode surface and so prevent signal loss due to adsorption of oxidation products; at a gold electrode the most positive potential (usually E2) removes the electrode's outer layer, exposing a fresh surface. Pulsed techniques may also be used with other electrode substrates, including carbon-based planar and micro-electrodes, and also carbon-based sensors. At these surfaces, pulses (anodic and/or cathodic) may be employed to electrochemically 'clean' the electrode by removing adsorbed materials rather than the outer surface of the electrode substrade itself.

Anderson asked Lingeman if his derivatizing agents could be used to label proteins. **Reply**: yes, but reproducibility is a problem.- Hardly so (**comment by Anderson**) in the case of labelling antibodies to be used as analytical reagents, since labelled is separable from unlabelled Ab; what matters is susceptibility to hydrolysis in aqueous buffer. **Response by Lingeman.**- There is stability for 1-2 days, and for longer storage one could keep at -20°. **Answer to H.M. Hill**: in general the derivatizing agents are purchasable, possibly from certain U.S. universities in some cases.

M. Bayliss asked Lingeman whether there are any means for exciting with a 'standard laser' (>400 nm) a compound excitable in the ~250-400 nm range. **Reply**: this can be achieved by 2-photon excitation (UV laser) or, with some compounds, by a pH change. **Remark by D. Perrett**: as the final derivatives are anionic, CE separations could be faster if reverse- rather than normal-phase. **Reply.**- Yes, if only anionic solutes have to be separated; but normally not so in the derivatization context where neutral/anionic/cationic solutes have to be separated and little time is saved. The applicability of this approach is also limited, with bio-samples, by matrix interferences. Moreover reversed-polarity CE is less robust because a quaternary ammonium surfactant should be added to reverse the polarity and this significantly reduces the stability of the system.

Bevan, answering R. Venn.- Freezing packed CE capillaries or columns seems to cause no damage to the particles despite the volume expansion.

Citations contributed by Senior Editor

For journals, Chr denotes Chromatography

'A Practical Guide to the Care, Maintenance and Troubleshooting of Capillary Gas Chromatography Systems': Rood, D. (1995), 2nd edn., Huthig, Heidelberg, 220 pp.

HPLC conditions and capabilities

"Advantages of using TLC as a pilot technique for HPLC".- Rozylo, J.K. (1994) J. Liq. Chr **17**, 3641-3653. Optimization software packages: comparison of ICOS and DIAMOND, and pitfalls [cf. 'Expert systems' Index entry in Vol. 23].- Bowman, P.B. et al. (1993) J. Pharm. Biomed. Anal. **11**, 1295-1301. Graphitic carbon, with optimized eluant and UV wavelength, for separating urinary drugs and glucuronides (zidovudine, chloramphenicol, β-oestradiol): Ayrton, J. et al. (1995) J. Chr **B 667**, 173-178. The effectiveness (isocratic; 45 min) for serum atenolol of bare silica, through ion-exchange and silanol mechanisms; liquid-liquid extraction [title misleading] or ethylsilane SPE at outset.- Simmons, B.R. & Stewart, J.T. (1995) Anal. Lett. **28**, 2017-2026. "Comparison of silica-, alumina- and polymer-based stationary phases...": a study that embraced drugs; the PRP type was least suitable.- Jedrejewski, P.T. & Taylor, L.T. (1995) J. Chr Sci. **33**, 438-445.

Tailing suppressors (for β-adrenergics): Hamoir, T. et al. (1994) J. Chr Sci **32**, 14-20. HPLC-fluor with three retention systems (switching tactics) was used for neopterin in serum: proteins were separated by size exclusion; analyte isolation was adsorptive and resolution was achieved by ion-exchange.- Huber, J.F.K. & Lamprecht, G. (1995) J. Chr **B 666**, 223-232.

For individual corticoids (some synthetic) in urine, with a C-18 column, optimization of the aqueous acetonitrile mobile phase and of UV detection wavelength was investigated; initial isolation was by SPE (PRP) or solvent extraction (NaCl obviated emulsion formation).- Santos-Montes, A. et al. (1994) J. Chr **B 652**, 83-89.

Supercritical fluid chromatography (SFC)

Artemisinin, solvent-extracted from whole blood, was applied (some variability) to a packed RP (CN) microbore column and subjected to SFC-ECD with CO_2.- Mount, D.L. et al. (1995) J. Chr **A 666**, 183-187. Certain non-drug analytes have been run with CO_2 on a capillary column, coated with DB-5, which was linked both to FID and to an MS (CI or EI; SIM) detector.- Buecheri, T. et al. (1994) HRC **17**, 765-769.

In J. Chr (1995) **B.- 664**, 317-327 (Gonzalez, M.L. et al.): **GC** injection mode to preserve cocaine and metabolites; **669**, 408-412 (Otsubo, K. et al.): detecting benzodiazepines by **TLC.**

Separations with chiral stationary phases (cf. #E-5 & #ABC)

J. Haginaka et al. prepared OVM columns and directly loaded serum containing (e.g.) ketoprofen [(1993) J. Chr **B 620**, 199-204]; amongst separated OVM 'variants' they found only one was chirally effective [(1995) Anal. Chem. **67**, 2539-2547]. For methadone, CN and AGP columns were linked in that order [title unclear]: Kristensen, K. et al. (1994) J. Chr **A 666**, 283-287. - Cf. 289-293 (Quaglia, M.G. et al.; Chiracel OD for a pure β-blocker) and 421-432 (Küsters, E. et al.: enantiomeric sulphoxide analytes; cellulose tribenzoate - Chiracel-OB - surpassed other cellulose-based CSP's). A mini-review [in French] of protein-type CSP's (NSAID's etc.): Felix, G. (1995) Analusis **23**, M20-M22; cf. M22-M26: a 'system' for CSP choice.

Chiral GC using a modified α-cyclodextrin phase was achieved (with derivatized ephedrine, phenamium and metabolites) in the effluent stream (switching) from a conventional GC run; initial liquid-liquid extraction was thereby simplified.- Wang, X. et al. (1994) J. Chr. **B 653**, 98-101.

Note on a 'myth'.- The idea that the *R*-isomer of thalidomide is safe is false - Simonyi, M. (1995) Chem. Br. **31**, 942, citing art. in Nature (1967) **215**, 296 (teratogenicity) and Eriksson, T. et al. (1995) Chirality **7**, 44 (racemization).

Capillary electrophoresis (CE; first example is chiral)

Enantioselective separation was achieved with pure drugs, including clenbuterol (best), amphetamine and selegiline, by pH 2.5 CE with β-cyclodextrin but preferably no modifier in the medium.- Ingelse, B.A. et al. (1995) HRC **18**, 348-352.

MECC-UV after deproteinization can separate 9 β-blockers in serum.- Lukkari, P. et al. (1994) J. Chr **A 674**, 241-246. MECC-UV with a bile salt was effective, after solvent extraction, for diverse drugs (e.g. antipsychotics) added to urine.- Aumatell, A. & Wells, R.J. (1995) J. Chr **B 669**, 331-344.

MECC-fluor (laser) has been explored for bioanalysis by IA, using theophylline-containing serum exposed to 'FPIA' (Abbott) reagents.- Steinmann, L. et al. (1995) Electrophoresis **16**, 1912-1916. CE detection by MS(ESP) has been explored with non-aqueous (methanolic) media containing a surfactant, in a study focused on tamoxifen metabolism in hepatocytes.- Lu, W. et al. (1996) Anal. Chem. **68**, 668-674.

Derivatization for HPLC or CE

Reviews in a J. Chr issue (1994; **B 659**; 241 pp.) include surveys by I.S. Krull et al. of derivatization strategies (pre- or post-column; pp. 1-17) and polymer-linked reagents (19-50). For urinary glucuronides (steroidal), derivatization using a substituted hydrazide allows analysis by HPLC-fluor (Hirose, T. et al. (1994) J. Chr **B 654**, 171-176. [The 'Derivatization' index entry in past vols. of this series may be helpful.]

Section #F

DATA ACQUISITION AND VALIDITY IN RELATION TO DRUG APPRAISAL

The suffix '**n**' as in #F-4n distinguishes 'NOTES' from main articles. COMMENTS (starting on p. 336) comprise Forum discussion remarks and Editor's 'annotations'.

#F-1

HAEMOLYTIC PLASMA: UNEXPECTED PROBLEMS WITH A VALIDATED ASSAY IN ROUTINE ANALYSIS

P. Heizmann†, R. Gora, J. Körner & S. Maier

Department of Drug Metabolism and Pharmacokinetics,
F. Hoffmann-La Roche Ltd., CH-4002 Basel, Switzerland

In relation to 'assay robustness', an example is now given of a 'validated' HPLC assay for a particular drug where unexpected problems arose with the routine liquid-liquid extraction procedure when analyzing samples from a toxicology study. The i.s. peak heights showed unacceptably large variations within each analytical series, casting suspicion on the results. It turned out that the effects arose from two factors: (1) the degree of haemolysis in the plasma sample▫, and (2) the peroxide content of the THF component in the solvent used to re-constitute and store the dry sample extract pending HPLC injection. The investigation of the strange effects and the final (simple) remedy are now described.*

During the development of an analytical procedure, various performance criteria have to be evaluated [1-4] before the 'validated' assay can be successfully used in routine analysis; but, even during subsequent routine application, unexpected problems may arise in a validated and seemingly well established procedure. Revalidation experiments and methodological changes may then be needed, as in the following case history where such unexpected and time-consuming problems arose.

The drug (tolcapone, Ro 40-7592), under clinical investigation for Parkinson's disease, is an orally active, selective inhibitor of catechol-*O*-methyltransferase (COMT), important in the inactivation of catecholamine neurotransmitters such as dopamine. COMT also methylates dopamine's precursor, L-dopa,

†addressee for any correspondence
**Abbreviations*.- i.s., internal standard; THF, tetrahydrofuran; LLOQ, lower limit of quantification.
▫This aspect (but not the later THF study) was considered, for the drug concerned, by Heizmann & Gora in Vol. 23 (p. 365), this series (eds. & publisher as now; 1994); assay details appear therein, with chromatograms.-*Ed.*

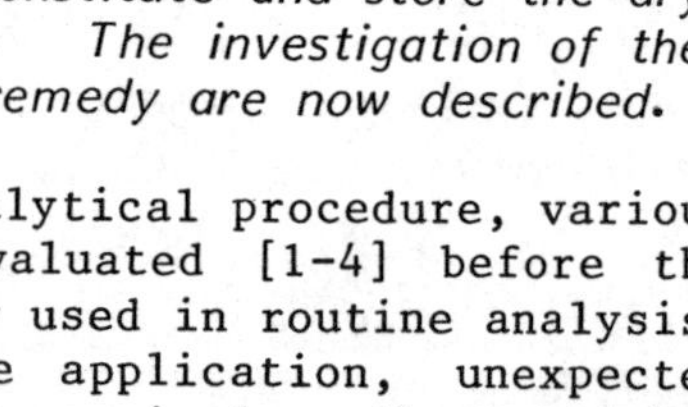

parent drug: R_1=OH, R_2=CH_3
metabolite: R_1=OCH_3, R_2=CH_3
i.s., R_1=OH, R_2=Cl

as used for the symptomatic treatment of Parkinson's disease. Co-administered with L-dopa, tolcapone increases its bioavailability and duration of action.

ANALYTICAL PROCEDURES: see Panel

For the drug and its metabolite [structures on previous p.] in animal and human plasma samples, three HPLC assays had been developed. In Assay **3** the i.s. served in measuring the drug, but external standardization had to be used for the metabolite.

Assay 1 for the parent drug, used in early dose-finding studies in man, entailed liquid-liquid extraction (pH 2 phosphate buffer). R_T typically: drug 7.2; metabolite 8.6; i.s. 10.2

sample work-up	: • extraction with n-butylchloride / ethylacetate (at pH 2) • evaporation, reconstitution in mobile phase • RP-HPLC
column	: 125 x 4 mm, Hypersil ODS, 5 µm
mobile phase	: methanol / buffer pH 2 / THF (55 / 45 / 5)
detection	: electrochemically (coulometric)
sensitivity (LLOQ)	: 2.5 ng/ml (parent drug)

Assay 2 (drug & metabolite) was developed when the sensitivity demand shrank; initial protein precipitation, and UV instead of EC detection.

sample work-up	: • acetonitrile protein precipitation • dilution with buffer pH 2 (1:1) • RP-HPLC
column	: as for assay **1**
mobile phase	: as for assay **1**
detection	: UV (**270 nm**)
sensitivity	: 50 ng/ml (parent drug, metabolite)

Assay 3 (drug & metabolite) had extraction as in **1** and detection as in **2**.

sample work-up	: as for assay **1**
chromatography	: as for assay **2**
detection	: UV
sensitivity	: 100 ng/ml (parent drug, metabolite)

VARIABILITY PROBLEMS

Assay **3**, with liquid-liquid extraction, was in routine use for toxicokinetic studies. When analyzing samples from a drug toxicity study, large variations in i.s. peak height were observed, as illustrated in Table 1 where i.s. heights in the same analytical series on two consecutive days are tabulated. The i.s. was added in fixed amount to each sample, so heights should have stayed constant within an analytical series. The variations were very large in extracts of the unknown samples and lower in the calibration samples. However, variation was acceptable for the parent-drug duplicate values (calculated from peak height ratios, drug/i.s.).

Table 1. Test samples (**1-12**) analyzed on two consecutive days: peak heights for i.s. (fixed amount added) and µg/ml found for the drug (<LL = below the 0.1 µg/ml LLOQ).

	Pk. ht., i.s.		Drug concn.	
Day:	1	2	1	2
1	1080	750	<LL	<LL
2	1220	990	10.7	11.1
3	1180	570	5.15	5.69
4	1030	750	0.78	0.85
5	950	570	0.29	0.27
6	290	25	<LL	<LL
7	1050	940	<LL	<LL
8	780	840	2.51	2.43
9	610	320	5.55	6.39
10	790	450	1.69	1.76
11	370	780	<LL	0.17
12	980	930	<LL	<LL
c1	1120	Day 1 values for calib. stds.: mean 1260; C.V. 13%		
c2	1220			
c3	1470			
c4	1090			
c5	1380			

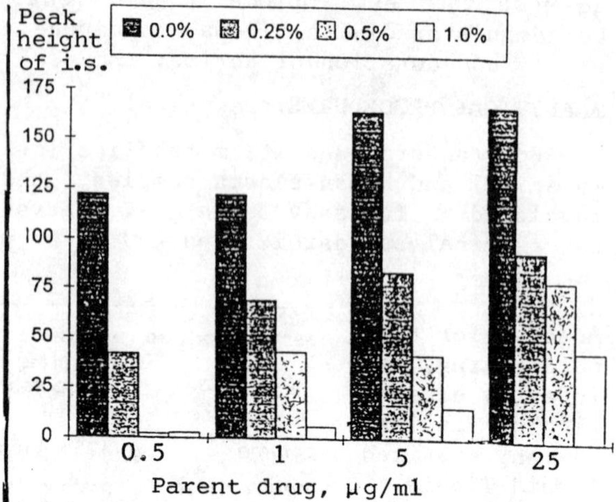

Fig. 1. Influence of haemolysis on the extraction procedure: i.s. recovery (as peak height) from human plasma with varying 'degrees' of haemolysis, 0.25% (slight redness) to 1.0% (distinct red).- See text, also (first p.; other results) ¤.

The unsatisfactory situation could not be overcome at that time, and analysis was continued but with duplication for every sample. Later, for further animal toxicokinetic work, human and rat spiked QC's were analyzed and the following i.s. peak heights found ['0' = no peak found; *in italics:* mean peak height and drug/i.s. height ratio]:

human, h 0.5 (µg/ml): 1540, 1650, 1610, 1150; *1490, ratio 0.18*
" h 5.0 ("): 330, 1310, 1290, 1490; *1360, ratio 1.77*
rat, r 0.5: 0, 150, 0 0; r 5.0: 180, 110; 150; 70; *130; ratio 1.96*

As drug and i.s. extraction was, surprisingly, abortive in most r 0.5 samples and i.s. peak height drastically reduced in r 5.0 and as the plasma used for spiking was very haemolytic, it was thought that haemolysis might account for these assay findings.

Results in Assay 3 with simulated haemolysis (Fig. 1)

Small amounts of deep-frozen and hence entirely haemolyzed whole blood were added to non-haemolytic plasma to give a blood content of 0.25, 0.5 and 1.0%. The resulting 'haemolytic' plasma was extracted after spiking with parent drug and metabolite and i.s. in fixed amount such that a constant peak height would be expected. As shown in the Fig. 1 plot of i.s. peak height *vs.* drug concentration, the higher the concentration the higher was the i.s. peak height for that haemolysis level; moreover, for a given concentration the greater the haemolysis the lower were the peak heights for both i.s. and (not shown; cf. ¤) the drug. Metabolite peak heights were unaffected by haemolysis.

¤ is on 1st p., footnote

Fig. 2. Influence of haemolysis in the protein precipitation procedure: recovery of i.s. and parent drug (as peak height) from non-haemolytic and haemolytic human plasma.

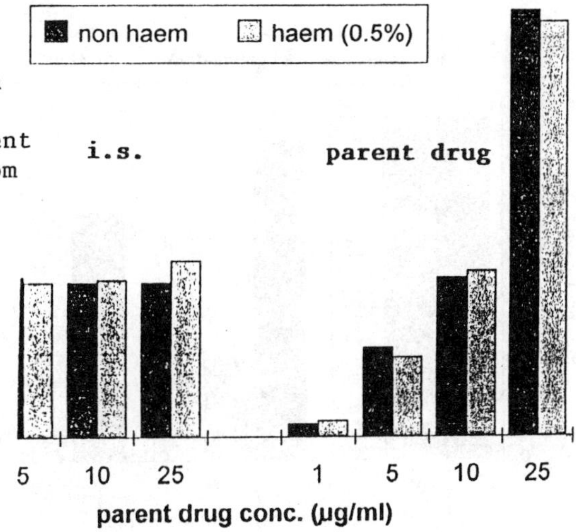

The question arose as to what extent the precipitation procedure (**Assay 2**) might be influenced by haemolysis, as was similarly investigated. Non-haemolytic and haemolytic plasmas gave equal peak heights for the i.s.; likewise for the parent drug, the increase with rising spike concentration being linear (Fig. 2). In view of the results with the extraction procedure it was discontinued and the protein precipitation procedure successfully revived for routine use.

EFFECTS FROM MOBILE-PHASE TETRAHYDROFURAN

At a late stage the high sensitivity of the early extractive assay (**1**; coulometric detection) was again needed, for determining the free (unbound) fraction of the parent drug in plasma samples from a clinical study. The assay was set up again and, as feared, the problems with liquid-liquid extraction returned.

With his set of spiked calibration standards, the analyst concerned noticed, for both parent drug and i.s., a peak-height decrease with time spent in the final injection solution (the mobile phase). Several investigations pointed to THF, present in the mobile phase, as the culprit. With a different (fresh) batch the effects were less.

Instability of the drug under oxidizing conditions

Catechol structures (as in the parent drug and i.s.) are known to be prone to oxidation*, as was confirmed directly by the following experiments. A methanolic solution of the drug, metabolite and i.s. was evaporated to dryness, reconstituted in dilute hydrogen peroxide, and analyzed by HPLC. Peak heights decreased as the hydrogen peroxide concentration was raised (Fig. 3), even the metabolite being affected to some extent.

*considered on p. 99 of Vol. 22 (1992) by R.J. Simmonds *et al.*-Ed.

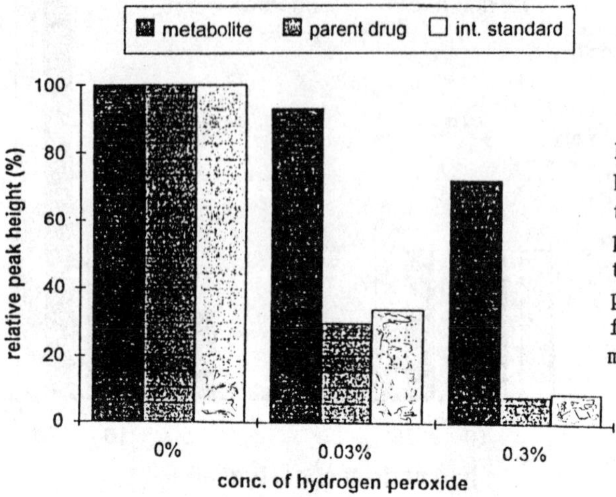

Fig. 3. Influence of hydrogen peroxide at two concentrations on peak heights relative to control with none present (= 100% value), for parent drug, metabolite and i.s.

Similar findings were observed under the influence of air: the longer the dry evaporation residues were exposed to air, the greater were the peak height decreases for the parent drug and i.s. (~25% within 3 h); again the metabolite was less affected.

Peroxides in the THF

Peroxides can be formed in unstabilized THF,† and if present in the final solvent could react with the catechol moieties and so decrease peak heights, as was investigated. The evaporation residues from extracts of spiked haemolytic and non-haemolytic plasma samples were re-dissolved in solvent without and with THF and injected immediately ('100%' value) and again after storage for 1-2 h; peroxides had been formed artificially in the THF by stirring it for 24 h exposed to air, their presence being indicated by colour reaction with KI-starch paper.

In solvent without THF, storage entailed no peak-height reductions, whereas with peroxide-containing THF present the peak heights decreased for haemolytic plasma - but not for non-haemolytic plasma, nor was the metabolite affected (Fig. 4). Evidently the THF (or peroxides therein) in the final solvent was one factor that contributed to the observed instability. But only the combination of haemolytic sample *plus* peroxide-containing THF caused losses, obviated by omitting THF from the final solvent (innocuous in the mobile phase, each run being brief).

To investigate whether parent drug and i.s. were indeed affected in parallel, calibration samples set up in artificially prepared strongly haemolytic plasma were extracted. The final solutions of every sample (THF present) were injected, immediately after dissolution and again after storage for 1 h.

†Stabilizer example (Merck; not in HPLC grade): 2,6-di-tert.-butyl-4-methylphenol. Distinguishing 'good' THF was not achieved.

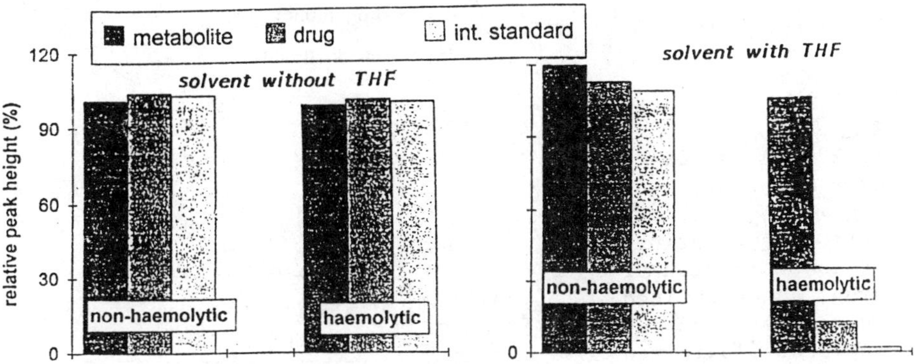

Fig. 4. Influence of peroxide-treated THF on the decreases in peak height during pre-HPLC storage (no-storage value = 100) for extracts from non-haemolytic and haemolytic (1%) plasma samples containing drug, metabolite and i.s. (each 1 µg/ml).

The drug gave post-storage ('found') heights much lower than the immediate ('expected') values (Fig. 5). Peak heights for the i.s. varied widely; but the drug/i.s. peak-height ratios correlated linearly with the spiked concentrations (Fig. 5). Thus the i.s. had compensated for the fall in drug peak height. The metabolite peak heights were unaffected (not shown). This explains why, in extraction-based assays, internal standardization could be used for the parent drug but not the metabolite.

INFLUENCE OF Fe^{2+} or Fe^{3+} IONS

One conceivable cause of the effects seen with haemolytic samples was haem-derived Fe(II) or, formed post-release, Fe(III). Iron ions were identifiable as red-coloured Fe(III)-complexes with KSCN in the solvent extract (drying-down residue dissolved in 3 N HNO_3) of a strongly haemolytic plasma, but not with a seemingly non-haemolytic one, nor in the dried-down supernatant from haemolytic plasma after acetonitrile precipitation.

This avenue was pursued by adding small amounts of Fe(II) salts to non-haemolytic plasma. The samples were acidified and submitted to the standard extraction procedure. The final THF-containing solutions were stored for 1 h and then injected for HPLC, with the results shown in Fig. 6. The peak heights for the drug and i.s. (not metabolite) underwent, during storage, greater reductions (25-40%) if Fe(II) ions had been added. The same effect was found after adding Fe powder, which gave traces

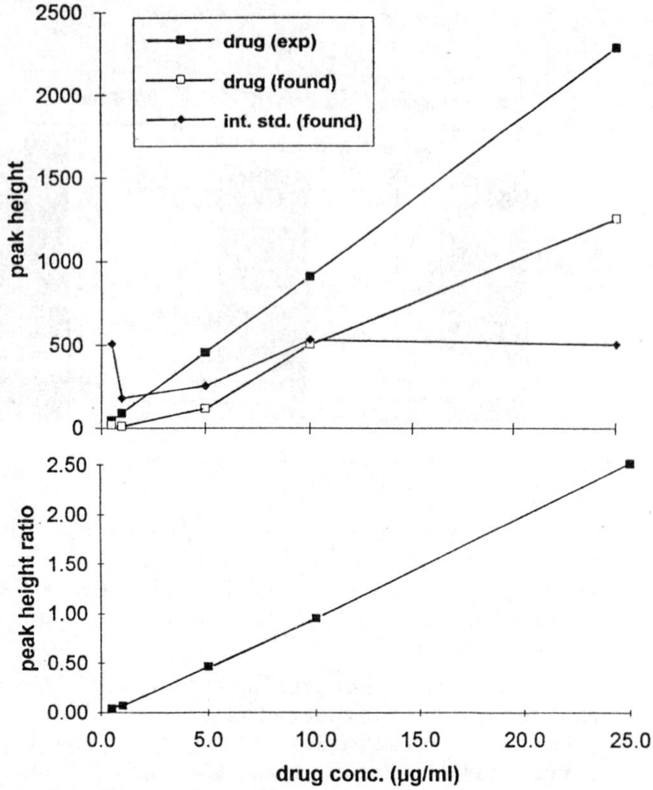

Fig. 5. Peak heights for drug and i.s., and *(lower part)* their ratio, in relation to concentration of drug, as determined in re-dissolved residues ('oxidized' THF present) from extracts of very haemolytic (1%) plasma:- re-assay values ('found') after 1-h storage and, for the drug, expected values ('exp'; from the immediate assay) if no fall during storage.

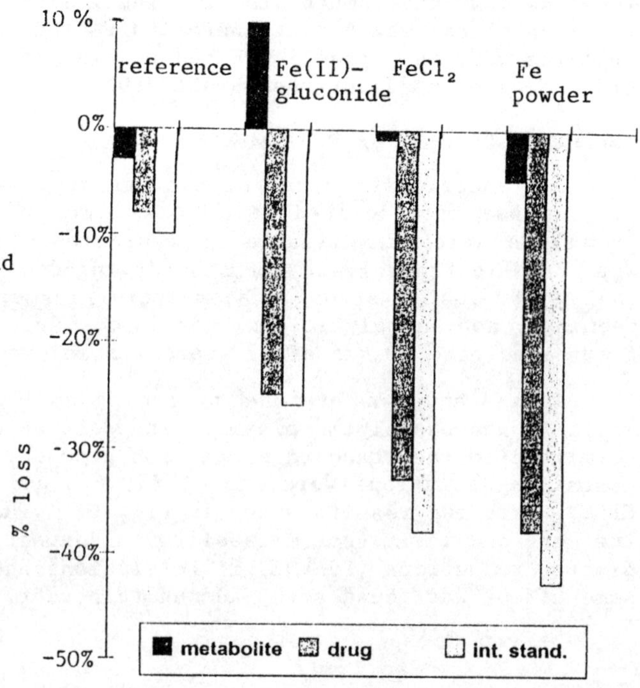

Fig. 6. Changes with 1-h final storage (THF present) in the peak heights for drug, metabolite and i.s., added (each 1 µg/ml) to non-haemolytic plasma with Fe additions also (see text for Fe powder).

of ionic Fe by dissolution under the acidic conditions used. The results support our theory that indeed a role is played by the presence of ionic iron in haemolyzed plasma.

The mechanism of the presumed interaction is not fully understood. A possible explanation may relate to the known formation of chelates between Fe(III) ions and catechols, the parent drug and i.s. being in this class - not the case for the metabolite, which is analytically unaffected.

SUMMARY OF THE FINDINGS

The parent drug and i.s. are very sensitive to oxidation. Two factors influenced their stability in the final injection solvent of the extractive procedure: the extent to which the plasma sample was haemolyzed (liberating ionic iron) and the peroxide content of the THF in the final solvent. Where *both* factors were strongly operative, significant reductions in drug and i.s. peak heights were observed, as in the Table 1 example.

Parent drug and i.s. were generally influenced in parallel. However, it could not be guaranteed that the i.s. would fully compensate for the parent drug peak-height reduction in all cases. The problem was avoided simply by omitting the THF in the final re-dissolution step of the extraction assay, or by using the protein-precipitation assay if sensitive enough for the studies concerned. The mobile-phase THF caused no trouble.

CONCLUDING COMMENTS

During the development of an analytical procedure different assay performance criteria must be evaluated. Having done this it is at the discretion of the analyst when to regard an assay procedure as fully validated and write the final method report. This should normally await the analysis of an adequate number of *in vivo* samples, as only by analyzing study samples can assay robustness be proved. As shown above, even with a validated, seemingly well established, procedure unexpected problems may arise that call for time-consuming revalidation investigations. Analysts must realize that developing an assay is a sustained process, maybe final only after a long routine assay period.

References

1. Buick, A.R., Doig, M.V., Jeal, S.C., Land, G.S. & McDowall, R.D. (1990) *J. Pharm. Biomed. Anal. 8,* 629-637.
2. Shah, V.P., Midha, K.K., Dighe, S., McGilveray, J.J., Skelly, J.P., Yakobi, A., Layoff, T., Viswanathan, C.T., Cook C.E., McDowall, R.D., Pittmann, K.A. & Spector, S. (1992) *J. Pharm. Sci. 81,* 309-312.
3. Hartmann, C., Massart, D.C. & McDowall, R.D. (1994) *J. Pharm. Biomed. Anal. 12,* 1337-1343.
4. Daghar, D., Burnett, P.E., Choc, M.G., Gallicano, K. & Hooper, J.W. (1995) *J. Pharm. Biomed. Anal. 13,* 89-97.

#F-2

A DATABASE FOR BIOANALYTICAL RESULTS IN DRUG DEVELOPMENT

D. Browne, P. Maude & L. Robotham†
Dept. of Biopharmaceutical Analysis, Corning Hazleton (Europe), Otley Road, Harrogate HG3 1PY, U.K.

Bioanalytical 'productivity' has escalated. The processing, evaluation and reporting of data may now be rate-limiting. We have developed a flexible database from a commercial information management system. Generated reports can include CAL and QC data, validation, study results (PK, toxicological, clinical) and audit comments, apt for clients' needs.*

Our Department analyzes, following GLP* principles, very numerous biosamples, largely by GC or HPLC and sometimes with MS detection. The results serve to help PK studies (bioavailability, bioequivalence), and also dose choices (toxicological, clincal). Samples may arise, with varying time-scales, from persons challenged with single or multiple formulations on one or more occasions. The reporting capabilities of many present-day MS and chromatography data handling systems are limited, often embracing only data from single samples or batches. To create tables for a long-running study, data are often manually compiled and maybe later reformatted for entry into other software packages to assist generation of PK parameters. The LIMS alternative for handling the data, whilst very apt for laboratories involved with QA and typically doing multiple analyses on each sample, is less apt for contract laboratories doing PK because samples usually undergo single analytical protocols.

We describe here an alternative approach which uses a commercially available 4th-generation-language (4GL) relational database management system (RDBMS) that can accept Data System (MS, Chromatography) data.

METHODOLOGY AND SYSTEM REQUIREMENTS

Samples are rendered suitable (extraction or other pre-treatment) for analysis by automated chromatographic methods, typically in batches of 40-50 unknowns plus CAL/QC samples. Data from GC or HPLC runs are captured and processed using a Multichrom #2.0 (Fisons Labsystems, Altrincham, U.K.) operating on two MicroVax #3800 mini-computers (Digital Equipment Corp., Maynard, MA, U.S.A.). The processed data are automatically

---

**Abbreviations*.- CAL, calibrant; GLP, Good Laboratory Practice; LIMS, Laboratory Information Management System; PK, pharmacokinetic(s); QA/QC, Quality Assurance/Control (sample).
See above for 4GL and RDBMS, and next p. for SQL.
†addressee (or Dr H.M. Hill) for any correspondence

transferred to a Novell 3.11 PC network using Pathworks #4.1, and moved to a restricted directory on a PC server (Compaq SystemPro XL # 486 DX50 network server with 32 mB RAM, 2 gB hard disk and 32-bit Network card). Chromatography data from SCIEX mass spectrometers are captured and processed using MacQuan #1.3 (Sciex Corp.) on Apple Macintosh Quandra 650 or 950 computers running an Apple #7.1.2 operating system. The processed data are manually transferred to the Novell network using an Apple file translator and moved to the same restricted directory on the PC server. Transferred data are automatically processed by a PC (486 SX IBM compatible with 8 mB RAM running Microsoft Windows #3.11) which sorts or parses the data according to a strict naming convention (discussed below). This PC also administers the within-Biobase audit logging procedures.

Users access the data *via* a security screen which requires the entry of user-name and password; security can be configured to user-study-task specificity. Using a Windows-compatible PC (#486 DX2), the study data are reviewed for inclusion as final data. Tables are generated using pre-defined Structured Query Language (SQL) commands, operating under Paradox for Windows #4.5 (Borland International), and may be printed out in defined Table layout. Each is time-stamped with a unique audit identification number. Each can also be exported as comma-separated values for further manipulation with commercial software.

PRINCIPLE OF OPERATION

'BioBase' operates in a client server relationship, the server providing security and audit support. A parser engine operating as a server processes, sorts and consolidates incoming data, working to a defined format. Audit information is event-driven and allows specific actions to be linked to key events, and can be used as an active rather than passive auditor.

Parser operation (Scheme 1).- The parser sorts the data according to the characters used in the sample name, which allows the data to be reported in separate Tables. Moreover, data from different analytical batches can be compiled and consolidated in summary Tables. This requires that a strict naming convention be used to identify samples, CAL's, QC's, analytes, etc.

Batch naming convention.- Numbering is in sequential order from 01 to 99. The batch no. is prefixed to identify the matrix and type of regime followed. Matrices include:

B	Blood	F	Faeces	P	Plasma	SA	Saliva	U	Urine

Regime: VAL = validation phase, B = routine batch analysis. Reinjected batches get the suffix 'I', and re-extracted or re-assayed (i.e. repeated) batches are denoted by the suffix 'R'.

Typical examples: PB18 PVAL05

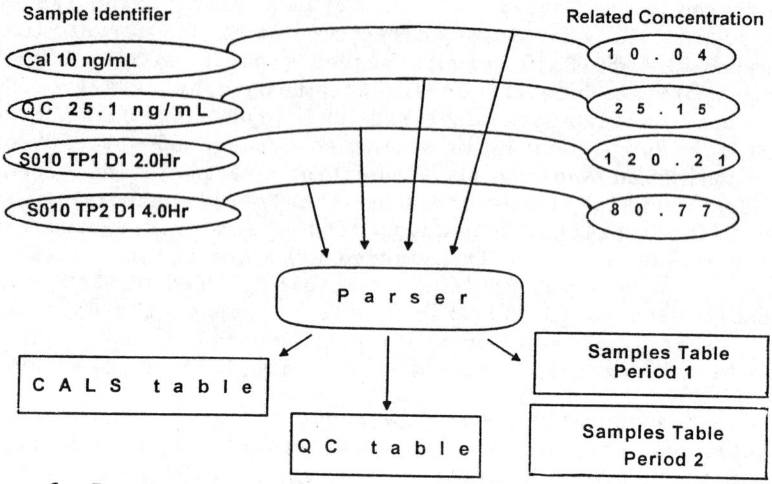

Scheme 1. Parser operation.

Sample naming convention.- CAL's and QC's are denoted:
CAL{nominal concentration}{units}
QC{theoretical concentration}{units}
and clinical study samples:
S{subject no.}TP{treatment period}sampling phase
with possible further definition of the sampling phase:
Occasion - phase within the study; e.g. OccA
Dose - formulation; e.g. DoseB
Day - day no. of dosing, e.g. Day 1
Hour - decimal hours from dosing, e.g. 1.0 Hr
Minute - decimal minutes from dosing. e.g. A B ......

Toxicology study samples.- In the convention used for clinical study samples, Animal no. replaces Subject no., with the 'S' prefix dropped, and there may be group/sex identifiers.
Examples: CAL 100 pg/ml S010 TP1 Dose A D1 2 Hr 30 mins B
QC 10.5 ng/ml GP4 M 1256 TP 1 D1 4Hr

FINAL REPORTED RESULTS (Verification: Scheme 2)

Sample analyses may be repeated during the course of the study for many reasons, e.g. instrument problems, poor chromatography, or other causes of anomalous results. Any duplicate results are highlighted prior to compilation of the final data Tables. The study director will evaluate the replicate results during the data-review phase. The replicate value(s) will then be flagged for exclusion, and marked with a defined error code. This permits tracking of data for GLP purposes. Data can be reported in a variety of defined formats and also *ad hoc* reports produced using a commercial report generator. PK profiles can be viewed and printed. Reports include CAL's and QC's, validation, clinical and toxicology data and audit comments.

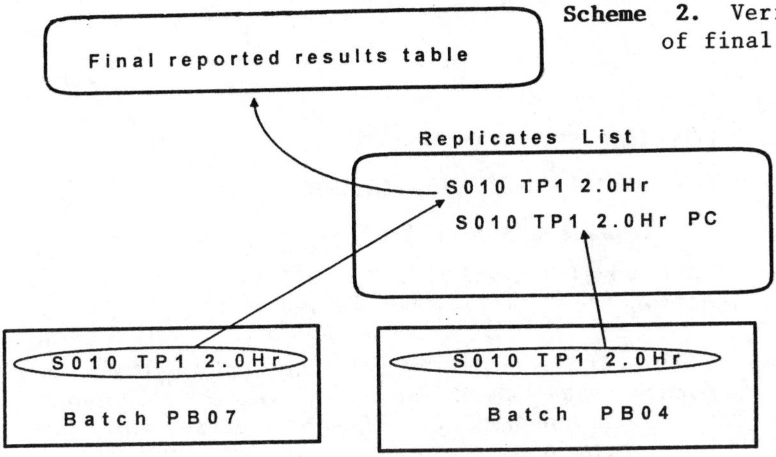

Scheme 2. Verification of final reported results.

STUDY DESIGNS

Depending on the study concerned, operation can be in free-study mode where the data design defines the data structure, or in a protocol-driven mode where parameters have to be defined, data are added in batch mode, and any batch not conforming to the defined protocol is rejected. A wide spectrum of study designs can be accommodated with the above naming convention, ranging from single-analyte plasma profiles for 12 subjects over an 8-h period to multiple groups with several doses (or formulations) on a number of occasions and measuring plasma and urine analyte concentrations over several days. Animal studies can also be accommodated as either single- or mixed-sex groups, permitting the easy tabulation of groups of results.

CONCLUSION

A database management system is an effective tool for tabulation of data generated by diverse types of analytical equipment operating on different hardware platforms. Tables of interim or final results can be generated in a rapid GLP-compliant fashion, allowing fast turn-round of results.

The authors are grateful to colleagues in the Department for their assistance and input during the development and testing of the system's applicability.

#F-3

CURRENT VALIDATION AND GLP ISSUES WITH RESPECT TO BIOANALYTICAL METHODS

H.M. Hill†, K. Tennant & S.A. White

Corning Hazleton (Europe), Dept. of Biopharmaceutical Analysis, Harrogate, N. Yorks. HG3 1PY, U.K.

Guidelines for the validation of bioanalytical methods and their routine application have been documented in the 'Consensus' report of the 1990 Washington meeting. They have subsequently been accepted by the international pharmaceutical industry and are the subject of constant review through regular scientific meetings and published scientific critiques. These critiques largely question the statistical limitations of the mainly empirical Washington Consensus guidelines concerning the routine analysis and acceptance of data.

While the application of these guidelines is, in general, done in accordance with GLP principles, only pre-clinical studies can formally claim compliance with GLP. However, in the U.K. the DoH* is taking the initiative in preparing guidance for the application of GLP to analytical studies.*

In the late 1980's and early 1990's a number of meetings and publications [1-4] considered criteria for the validation of bioanalytical methods which were to be used to generate data for determining bioavailability and bioequivalence as well as definitive pharmacokinetic parameters. However, it was not till the December 1990 meeting held at Crystal City, Arlington, Washington D.C. (organized by AAPS, FIPS, HPB, AOAC and FDA*; variously described by one or more of these names), hereinafter referred to as the Washington meeting, that generally accepted guidelines were agreed. The consensus appeared in a variety of journals [e.g. 5]. Since then the guidelines have been reviewed

---

†addressee for any correspondence

**Abbreviations include:* AQL, acceptable quality level; GLP, Good Laboratory Practice; cGMP, current Good Manufacturing Practice [*not* cyclic 5'-GMP!]; i.v., intravenous; LLOQ, lower limit of quantification; LOD, limit of detection; QC, quality control. ORGANIZATIONS etc.:- AAPS, American Assocn. of Pharmaceutical Scientists; AOAC, Assocn. of Official Analytical Chemists; CFR, Code of Federal Regulation; CPMP, Committee for Proprietary Medicinal Products; DoH, Dept. of Health; FDA, Food and Drug Administration; FIPS, Fédération Internationale Pharmaceutique; HPB, Health Protection Branch (Canada); ICH, Internat'l Conf. on Harmonization; ISO, Internat'l Standards Organization; MCA, Medicines Control Agency; (A)NDA, (Abbreviated) New Drug Applic'n.

at a number of international meetings and symposia where the regulatory bodies have been significantly represented: notably Bad Homburg 1992 and Bio '94 in Munich and latterly review meetings at the annual (November) AAPS meetings held 1994 in San Diego and 1995 in Miami.

Not all of these meetings have been formally documented; hence opinions expressed at them can at best only be regarded as advisory. This is at complete variance with the requirements for validation of analytical methods used in the analysis of pharmaceutical products, be it for regulatory stability studies or QC release. These methodologies have been aboard the ICH process, ultimately resulting in an ICH document [6] which describes validation requirements for these methods; subsequently a 'how to' extension document [7] emerged, which provides guidance as to how these parameters should be evaluated. It should be emphasized that this document has not yet been finalized and at the time of writing† is subject to change.

The guidelines developed by the ICH process have been largely incorporated into the CPMP guidelines [8]. The guidelines also attempt to draw bioanalysis into the criteria. While the Washington guidelines have not *per se* been incorporated into regulatory guidelines, it would be true to say that they are a reference/pivotal guidance document of many regulatory bodies including the FDA.

The position reached

Perhaps an over-riding theme of post-Washington discussions has been the number of statistical critiques of the consensus guidelines which were in general empirically derived. As a consequence the evolving consensus necessitates the reinterpretation of guidelines on a regular basis. Most recently a meeting in Japan, Bio 96, will discuss validation criteria with respect largely to bioavailability/bioequivalence studies.

Since their publication in 1992, the Washington guidelines have gained universal acceptance throughout the pharmaceutical industry. While they provide guidance as to the specific validation requirements - both pre-analysis and during the sample analysis - they do not provide guidance on the appropriate regulatory requirements. Thus pre-clinical studies are universally recognized to be subject to GLP [9]; indeed, this aspect has been reinforced by the ICH guidelines on Toxicokinetics [10]. However, confusion exists in some quarters as to the application of GLP to other studies. Thus, while the analysis of samples from clinical studies may be appropriately carried out according to GLP principles, it is not appropriate to claim GLP compliance. Indeed, referring particularly to NDA and ANDA studies, where the marketed product is subject to evaluation

†mid-April 1996

through bioavailability/bioequivalence studies under the terms of a CFR Code [11] for human *in vivo* bioequivalence studies, the most appropriate guidelines would appear to be cGMP.

The U.K. DoH Monitoring Authority has recognized the need for additional guidance when applying GLP to the analytical laboratory, and are currently working with industry to produce an advisory document. The final draft is currently (mid-April) under review, and it is hoped that the document will appear later in 1996.

Regardless of these apparent anomalies in the application of GLP, interpretation of validation criteria and procedures undergoes continuing update and re-evaluation, consistent with changing technologies and scientific practice. The parameters most widely accepted as being indicative of method performance include: Response function/linearity, Accuracy, Precision, Selectivity/specificity, Recovery, Sensitivity and Stability. These are now broadly accepted by all regulatory authorities and investigators in the bioanalytical field. However, there are still areas of discussion or contention in terms of what criteria should be used to determine acceptability. The latest interpretation of analytical validation is discussed below in terms of current regulatory requirements in respect of chromatographic techniques.

PRE-VALIDATION ISSUES

Prior to 'validating' an analytical method it is essential that a number of contentious issues be evaluated prior to the validation program and that some criteria be established to indicate that the method developed is likely to be capable of being validated.

Besides establishing such criteria it is imperative that objective criteria be established as to the acceptability of the validation data. Thus a validation plan/procedure/protocol should embrace:

- parameters to be evaluated;
- procedure for evaluating them;
- results derived from the experiments;
- acceptance criteria for the validation parameters.

Last but not least, the documentation of the validated method must include 'sensitive/critical' procedures which may impact on the validation criteria and development of system-suitability criteria. Validation should always be regarded as a continuing process [12]. Changes in the nature of the sample and/or dose may call for revalidation prior to or during sample analysis.

While it is generally accepted that the choice of calibration function can significantly impact upon the accuracy and to some extent precision of the assay, procedures used to determine and evaluate the response function are still much debated.

Assay robustness can hardly be evaluated until the validation process is complete. However, it would seem pointless to embark upon the validation unless it is likely that it will succeed. Hence it is proposed that as a basis for continuation a calibration curve be evaluated along with 6 QC samples, 2 at each of three concentrations corresponding to low (Lo), medium (Me) and high (Hi). A *go/no go* decision should be made on the basis of the acceptability of the calibration function and the variability and accuracy of the QC's.

VALIDATION

Response function.- Washington guidelines recommend use of the simplest function to define the response-concentration relationship. This is usually determined by plotting residuals of a number of calibration curves obtained in the validation process, to determine whether the most appropriate model has been used [13]. Weighting is sometimes necessary to improve the fit. Empirically, the larger the dynamic range the greater the weighting needed, e.g. $1/x$ ($1/y$) or $1/x^2$ ($1/y^2$). (In ref. [2], a notable validation art. by D. Dell, p. 9, concludes with a survey of weighting practices [in 1989].- *Ed.*) In general, weighting is essential for quadratic calibration functions [13]. For linear functions the slope of a plot of log concn. *vs.* log-(variance of replicate determinations at each concn.) provides an objective assessment of the degree of weighting required.

Pragmatically it has been suggested that perhaps the most complex calibration, the log quadratic function usually ascribed to J.G. Wagner, could be classified as a universal function - hence $\log y = a \log x^2 + b \log x + c$ can reduce to a simple linear function if the response-concn. relationship demands [13]. Whatever criteria are used, they must be appropriate.- A well defined standard curve is crucial for assay accuracy.

Accuracy (closeness of determined value to true value).- According to the Washington meeting, accuracy and precision should be determined using a minimum of 5 determinations per concentration. The mean should be within ±15% of the actual value except at LLOQ where it should not deviate by >±20%.

Precision has mainly a statistical definition, usually related to the C.V. (R.S.D.; S.D. expressed as % of the mean of several determinations). However, the 'determinations' used to determine the mean can affect the definition of the derived values. Thus it is common to regard precision derived on a single occasion (usually same day or run) by the same analyst on the same instrumentation as 'repeatability', while regarding similar data generated on different occasions usually with the same analyst and equipment as 'intermediate precision'.

'Reproducibility' on the other hand, frequently used interchangeably with 'precision, should be used only where determinations were made on different occasions with different analysts and equipment. These ISO-specified criteria were integrated into both ICH and CPMP guidelines. Acceptable precision as defined in the Washington guideline is ±15% at all concentrations except the LLOQ where it can be ±20%. This does not take account of differences between repeatability and intermediate precision. Thus, if repeatability is ±15% then it is unlikely that intermediate precision will meet the ±15% acceptance criteria. Hence empirically [14] repeatability should be ±10% or less to 'achieve' acceptable intermediate precision.

Sensitivity is usually defined as the lowest concentration on the standard curve which meets the Washington criteria (see Accuracy and Precision) and suffices to establish the kinetic profile. This *lower* limit of quantitation (LLOQ) may not be the *lowest* limit attainable.

In addition, the LOD (detection limit) may also be ascertained, usually in situations where it may not be possible to quantitate low levels of an analyte but where some idea is required of presence or absence. Thus in pharmaceutical terms it may be necessary to prove non-absorbance of a topical agent, and invoking the LOD criterion is appropriate. Based on IUPAC [15] definitions and refined by the American Chemical Societies Standing Committee on Environmental Improvement [16], the LLOQ and LOD can be derived from the same base parameters. Using the measure of S.D. for a number of determinations these parameters may be determined as follows:

LLOQ = Mean ±10 S.D. LOD = Mean ±3 S.D.

where Mean is the mean blank signal from several determinations. *Note:* these definitions are different from those proposed in the Washington guidelines and may produce different results.

Specificity.- A specific assay is one in which the chromatographic peaks are quantifiable and uniquely identifiable, as in LC-MS-MS. Washington guidelines suggest that the specificity of the assay methodology be established using six independent sources of the same matrix. This can only evaluate the possibility of interference from endogenous sources. In addition it is essential to provide some evidence of peak purity in respect of known and/or potential metabolites and degradation products.

While LC-MS-MS methodologies represent one of the highest forms of specific assay, the detector's actual response to the analyte can be significantly modified by co-eluting components which in themselves may be 'invisible' to the MS detector under conditions specific to the analyte of interest. This matrix effect can be evaluated by determining the analyte's response in the same matrix from a range of different individuals.

FOLLOW-UP: VALIDATION REINFORCEMENT AND ROUTINE ASSAYS

Recovery and Stability.- The parameters thus far can and must be evaluated in each batch of samples that is run after validation, to ensure that method performance and characteristics accord with the validated method. In addition there are several parameters that should be evaluated as part of the validation, to ensure that the method is optimized and the samples are suitable for analysis. These include recovery and stability. Whatever criteria are used to determine stability, it is essential to ensure that the analytical method is not compromised by analyte loss during storage, sample processing or analysis. While a knowledge of analyte recovery is useful in understanding sample processing, low recovery may not itself compromise the integrity of the assay. Indeed it may provide a useful tool in improving the sensitivity and/or robustness of the assay, in that low recovery may compromise the assay's sensitivity and overall precision. 'Recovery' as used in this context relates to the absolute amount of analyte recovered during the method. It is sometimes used to compare the difference between spiked (nominal) and derived values, and in this usage is analogous to 'bias'.

Procedures in routine assays

Once validated the assay may be applied to the routine analysis of biological samples. Usually these are analyzed in batches whose size is determined by the pharmacokinetic study design and the robustness of the analytical system. Even though the method counts as valid the analyst must review and evaluate the results in order to ensure that the method is performing appropriately. In some cases additional validation, re-validation or cross-validation may be needed. Thus method validation should be considered an evolving process [12]. In the Washington guidelines it is recommended that 6 QC samples be run with each batch, viz. two samples at 3 concentrations - low (2 or 3 times the LLOQ), medium (20%-50% of highest standard) and high (~80% of highest standard). While suggested concentrations have not been included within the guidelines, Dighe & Adams of the FDA [17] have suggested those cited here.

Thus each analytical batch should consist of a range of standards used to define the response function, a total of 6 QC's, a blank from the same matrix as used to prepare the calibration standards, a reagent blank and appropriate system-suitability samples designed to document the system's ability to quantify the unknown samples. Perhaps the most important parameters are sensitivity and specificity. Thus prior to batch start-up two samples of blank matrix processed as for the unknowns are run - one with LLOQ spiking and one unspiked.

These will provide information both on the sensitivity and specificity of the system, while calibration standards and QC's interspersed through the batch monitor the relative R_T for analyte and (if used) internal standard.

Standard curve.- The number of standards used to define the response function (the blank never counts) should be 5-8 (Washington). While an FDA guide [18] proposes 6-10, more may be required for non-linear/complex response relationships. In general there is little guidance as to their distribution over the concentration range of the response function, although some of the statistical implications and practicalities have been discussed elsewhere [19]. Empirically, however, there must be QC's at all concentrations to allow evaluation; thus there must be more than one standard above the high QC.- Otherwise loss of the highest standard would preclude evaluation of the high QC's since extrapolation for any reason is unacceptable. Likewise the low QC must have two standards below it.

Batch order is not defined within the guidelines, and indeed the variety of arrangements used testifies to the potential pitfalls and advantages of the different systems. These range from (1) randomization of all samples (unknowns treated the same as QC's and calibration standards) to (2) placing the calibration standards in ascending order of concentration with QC's randomly or evenly distributed throughout the batch, while samples are placed in profile order. Although (1) provides a truly non-biased configuration, if the slightest degree of carryover exists this could significantly impact on the lower concentrations, with dramatic effects on the interpretation of the final stage of half-life assessment. While carry-over can and should be evaluated prior to and during validation, the number of factors causing it are so numerous and varied that it is an ever-present problem. Vigilance post-validation is vital.

Arrangement (2) miminizes the potential effects of carry-over but can lead to systematic bias in the results. So there appears to be no ideal/recommended procedure; all appear to be acceptable if their relative merits are known and understood.

Batch acceptability.- The use of acceptance criteria based on QC criteria is well documented and has been subjected to a number of critiques [20-22]. However, it is essential that the more qualitative aspects of chromatography not be forgotten; indeed such criteria may point to problems at an early stage. Among these are reproducibility of chromatographic profiles, including peak shapes and consistency; it follows that there should be evaluation of the appropriateness of baseline assignment and choice between peak height and area [23]. Not until these parameters have been rationally and critically evaluated should the response function be determined and subsequently the unknowns and QC's.

Prior to judging batch acceptability the LLOQ should be verified; thus the LLOQ and blank should be evaluated first; if the blank response is >20% that of the LLOQ, the LLOQ may well be unsustainable throughout the batch. Ideally there should be an LLOQ sample (system suitability) at the batch start and finish to settle this. If the LLOQ is not confirmed, the next available standard that fulfils these criteria should be used.

In respect of acceptability of the batch based on QC's, the Washington guidelines suggest that 4 out of 6 QC's (including at least one at each concentration) should be within ±20% of the nominal - often referred to as the '4:5:20 rule'. It has generally been considered that the 20% acceptance level at all concentrations represents a mere minimum (baseline) for acceptability; thus a range of additional individual criteria has been established. Perhaps the four most widely used are 10:10:10, 20:15:10, 20:15:15 and 15:10:10, the 3 values being for the low, medium and high QC respectively. While the Washington criteria intuitively appear somewhat lax, this is not necessarily the case when practicalities are taken into account; indeed the converse may be reckoned true. Thus Hartmann & co-authors [20] and other workers [21, 22] have shown that while significant amounts of 'acceptable' data may be discarded, it is unlikely that 'bad' data may be accepted. The term 'accepted' refers to validation criteria whereby an assay may be deemed acceptable provided that accuracy and precision are within ±15% of nominal and have C.V. ±15%. Thus if the 15% QC criterion is applied an assay which has a 'bias' of -15% and a C.V. of ±15% would entail 50% of acceptable batches being rejected, even for assays whose analytical performance is well within the Washington criteria; e.g. with C.V. 10% and bias 10%, the false rejection rate is still 12% [21]. Kringle [21] summarized the problem thus.- "For assays with moderate (or greater) inherent bias and variability, the false rejection rate is too high, requiring the unnecessary repetition/rejection of runs"; and "For assays with low bias and variability, the false rejection rate is low, although the power to detect important deteriorations in assay performance is compromised".

Further acceptability considerations.- Kringle proposes the use of a Shewhart or multivariate statistical QC procedure wherein the rate of false rejection is fixed - and is acceptably small based on the statistical power of the procedure. Thus further tightening of these criteria, e.g. 4:6:10 or 4:6:20 rather than 10:10:10, could result in the rejection of more 'acceptable' data.

More recently Selinger [22] has proposed the use of inspection of variables as an acceptance criterion. This is based on MIL-STD-414 which has since been absorbed into ISO 3591 and relies on the use of an AQL (acceptable quality level) which although derived mathematically must be set empirically between the producer (bioanalyst) and the consumer/user (pharmacokineticist or regulator). The AQL is derived from two sets of tables, based on batch size and no. of QC's (usually 6 as in the Washington guidelines); as a pre-requisite the QC accuracy and mean and S.D. for all QC's in the batch must be calculated. The option is available to use, instead of the nominal concentration, the observed mean - calculable after assaying 6 or more QC's from the same batch. The AQL provides an indication of % 'defective' results that can be considered acceptable as part of the process. Thus an AQL of 6.5% with specifications set as ±20% or ±15% (i.e. consistent with the Washington guidelines) would give C.V. 15% and 12% respectively.

In general the alternatives to the 4:6:20 guidelines are complex and more difficult to implement; as such their acceptance by bioanalysts is debatable. However if they can be incorporated into a computerized data evaluation process their chances of acceptance might be improved. User-friendliness is a major factor in the implementation of such systems. On this basis, use of confidence limits may thus be more appropriate, as suggested within the Washington guidelines. To date few workers appear to have invoked alternative acceptance criteria on a routine basis [D. Dell, pers. comm.].

Acceptability of individual results

Although using the above criteria a batch may be deemed acceptable, it is still not uncommon to see individual samples that may be flawed† - usually either due to sample processing/ analytical reasons or manifest as a pharmacokinetic anomaly although apparently otherwise acceptable. In either case it is essential to establish *a priori* criteria for rejecting and hence repeating the analysis. This is particularly important for what may be regarded as anomalous results since reviewers including regulatory authorities may rightly view repeating these samples with suspicion. In such cases, in our laboratory we have developed common-sense criteria to suggest where a repeat is needed. However, in order to ensure consistency and a degree of objectivity we define anomalous results as those samples which provide an unexpected measurable result (e.g. a pre-dose sample) or, conversely, have a result below the LLOQ where a measurable result might have been expected. Besides, for single oral doses a value may be considered for repeat analysis if either of the adjacent values is 100% higher or lower than that value. Similarly for i.v. doses, if during the distribution phase a beyond-the-peak sample is 20% or more above that for the preceding timepoint, then it should be considered for repeat. Such repeats should be carried out in duplicate, and where appropriate one or more of the values adjacent to the 'suspect'/'anomalous' value should also be repeated.

†The Index entry 'Outliers' in past vols. is pertinent.-*Ed.*

Although repeat criteria have not been widely documented there are several 'logic trees' designed to provide acceptance criteria for repeat analyses - some complex and all-embracing, covering every eventuality; others are minimalist and hopefully pragmatic in their implementation. Lang & Bolton [24] provide a flow chart for evaluating repeats. Whilst inherently complex and comprehensive, it can be reduced to simpler formats bearing in mind that this reductionist approach carries a higher risk.

While all repeat assays are done in duplicate, evaluation of the results differs depending upon whether further repeat analyses are possible. Thus, if not and if the duplicates and presumably the original results agree within 15%, the original is accepted. If they differ by >15% but <30% the mean of the original and re-assay is reported, while if >30% no result is reportable. A minimalist approach would be to take the median of all 3 values (original and duplicate repeats) regardless of variability, although this can be modified by incorporating a rejection criterion - that if all values differ by >30% then no result is possible.

While additional repeats may be feasible, the complexity of the logic tree required makes a decision as to which result to use increasingly impractical. If significant numbers of repeats fall outside the 30% acceptance level, the analyst should seriously consider the robustness/validity of the assay before embarking on an unending series of repeats.

Thus while criteria for characterizing and validating a bioanalytical method are well accepted, criteria for acceptability of analytical results are still an issue. Nevertheless the issues have been widely discussed and provide a more than adequate basis for bioanalysts to develop their own criteria. The overriding criteria are those of consistency and balance between the excessive rejection of good data and excessive acceptance of doubtful data as an expediency for maximizing throughput. As such these criteria should be derived and agreed by both the analyst and the consumer.

Concluding comments.- Debate continues on ways to quantitate 'stability' and criteria, whereas there is fair agreement on criteria for characterizing an analytical method and judging acceptance at batch and individual-result level. Acceptance largely revolves on the '4:6:20 rule'; but there is growing diversity of approaches. These should, to win consensus, be simple, rapid and user-friendly as well as scientifically acceptable; the onus is on the scientific community, as it is unlikely that regulatory authorities will provide definitive guidance in this area.

ADDRESSES (cf. refs. overleaf)- MCA, EuroDirect Office, Info. Centre, 1 Nine Elms Lane, London SW8 5NQ: for ICH documents - & CPMP, also procurable from Commission of the European Communities, Director-General III, Rue de la Loi, B-1049 Brussels.

References [Abbreviations on title p. of art., *; procurement:† H.M.H. can advise; RSC = Royal Soc. of Chem., Cambridge]

1. Aarons, L., Toon, S. & Rowland, M. (1987) *J. Pharmac. Meths. 17,* 337-346.
2. Phillips, L.J., Alexander, J. & Hill, H.M. (1990) in *Analysis for Drugs and Metabolites Including Anti-infective Agents* [Vol. 20, this series](Reid, E. & Wilson, I.D. eds.), RSC, pp. 23-26.
3. Buick, A.R., Doig, M.V., Jeal, S.C., Land, G.S. & McDowall, R.D. (1990) *J. Pharm. Biomed. Anal. 8,* 629-637.
4. Karnes, H.T., Shiu, G. & Shah, V.P. (1991) *Pharm. Res. 8,* 421-426.
5. Shah, V.P. *& 11 co-authors* (1992) *Pharm. Res. 9,* 588-592.
6. ICH (CPMP approval: Nov. 1994) Tripartite Guideline: *Validation of Analytical Procedures.*
7. ICH (draft: 29 Nov. 1995) Topic Q2B, Step 2: *Validation of Analytical Methods: Methodology.* MCA Euro Direct Pubn. 281/95.
8. CPMP (19xx) Analytical Validation [1995 & earlier up-datings].
9. OECD (1982) [ISBN 9264 123679] Final Rept. of Expert Groups on GLP: *Good Laboratory Practice in the Testing of Chemicals.* Conforms with Directives 876/18/EEC & 88/320/EEC; GLP principles adopted by DoH (London) GLP Monitoring Unit.
10. CPMP (approved Nov. 1994) Toxicokinetics. *The Assessment of Systemic Exposure in Toxicity Studies.* MCA [as for7.] 5081/94 (ICH1).
11. CFR Part 320 [Federal Register 21] *Bioavailability and Bioequivalence Requirements.* [Is in July 10, 1989, issue.]
12. Centre for Drug Evaluation & Research, CDER (1995) Reviewer Guidance: *Validation of Chromatographic Methods.*
13. Hill, H.M., Causey, A.G., Lessard. D., Selinger, K. & Herman, J. (1992) in *Bioanalytical Approaches for Drugs .....* [Vol. 22, this series; eds. as in 2.], RSC, pp. 111-118.
14. Horwitz, W., Kamps, L.R. & Boyer, K.W. (1980) *J. Assoc. Off. Anal. Chem. 63,* 1344-1354.
15. IUPAC (1978) Nomenclature, Symbols, Units and their Usage in Spectrochemical Analysis 2: *Spectrochim. Acta 33,* 241-245.
16. Keith, H.K., Crummett, W., Deegan, J., Libby, R.A., Taylor, J.K. & Wentler, G. (1983) *Anal. Chem. 55,* 2210-2218.
17. Dighe, S.V. & Adams, W.P. (1991) *Bioequivalence: A United States Regulatory Perspective....* [Vol. 48 in series Drugs & the Pharmaceutical Sciences; ed. Welling, P.G. .....].
18. FDA (1991) Compliance Program Guidance Manual, Chap. 48. In vivo *Bioequivalence.* [FOI Services Inc., Washington, a source.]
19. Leppard, J.P. (1981) in *Trace Organic Sample Handling* [Vol. 10, this series] (Reid, E., ed.), Horwood, Chichester, pp. 320-335. [Is O/P. Corrigenda in Vol. 12 (1983; ed. E. Reid; Plenum.]
20. Hartmann, C., Massart, D.L. & McDowall, R.D. (1994) *J. Pharm. Biomed. Anal. 12,* 1337-1343.
21. Kringle, R.O. (1994) *Pharm. Res. 11,* 556-560.
22. Selinger, K. (1996) *J. Pharm. Biomed. Anal.,* in press [pers. comm.].
23. Dyson, N. (1990) *Chromatographic Integration Methods* [RSC Chromatography Monographs, ed. Smith, R.L.], RSC.
24. Lang, R.J. & Bolton, S.M. (1994) in *Pharmaceutical and Biomedical Applications of Liquid Chromatography* (Riley, C.M., Lough, W.J. & Wainer, I.W., eds.), Pergamon/Elsevier, Oxford, pp. 345-367.

†some guidance at foot of p. 325

#F-4n

A Note on

POST-VALIDATION PROBLEMS WITH INCURRED SAMPLES*

†M. Bentley, ¤J. Borlak & K. Tennant

Corning Hazleton (Europe), Dept. of Biopharmaceutical Analysis, Harrogate, N. Yorks. HG3 1PY, U.K.

¤Kali-Chemie Pharma GmbH, Hans-Boeckler-Alle 20, D-30173 Hannover, Germany

A validation program is a set of procedures to evaluate and define the critical parameters of analytical methods: specificity, sensitivity, precision, accuracy, response function, recovery, stability and between-batch variation [1]. Regulatory bodies, particularly the FDA, recognise that "validation should be ongoing in the form of revalidation with method changes" [2].

Validation programs can therefore be considered a 'best estimate' of assay performance, as the experiments are well defined and use good-quality matrices. They do not take into account the variable nature of incurred *(ex vivo)* samples due to factors such as unstable metabolites, haemolysis and endogenous matrix interferants that arise from disease states as well as particular anticoagulants, sample ageing and co-administered drugs. Only with the analysis of many 'real' samples may the requisite critical parameters emerge for ensuring assay robustness over lengthy periods of use with changes such as different analysts, instruments or columns. The following assay entailed solvent extraction and HPLC with electrochemical (EC) detection.

tedisamil (an anti-protozoal)

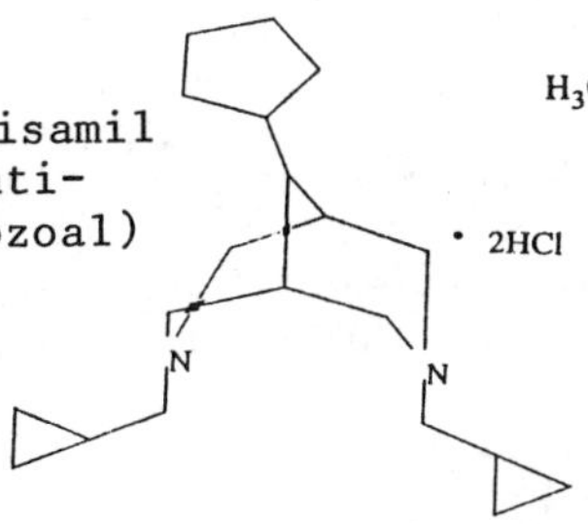

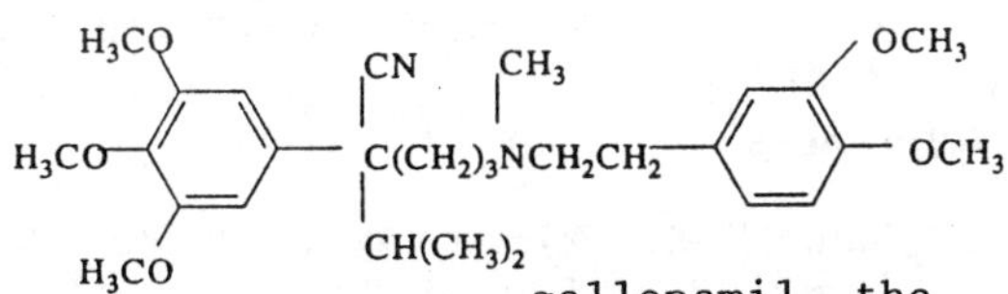

gallopamil, the internal standard (i.s.)

Analytical procedure for tedisamil

To plasma or serum (0.1-0.5 ml) are added 1 M NaOH (50 µl), i.s. and t-butyl methyl ether (3 ml). After mixing and centrifugation, the aqueous layer is frozen using an acetone/dry ice bath (so getting good phase separation without centrifugation).

*The term 'incurred' connotes 'real' study samples received.
†addressee for any correspondence

The organic layer is transferred to a clean tube containing 50 mM phosphoric acid (200 µl) and mixed. The upper aqueous layer is transferred to vials and ~15 µl aliquots are submitted to HPLC-EC (analytical cell 650 mV, guard cell 850 mV).

The packing for the pre-column and main column (200 × 4 mm) is Nucleosil 100 5-CN, and the mobile phase is ~10 mM ammonium acetate/acetonitrile/methanol (2800:1000:125 by vol.) adjusted to pH 3, with 0.8 ml/min flow rate. The R_T is ~5.2 min for tedisamil and ~8.0 min for the i.s.

Post-validation issues

Chromatography.-Following the validation and analysis of many samples, two precautions were found to be vital for assay robustness. Mobile-phase pH had to be rigorously controlled since even small changes (<0.5 pH units) gave R_T's which were 4-fold greater than those validated and mentioned above. Each time the assay was set up the 'system' required conditioning by injection of plasma extracts (for good sensitivity and peak shape).

Plasma collection.- After validation of the assay in mouse plasma using EDTA as anticoagulant, some samples were supplied with lithium heparin as anticoagulant. A cross-validation was done to compare the two anticoagulants and gave results that differed by 20%; therefore full validation had to be done. The explanation may be binding of tedisamil to heparin.

Species-specific stability issues.- Following successful validation with several matrices, problems arose with rat serum and plasma: the i.s. was not stable with respect to time in the extracts in the autosampler. The addition of 100 µl methanol to the matrix prior to the extraction prevented this phenomenon. The merit of matrix validation embracing each species is clear. Although experiments have not been done to determine the exact mechanism of the phenomenon encountered, it is thought that the haemolysis which is more evident in rat samples leads to higher plasma/serum levels of an interfering enzyme that part-survives the extraction.

References

1. Shah, V.P., Midha, K.K., Dighe, S. & *9 co-authors* (1992) *Pharm. Res. 9,* 588-592.
2. Food & Drug Administration (1994) *Reviewer Guidance. Validation of Chromatographic Methods.* CDER, 94. Washington, D.C.

#F-5n

A Note on

HIGH THROUGHPUT DETERMINATION OF DRUG-PROTEIN BINDING USING IMMOBILIZED HSA AND LC-MS

Philip R. Tiller†, Ian M. Mutton, Stephen F. Lane & Christopher D. Bevan

Structural Chemistry Dept., Glaxo Wellcome plc
Medicines Research Centre, Stevenage, Herts. SG1 2NY, U.K.

Drug-binding characteristics are important [1, 2] and have been extensively studied with human serum albumin (HSA), which comprises ~5% of the plasma weight and ~85% of total protein. As recently reviewed [3], drugs appear to bind to HSA *via* two main sites, characterized by 'typical' ligands - e.g. warfarin for site I and indoles or benzodiazepines for site II.

Immobilized serum albumins can be used to determine drug binding [4, 5], with good correlation between conventionally measured binding and binding to a column expressed as k'/(k'+1) [1]. Binding to HSA columns has hitherto been studied using UV detection. The utility of MS detection has now been examined [6]. It could offer two major advantages over UV: (1) its better sensitivity might allow the simultaneous examination of several molecular species with minimal between-drug competition for binding sites, through injecting onto the column at greater dilution than is possible using UV detection; (2) because MS is very selective, drug-protein binding can be determined even if different drugs have the same k' value since each would afford a unique molecular ion (but isomers would need separate binding determinations).

The present approach could be important in drug discovery through quickly ranking the protein-binding affinities of a series of compounds early in the research process.

Determining k'.- MS spectra were acquired on a TSQ-700 instrument using the APcI source (vaporization at 450°, capillary at 200°), scanning from 100 to 600 amu in 1.0 sec. The 5 µl on-column injections of each compound (50 ng; glucose used as t_o marker) were from 10 µg/ml solutions in the eluent - 50 mM ammonium acetate with 4% (v/v) n-propanol, run at 0.8 ml/min in an HP 1050 system. An HSA column (50 × 4.6 mm; used at 37°) that had not been exposed to acetonitrile (which could damage the protein) was obtained from Shandon HPLC (Runcorn).

†present address: c/o Finnigan MAT, 355 River Oases Parkway, San Jose, CA 95134, U.S.A.

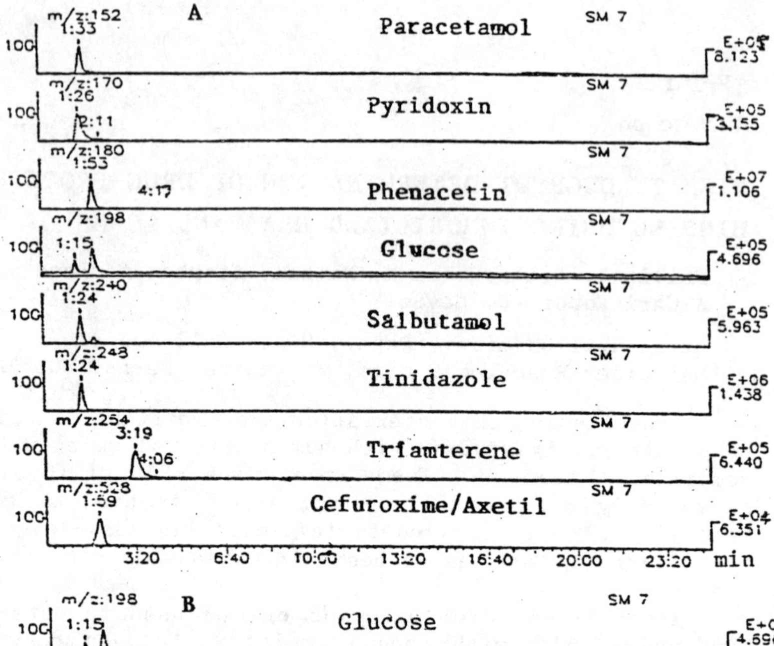

Fig. 1, A & B. Sets of chromatograms, each for a pure analyte giving m/z as stated. The t_o found was 1 min 15 sec ('1:15') net; the void vol. was 27 sec for the system up to the column.

RESULTS AND DISCUSSION

Fig. 1 shows mass chromatograms which furnished retention times (t_r) for each of a range of drugs having known binding values as given in Table 1 along with the times. The penultimate column tabulates k'/(k'+1) values for the drugs chromatographed individually. As shown in the right-hand column, these values were virtually identical when a mixture of the drugs was run (correlation coefficient 0.999).

Table 1. Values for the drugs analyzed: % binding to HSA as established (in-house if no ref. given) by ultracentrifugation or ultrafiltration, t_r as min:sec but in sec for calculating $(t_r - t_o)/(t_o - 27) = k'$, tabulated for the drug run alone; thence $k'/(k'+1)$ for the drug run alone or in admixture.

Drug	mol. wt.	% binding	t_r	k'	$k'/(k'+1)$	$k'/(k'+1)$ mixture
Glucose	180	0	1:15	-	-	
Salbutamol *	239	7.5	1:24	0.187	0.157	0.157
Pyridoxin	169	10 ¶	1:26	0.229	0.186	0.186
Tinidazole	247	12 ¶	1:24	0.187	0.157	0.157
Paracetamol	151	24 ¶	1:33	0.375	0.273	0.273
Phenacetin	179	33 §	1:53	0.792	0.442	0.442
Cefuroxime-Axetil	510	40 §	1:59	0.916	0.478	0.478
Triamterene *	253	57 ¶	3:19	2.583	0.721	0.721
Alprazolam	308	70 §	4:43	4.333	0.812	0.812
Ondansetron	293	74	8:14	8.729	0.897	0.897
Quinidine	324	90 §	7:19	7.583	0.883	0.883
Lorazepam *	322	90 §	7:10	7.396	0.881	0.881
Quinine	324	90 §	7:19	7.583	0.883	0.883
Salmeterol	415	95	7:37	7.958	0.888	0.888
R-Temazepam	300	97 §	4:42	4.312	0.812	0.812
S-Temazepam	300	97 §	6:19	6.333	0.863	0.860
Diazepam	284	98 §	7:56	8.354	0.893	0.893
R-Warfarin *	308	99 §	15:42	18.062	0.947	0.947
S-Warfarin *	308	99 §	20:54	24.562	0.961	0.951

*used as a binding calibrant
¶Value from Clarke's 'Isolation and Identification of Drugs' (Pharmaceutical Press, 1986)
§Clin. Pharmacokinet. **15**, 254-282 (1988)

As shown in Fig. 2, HSA $k'/(k'+1)$ values correlated well (0.94) with protein binding values although these came from diverse sources. Most investigators in this area concentrate on drugs that exhibit >60% protein binding, reckoning that only if this is high will an effect on the drug's clinical activity be pronounced [5]. We aimed to develop an approach applicable to a series of compounds whatever their degree of binding. We observed that several of the drugs were co-eluting (salbutamol/tinidazole/pyridoxin, quinine/quinidine/lorazepam, ondansetron/diazepam, alprazolam/temazepam); but this had no effect on the determination of their k' values by MS.

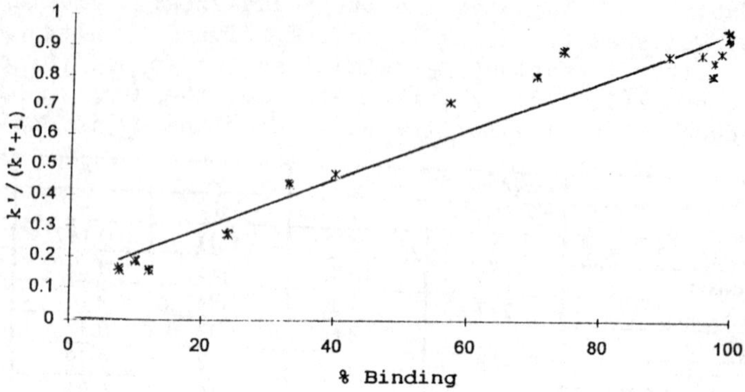

Fig. 2. Regression of % binding against k'/(k'+1).

For a set of unknown compounds the approach is made thus.- (1) Prepare a mixture, each 10 ng/μl, along with glucose as the t_o marker and including *R,S*-warfarin, lorazepam, triamterene and salbutamol. (2) Using the values for these four compounds, plot k'/(k'+1) against % binding. (3) Use the line obtained to get % binding values for the unknowns.

MS data gained in the full-scan mode provided ample sensitivity at the dilutions used; but these could, if necessary, be greater through employing SIM to get even better sensitivity.

References

1. Noctor, T.A.G., Diaz-Perez, M.J. & Wainer, I.W. (1993) *J. Pharm. Sci. 82,* 675-682.
2. Milton, A.D., Feick, A. & Salama, Z.B. (1994) in *Biofluid and Tissue Analysis for Drugs including Hypolipidaemics* [Vol. 23, this series; Eds. and publisher as for present vol.], pp. 51-57.
3. Wainer, I.W. (1994) *J. Chromatog. A 666,* 221-234.
4. Lammers, N., de Bree, H., Groen, C.P., Ruijten, H.M. & de Jong, B.J.: [a] (1989) *J. Chromatog. 496,* 291-300; [b] (1988) in *Bioanalysis of Drugs and Metabolites, especially Anti-inflammatory and Cardiovascular* (Reid, E. *et al.,* eds.), Plenum Press, New York, pp. 301-303.
5. Noctor, T.A.G., Wainer, I.W. & Hage, D.S. (1992) *J. Chromatog. B 577,* 305-315.
6. Authors as in present art. (1995) *Rapid Comm. Mass Spectrom. 9,* 261- 263.

#F-6n

A Note on

VALIDATION OF ENDOGENOUS ANALYTES USED AS PHARMACODYNAMIC MARKERS

J.E.C. Burnett, T. Thompson & H.M. Hill

Corning Hazleton Europe, Otley Road, Harrogate HG3 1PY, U.K.

Many endogenous compounds serve as PD* markers in drug treatment, for monitoring and/or measuring therapeutic effects. A changed level may accompany or actually mediate the effect. In bringing drugs from the pre-clinical stage to the market, it is becoming more common to assay plasma and urine for selected endogenous analytes[ᵒ] to fill out the drug-action picture. Drug development is helped by understanding of PK and PD parameters, thence appropriately tailoring formulation and delivery system.

By consensus (Dec. 1990 meeting in Washington, DC) [1], PK regulatory requirements should hold too for PD analyses. Their validation embraces: (#A†) specificity [imperilled in IA's by cross-reactivity (#B)]; (#C) sensitivity; (#D) precision; (#E) accuracy; (#F) recovery; (#G) stability. Calibration and QC standards should ideally be prepared in the matrix medium free of the endogenous analyte concerned - which is often hard to achieve (#H). In suggesting that innovative PK and PD studies in Phase I could streamline Phases II and III, Colburn *et al.* [2] remarked "the FDA is a regulatory agency that is not only open to consideration of novel approaches by sponsors of investigational new drug applications (IND) but actively encourages innovation"; FDA staff might, through pre-IND meetings, become helpfully involved in the design of Phase I trials.

Analyte availability (#I) for requisite characterization or reference material may be a problem, if entailing purification from bio-matrices that furnish only very small amounts, as with proteins, peptides or (#J) urinary 'collagen cross-links'. Hence multi-centre studies that involve different analytical laboratories must be interpreted cautiously. Such constraints together with the pertinence of the accuracy data should be emphasized and discussed during reporting.

**Abbreviations include:* DHT, dihydrotestosterone; PD, pharmacodynamic; PK, pharmacokinetic; (R)IA, (radio-) immunoassay.

†The Senior Ed. has shrunk the text to 'generalizations' and, alluding to his #A #B etc., authors' examples (no refs.). The PD analyses theme is due to be explored at the 1997 Forum and in Vol. 25. In Vol. 20 and elsewhere in this book series there is pertinent validation guidance (consult Indexes for #A etc.).

ᵒany quantifiable substance; includes enzymes as well as compounds

Reliability is questionable where there seems to be only 10-20% change in marker level (especially if downwards), unless 2-5% precision and accuracy are attained, intra- and inter-day (#D, #E); this calls for adequate sensitivity, with precision no worse than 15-20% at the LLOQ (#C). Recovery and stability assessment (#F, #G) should take account of any uncertainties in analyte identity or characterization (cf. #I).

MATRIX ISSUES (#H), WITH IN-HOUSE EXAMPLES

If an increase in the PD marker is expected, calibration standards and QC samples prepared by spiking the plasma allow conventional validation. If a decrease, the problem that the calibration range is below background levels calls for stratagems, such as 'stripping' plasma, that are now exemplified.

DHT, normally 0.25-1.0 ng/ml plasma and reckoned to rise in benign prostatic hypertrophy, is formed from testosterone by the action of an α-reductase, drug-blockage of which can lower DHT 10-fold such that the calibration range's LLOQ has to be 25 pg/ml. Hence before spiking and RIA the plasma is stripped of DHT, by charcoal (50 mg/ml, overnight). #B needs checking.

α-Sympathomimetics, given illegally to food animals to promote growth, are measurable by GC-MS in liver homogenates and urine down to 10 ng/kg or /L after removing interferants by immunoaffinity on a Sepharose column with immobilized IgG Ab's.

'Collagen cross-links', pyridoline and deoxypyridoline (#J), are normal urinary constituents that can rise with bone-degrading diseases and fall with treatment to near-normal. Not being amenable to chemical synthesis, they are prepared from human *post-mortem* tibia, by extensive hydrolysis and prolonged extraction and purification. Characterization entails comparison with established reference samples held in small amount by recognized laboratories. Dilution of urine rich in cross-links (Paget's disease) can serve to get secondary standards.

Vitamin A (retinol) levels may be informative in drug treatments that could impair its absorption, or in Vitamin A treatment, e.g. for bowel-cancer prophylaxis. Throughout LC-MS assays now performed, on plasma with ascorbic acid present at the outset to limit oxidation, natural light was excluded, and sample preparation was minimal so as to limit Vitamin A degradation, particularly when evaporating under N_2. As solvents are especially prone to cause such oxidation, an aqueous column switching technique preceding LC-MS was developed successfully.

Matrix free of Vitamin A being unobtainable, the calibration curve was prepared with plasma spiked at different levels, and was back-extrapolated where there was the problem of subnormal levels. Precision was 5% at levels of 100-300 ng/ml but only 20-30% at <100 ng/ml.

Various biomolecules assayed by RIA.- Plasma and urine levels indicate the intracellular efficacy of drug therapies. Commercially available kits with calibration standards supplied in buffer were used; they include QC samples but where feasible these are prepared by spiking the analyte directly into control matrix. For cGMP both the plasma and the urine methods involve SPE clean-up prior to RIA. The calibration range is 0.5-6.4 nM for urine, but 0.025-5 nM for plasma where higher sensitivity is gained from an acetylation stage between SPE and the RIA.

For human growth hormone (hGH), gastrin, IGF and cortisol, kit assay entails direct, double-Ab RIA on serum or plasma and urine. The respective ranges for the plasma/serum calibration standards, prepared with authentic analyte, are 1-30 ng/ml (cf. normal range in adults: 0-7 ng/ml), 2-1200 pg/ml (0-90 pg/ml), 5-2000 ng/ml (0-200 ng/ml), and - for urine also - 27.6-1380 nM (a.m., 138-690 nM; p.m., 70-350 nM). Since analyte-free matrix was unobtainable, correction for background levels was made by analyzing multiple 'blank' samples. Any test samples that fell above the calibration range were diluted, further jeopardizing accuracy. Precision was generally 5-10% but 18% near the LLOQ.

SUMMARY AND CONCLUSIONS

To enable full use and correlation of the data, PD marker assays must be validated as for PK assays by the same rigorous protocols, carefully thought out in relation to issues such as:

(#I) establishing reference analyte identity and purity or, if not feasible, using 'real' QC's to compare inter-batch data;

(#H) preparation of calibration curves using stripped or else complete matrix, or buffer solutions, with standard additions;

(#G) analyte stability during validation, sampling and analysis.

Such issues and choice of approaches should always be investigated and fully discussed as they may bear on data reporting.

References

1. Shah, V.P., Midha, K.K., Dighe, S. *& 9 co-authors* (1992) *Pharmaceut. Res. 9,* 588-592.
2. Colburn, W.A., Williams, R.L. *& 5 co-authors* (1990) *J. Clin. Pharmacol. 30,* 210-222.

Comments on #F-1, P. Heizmann: HAEMOLYTIC PLASMA PROBLEMS
'Interpolation': #C-3 (spokesman: H. de Bree): ROBUSTNESS TESTING
#F-2, L. Robotham: A RESULTS DATABASE
#F-3, H.M. Hill: VALIDATION AND GLP ISSUES
#F-6n, J. Burnett: PHARMACODYNAMIC MARKERS

Heizmann, answering A. Gulaid.- We haven't tried antioxidants to prevent the oxidation - which couldn't be transforming parent drug to metabolite, since metabolism is not oxidative.

de Bree: robustness testing (#C-3; 56 factors for work-up, 15 for HPLC) takes ~3 months - **reply to C.A. James**, who also (others too) sought **opinions on co-medication aspects**, namely how best to check samples for drug interferences. **Responses.- de Bree**: in default of 'real' samples we spike at realistic levels (maybe metabolites too) into plasma ±analyte and take through the procedure, looking for effects on analyte peak or recovery; but for instrumental methods we don't look for subtle interactive effects. **Hill.**- Spiked-in authentic compounds (maybe unprocurable from the client; metabolites may be unknown or unavailable) may serve for screening, but we prefer to take real ('incurred') samples through the extraction procedure.

M.V. Doig asked Robotham if bar-code sample identifications had been linked to the database. **Reply**: no, but this could be a future development. **Reply to J.D. Gilbert**: the system, in an ASKI format, can indeed be used to acquire RIA data from gamma counters. **Reply to D. Dell**: for customers who want the data for PK interpretation, data can be exported in a format that can then be imported into a PK package. **Hill asked** about the database format's acceptability to a SAS program. **Reply**: the comma-separated-value (CSV) format would permit import to SAS and many other applications such as Lotus 123 and MS Excel. **Replies to other queries.**- No performance problems have been found with increasing size or complexity of data; projects are held in individual databases in separate directories and each user accesses the data from separate PC's. Whereas commercial LIMS's are ideal for QC facilities where many types of analyses are performed on single samples and reporting is focused on giving reports of certificate-of-analysis type, our database is designed to report many results in a standardized table format that is not generally available in commercial LIMS's.

Replies by Hill.- To A. Gulaid: (1) a high—medium—low spread of QC's and calibrants is generally appropriate; (2) for techniques such as HPLC with pre-extraction or RIA, the FDA doesn't seem to relax its accuracy and precision expectations. **To J.D. Gilbert**: in the FDA, as in other quarters, there are misgivings about the specificity of RIA's, whose credibility benefits from the desirable cross-validation against equally sensitive and intrinsically specific methods such as LC-MS-MS.

Hill, agreeing with R. Laplanche whose concern was with immunoassays: the definition of LLOQ, 20% bias and 20% C.V., seems apt for the upper limit too. **P. Heizmann commented** that the FDA's acceptance of a '15% criterion' for found values, e.g. QC's, illogically does not apply to calibration standards and so allow some extrapolation of the calibration curve. **Replies by Hill.- To M. Doig**: for accuracy limits - the FDA's preferred term being 'trueness', not 'bias'- ±15% around the nominal value is expected for all concentrations except LLOQ, needing only ±20%. **To E. Reid**: variability values based on n = 3 are prevalent but are to be deprecated; we prefer n = 4 (but - **remark by D. Stevenson** - a client might resist advocacy of n >3). To **C.A. James** concerning policies on part-repeating study samples and acceptability of results: repeats are not done for out-of-line analyses in PK studies. Whilst the FDA might hope for the whole study set to be repeated, so far they have been content as long as all data are reported.

Further replies by Hill.- To D. Dell: for the stipulated 'Characterization of test article' it suffices to quote the Batch no. and Analytical Certificate. **To H. de Bree**: whilst the FDA are very keen on test-article accountability they don't expect evidence of an alternative check of weighings for stock solutions and of confirmation of identity, but they may ask for it when inspecting a study. **To B. Meijer**: LOD needed (CPMP) for, say, a topical drug (risky systemic absorption?); irrelevant to PK studies.

C.A. James asked J. Burnett (reply: "None so far") whether problems had been encountered with differences in behaviour between 'stripped' and 'real' plasma. **Comment by S. Westwood:** 'real' urine samples show major differences in behaviour from 'stripped' or 'synthetic' urine and especially from buffer.

Citations contributed by Senior Editor For journals, Chr denotes Chromatography

Analyte lability example:- isoprostanes (compounds akin to prostaglandins; GC-MS analysis) in plasma suffer autoxidation, associated with hydrolysis of certain lipids, even at -20°.- Morrow, J.D. et al. (1990) Anal. Biochem. **184**, 1-10 [cited by Roberts II, L.J. et al. (1995) Biochem. Pharmacol. **51**, 1-9].

Protein binding methods review: Oravcová, J. et al. (1996) J. Chr **B 677**, 1-28; a CE enantioselective method: Ohara, T. et al. (1995) Anal. Chem. **67**, 3520-3525. Dialysis on-line with HPLC works well (for NSAID's) under certain conditions, notably conjoint treatment of the plasma with an organic solvent and a displacer: n-decanoic acid (better than ibuprofen) in acetonitrile was added.- Herraez Hernandez, R. (1995) J. Chr **B 666**, 127-137. In #ABC and elsewhere in this series (notably #F-5n herein) binding phenomena feature repeatedly [cf. 'Protein...' index entry]; Vol. 23 has a methods comparison (A.D. Milton). Update on HSA [art. #F-5n]: Hage, D.D. et al. (1995) J. Chr **A 693**, 23-32.

Assay robustness (ruggedness; cf. #C-3, also #F-3): three Plackett-Burman designs were investigated for an HPLC-based drug assay, with 2- or (better) 3-level testing of 6 factors:- mobile-phase pH, TEA concn., column temp., slope of gradient, detection wavelength and column choice.- Vander Heyden, Y. et al. (1995) Anal. Chim. Acta **316**, 15-26. [D.L. Massart is a co-author for this art.; likewise for ref. 2 in #C-3 and, in Vol. 23 (this series), an 'expert system' designed for SPE.]

Factor analysis for C-18 HPLC separation of various organic compounds, with water/methanol/THF gradient systems, has led to a model whereby retentions can be predicted for other analytes or solvents.- Lochmueller, C.H. et al. (1995) J. Chr Sci. **33**, 640-646.

Analyte Index

structured as in past Vols. [Cumulative Index in Vol. 22 - NOT Vol. 23; error on its spine!] - wherein some ***corrections, here denoted*** ᵅ*, could usefully be inserted; a generic name might be assignable to some putative drugs (e.g. DuP 697) by consulting* Merck Index *mid-1996 edition (CD-ROM available), or* Martindale *when a new edition appears.*

Almost all entries refer to bioanalysis of 'real' samples (as had been denoted by the prefix *r* in Vol. 22 & earlier vols.); **ch** denotes chiral investigation. For [1] [1], SEE OVERLEAF; likewise for the **Key to the chemical classification** based on some analytically relevant features e.g. an acid group (actual or, as with esters or lactones, potential). Prefix ᵅ may signify a classification re-think, relevant if consulting past volumes.

Therapeutic classes have some General Index entries and, as listed on p. 161, are the basis for the Assay Compendium #**ABC** which continues to p. 180. Pro-drugs are prefixed †.

ASSIGNMENT 'CATECHISM' based on chemical categories | † ¤ ch: see previous p.

Metabolites are generally not separately listed; the entry for the parent molecule is preceded by a superscript denoting study of Phase I metabolite(s), ^{1} (BOLD $^{\mathbf{1}}$ if N-desalkyl or desacyl). Only rarely were conjugates (Phase II) studied intact, prefix ^{2}.

Formulae for **parent molecules** as indexed (by generic name unless inapplicable) may appear on or following the cited page, except for Compendium entries (pp. 162-180): consult *Merck Index*.

'**Acidic**' assignment (to #Ia, #IIa or #IIIa) applies where the pKa is <6, so excluding phenols; conjugate acidity is ignored. 'Acidity' embraces, besides lactones, esters (hydrolysis being facile, and likely to occur *in vivo*) except where the molecule's *main* moiety does not house the acid group arising by fission.

Cyclic N, ineligible for an amine sub-category (#IIb"/#IIb'''), qualifies for #IIIy/#IIIz (and *may* be basic/UV-absorbing) *only* if *not* 'imide'/'amide' [-CO-N-C(O)-] such that #Iz may apply. (Unlike esters, **-CO-NHR** groups as in **pep**tides resist chemical hydrolysis; likewise sulphonamide groups.) In assigning a compound to #IIIz rather than #IIIy, cyclic O is disregarded.

The Scheme, devised in the GC heyday, is pertinent to liquid-phase operations in respects such as pH choice and amenability to fluorescent derivatization (difficult for secondary amines).

Amino group or non-imide cyclic N present?

- *NO: Category* I *(Compound non-basic: relevant to extraction)*
  - # Any acid group, free or (ester) potential?
    - *YES:* - see Ia *(May contain halo, etc., as for* Iz*)*
    - *NO:* # Any halogen, P or N (e.g. amide /imide/nitro) ? *(Relevant to GC detection)*
      - *NO* - see **Iy**
      - *YES* - see **Iz**
- *YES:* # Amino (non-cyclic) present ? (& maybe non-imide cyclic N)
  - *YES: Category* II *(Implications for derivatization and GC behaviour)*
    - # Any acid group, free or (ester) potential?
      - *YES:* see **IIa** *(Amphoteric/ zwitterionic, if not ester)*
      - *NO:* # Primary amino?
        - *YES* - see **IIb'**
        - *NO* # Secondary amino?
          - *YES* - see **IIb"**
          - *NO* i.e. tertiary /quaternary - see **IIb'''**
  - *NO: Category* III Cyclic N, not merely imide - *but not necessarily basic*
    - # Any acid group, free or (ester) potential?
      - *YES* - see **IIIa**
      - *NO:* # Any heteroatom besides N?
        - *NO* - see **IIIy**
        - *YES* - see **IIIz**

For SUMMARY see p. 344

=============

CATEGORY **II** *(amino, not in a ring)*

KEY, in brief (pp. 339 & 340: full version and the chemical classes)
Phase I metabolite also: [1] or, if des-N, [1]; ch, chiral study.
Affecting previous-vol. Index: ¤

SUMMARY OF CATEGORIES
Full version: p. 340 ('merit': p.339)

	I	II	III
Amino?	no	✓	no
Non-imide hetero-N?	no	maybe	✓
Acid or potential acid (not conjugate)?	✓ = Ia	✓= IIa	✓= IIIa
– no! (and not an ester)	Halo, P or N? – no: Iy – ✓ = Iz	Primary amino? ✓ = IIb' If no: 2^y = IIb" 3^y or 4^y = IIb"'	Hetero atom besides N? (not O) – no = IIIy, ✓ = IIIz

Only **parent compound** listed even if Phase I metabolite(s), [1] ([1] if N-des), also studied. Pro-drug, †; chiral study, ch. Previous-vol. Index affected: ᵅ. Page range 162-180: citation from Compendium (#**ABC**).

CUMULATIVE INDEX in Vol. #22 covers 5 vols., back to #14 (Plenum), and that in #12 covers the 4 earlier vols. (Ed. can answer queries)

Analyte Index corrections to previous vols.
besides those marked ᵅ above and in Vol. 23

Vol. 23: in #IIb" **amend** Dimetindene to Dimethindene; in #IIIa, **amend** the Lacidipine p. no. to 153;
Vol. 22: **delete** the #IIIy entry for Navelbine [#IIIa correct].

A NAME-SIMILARITY INDEXING NIGHTMARE:
In Vol. 22 (referring to its p.160): #IIb"' lists Oxcarbamazepine; #IIIy (under Benzodiazepines) lists Carbamazepine (& its epoxide) [Merck Index agrees]. Cf. (1) Oxcarbazepine, listed (#IIb"') in present vol.; (2) Merck Index listing of Carbamazine (#IIIy type) as a once-used synonym for Diethylcarbamazine (antihelminthic).

============

General Index

This Index deals mainly with features studied and with points of technique, indexed as in previous vols. so as to facilitate back-searching — likewise for the preceding Analyte Index (AnIx). Where an item in the 'Compendium' (Cm; pp. 161-180) is cited, on the actual Cm page the item may be asterisked, *.

In a page entry such as '17-', the '-' signifies coverage in depth/look at ensuing page(s) too.

================

AnIx denotes Analyte Index;
Cm denotes Compendium (#ABC)

AnIx = Analyte Index, Cm = Compendium

===============

Corrections (see also Analyte Index, p. 334 and ▫ alertings)

Vol. 23: on spine, **mask** the cumulative index phrase
Near foot of p.70, preceding GC-MS **insert** by
Formula, p.124 top-centre: COON **should read** COONa
Entry on p. 307 **to be** Ifosfamide

Vol. 22 [& see Vol. 23 p.420]
Foot of p.82 after acetate and : **delete** then [explained in present vol., p.187 footnote]
Footnote, p. 253: PBI, **not** HNI
General Index, in Immuno...FPIA entry: p. 164, **not** p. 165